HISTORICAL DICTIONARY

The historical dictionaries present essential information on a broad range of subjects, including American and world history, art, business, cities, countries, cultures, customs, film, global conflicts, international relations, literature, music, philosophy, religion, sports, and theater. Written by experts, all contain highly informative introductory essays of the topic and detailed chronologies that, in some cases, cover vast historical time periods but still manage to heavily feature more recent events.

Brief A–Z entries describe the main people, events, politics, social issues, institutions, and policies that make the topic unique, and entries are cross-referenced for ease of browsing. Extensive bibliographies are divided into several general subject areas, providing excellent access points for students, researchers, and anyone wanting to know more. Additionally, maps, photographs, and appendixes of supplemental information aid high school and college students doing term papers or introductory research projects. In short, the historical dictionaries are the perfect starting point for anyone looking to research in these fields.

HISTORICAL DICTIONARIES OF ASIA, OCEANIA, AND THE MIDDLE EAST

Jon Woronoff, Series Editor

Guam and Micronesia, by William Wuerch and Dirk Ballendorf. 1994.
Palestine, by Nafez Y. Nazzal and Laila A. Nazzal. 1997.
Lebanon, by As'ad AbuKhalil. 1998.
Azerbaijan, by Tadeusz Swietochowski and Brian C. Collins. 1999.
Papua New Guinea, Second Edition, by Ann Turner. 2001.
Cambodia, by Justin Corfield and Laura Summers. 2003.
Saudi Arabia, Second Edition, by J. E. Peterson. 2003.
Kyrgyzstan, by Rafis Abazov. 2004.
Turkmenistan, by Rafis Abazov. 2005.
Vietnam, Third Edition, by Bruce Lockhart and William J. Duiker. 2006.
India, Second Edition, by Surjit Mansingh. 2006.
Hong Kong SAR and the Macao SAR, by Ming K. Chan and Shiu-hing Lo. 2006.
Pakistan, Third Edition, by Shahid Javed Burki. 2006.
Iran, Second Edition, by John H. Lorentz. 2007.
Gulf Arab States, Second Edition, by Malcolm C. Peck. 2008.
Laos, Third Edition, by Martin Stuart-Fox. 2008.
Brunei Darussalam, Second Edition, by Jatswan S. Sidhu. 2010.
Mongolia, Third Edition, by Alan J. K. Sanders. 2010.
Bangladesh, Fourth Edition, by Syedur Rahman. 2010.
Polynesia, Third Edition, by Robert D. Craig. 2011.
Singapore, New Edition, by Justin Corfield. 2011.
East Timor, by Geoffrey C. Gunn. 2011.
Postwar Japan, by William D. Hoover. 2011.
Afghanistan, Fourth Edition, by Ludwig W. Adamec. 2012.
Philippines, Third Edition, by Artemio R. Guillermo. 2012.
Tibet, by John Powers and David Templeman. 2012.
Kazakhstan, by Didar Kassymova, Zhanat Kundakbayeva, and Ustina Markus. 2012.
Thailand, Third Edition, by Gerald W. Fry, Gayla S. Nieminen, and Harold E. Smith. 2013.
Syria, Third Edition, by David Commins and David W. Lesch. 2014.
Science and Technology in Modern China, by Lawrence R. Sullivan and Nancy Y. Liu. 2014.
Taiwan (Republic of China), Fourth Edition, by John F. Copper. 2014.
Australia, Fourth Edition, by Norman Abjorensen and James C. Docherty. 2015.
Republic of Korea, Third Edition, by James E. Hoare. 2015.
Indonesia, Third Edition, by Audrey Kahin. 2015.
Fiji, by Brij V. Lal. 2016.
People's Republic of China, Third Edition, by Lawrence R. Sullivan. 2016.
Israel, Third Edition, by Bernard Reich and David H. Goldberg. 2016.
New Zealand, Third Edition, by Janine Hayward and Richard Shaw. 2016.
Nepal, Second Edition, by Nanda R. Shrestha and Keshav Bhattarai. 2017.
Burma (Myanmar), Second Edition, by Donald M. Seekins. 2017.
Yemen, Third Edition, by Charles Schmitz and Robert D. Burrowes. 2017.
Chinese Economy, by Lawrence R. Sullivan. 2018.
Malaysia, Second Edition, by Ooi Keat Gin. 2018.
Tajikistan, Third Edition, by Kamoludin Abdullaev. 2018.
Postwar Japan, Second Edition, by William D. Hoover. 2019.

Iraq, Third Edition, by Beth K. Dougherty. 2019.
Democratic People's Republic of Korea, Second Edition, by James E. Hoare. 2019.
Chinese Environment, by Lawrence R. Sullivan and Nancy Liu-Sullivan. 2019.

Historical Dictionary of the Chinese Environment

Lawrence R. Sullivan
Nancy Liu-Sullivan

ROWMAN & LITTLEFIELD
Lanham • Boulder • New York • London

Published by Rowman & Littlefield
An imprint of The Rowman & Littlefield Publishing Group, Inc.
4501 Forbes Boulevard, Suite 200, Lanham, Maryland 20706
www.rowman.com

6 Tinworth Street, London SE11 5AL

British Library Cataloguing in Publication Information Available

Library of Congress Cataloging-in-Publication Data

Names: Sullivan, Lawrence R., author. | Liu-Sullivan, Nancy, 1963– author.
Title: Historical dictionary of the Chinese environment / Lawrence R. Sullivan, Nancy Liu-Sullivan.
Description: Lanham : Rowman & Littlefield, [2019] | Series: Historical dictionaries of Asia, Oceania, and the Middle East | Includes bibliographical references. | Summary: "Historical Dictionary of the Chinese Environment contains a chronology, an introduction, appendixes, and an extensive bibliography. The dictionary section has more than 200 cross-referenced entries on environmental degradation including air and water pollution, deforestation, desertification, and resource depletion"—Provided by publisher.
Identifiers: LCCN 2019018580 (print) | LCCN 2019980129 (ebook) | ISBN 9781538120354 (cloth) | ISBN 9781538120361 (ebook)
Subjects: LCSH: Environmental degradation—China—Dictionaries. | China—Environmental conditions—Dictionaries.
Classification: LCC GE160.C6 S85 2019 (print) | LCC GE160.C6 (ebook) | DDC 363.700951/03—dc23
LC record available at https://lccn.loc.gov/2019018580
LC ebook record available at https://lccn.loc.gov/2019980129

To Liang Congjie, Ma Jun, Qu Geping, and Yang Xin, environmental pathbreakers in the People's Republic of China, and to the memory of Professor Roderick MacFarquhar, Harvard University and Fairbank Center for Chinese Studies, eminent scholar and gentleman.

Contents

Editor's Foreword xi

Preface xiii

Reader's Note xv

Acronyms and Abbreviations xvii

Maps xxiii

Chronology xxix

Introduction 1

THE DICTIONARY 7

Glossary 305

Appendix A Chinese and Foreign Environmental Activists 307

Appendix B Chinese and Multinational Environmental Companies 309

Appendix C Environmental Laws and Regulations 311

Appendix D Environmental Nongovernmental Organizations (ENGOS) 313

Appendix E State Environmental Organizations 317

Appendix F National Nature Reserves and National Parks 319

Appendix G Wetland Nature Reserves and Parks 323

Bibliography 325

About the Authors 339

Editor's Foreword

The situation of China's environment is of importance to not only the People's Republic of China (PRC), but also the world as a whole, given the country's huge population, massive industrial production and creation of waste, and economic ramifications abroad through the import of natural resources and export of finished products. But it is not easy to explain, first, because China has gone through different periods—almost total disregard for the environment during the rule of Mao Zedong and Deng Xiaoping (1949–1992), and increasing and impressive efforts to save what remains of it, resolve old problems, and create a better environment since the 1990s. The other problem is that what is said, even by the government and its leaders, and embodied in national laws and regulations, is not always done. Thus, it is essential to follow the events on the ground, where there is increasing awareness among the people and the emergence of environmental nongovernmental organizations (ENGOs). That is the double task of *Historical Dictionary of the Chinese Environment*, carried out with numerous entries on such topics as the ecologically backward Great Leap Forward (1958–1960), production of garbage, radioactive waste, deforestation, desertification, and climate change, as well as ecological cities and villages and environmental awareness, often having two components—what is said and what is done, although certainly much more is being done now than before.

Given the remarkable reversal on ecological matters, the chronology is a good place to start, showing the frequent backward steps of the earlier "communist" regimes and the increasingly positive steps of subsequent "communist" regimes, from presidents Hu Jintao to Xi Jinping, the reference to communism again showing just how important it is to look beyond the terminology used. This crucial aspect becomes even more evident in the insightful introduction, giving some hope for a better trajectory into the future. The introduction does an excellent job of summing up the situation and providing a good basis for what follows. The dictionary provides entries on the various policies adopted, the people who implemented them—usually within the government but increasingly in the general population and sometimes even at the grassroots level—and the countless policies and programs that have been adopted and sometimes implemented and which bodies are responsible for their implementation. Even this large volume cannot cover all the ground, and further information can be found in the appendixes and bibliography.

This volume was written by two authors already familiar to readers of this series, Lawrence R. Sullivan and Nancy Liu-Sullivan. Dr. Lawrence Sullivan is professor emeritus of political science at Adelphi University and research associate at the East Asian Institute, Columbia University. Dr. Liu-Sullivan is a lecturer in the Department of Biology, College of Staten Island, and has conducted cancer research at the Sloan-Kettering Institute. Both have previously written about the ecology of China, namely in their translation of *China's Water Crisis*, by Ma Jun, and in several other historical dictionaries, as well as in articles and papers, *Historical Dictionary of the Chinese Economy* (2018), and *Historical Dictionary of Chinese Foreign Affairs* (2018), for Dr. Sullivan, and *Historical Dictionary of Science and Technology in Modern China* (2015), for Dr. Liu-Sullivan. This latest volume, by two people who know the subject and the country well, nicely rounds out our subcircle of books on science and technology in the People's Republic of China.

Jon Woronoff
Series Editor

Preface

The year is 2019, and planet Earth is obviously in trouble. Climate change is occurring *now* with melting ice caps and shrinking glaciers, leading to sea-level rise and storm surges, intense precipitation events—"bomb cyclones"—producing massive flooding, and widespread and long-term drought with rising global temperatures helping to spawn fierce wild fires and endangering wildlife and plant species: all major threats to the global economy especially agriculture.

The largest emitter of greenhouse gases (GHG) since 2007, and top polluter of the increasingly stressed Pacific Ocean, the People's Republic of China (PRC) is both a major contributor to environmental degradation and a leading contender to mitigate and stabilize global environmental conditions. Reviewing the history of the PRC from the periods of central economic planning (1953–1978) followed by the single-minded pursuit of economic growth and mass consumption beginning in 1978–1979, to the adoption of a more balanced approach stressing environmental protection and restoration beginning in the 1990s, *Historical Dictionary of the Chinese Environment* documents both the enormous damage to the country's natural environment and the dramatic attempts by the Chinese government and environmental nongovernmental organizations (ENGOs) at environmental amelioration and restoration. Major entries describing environmental degradation include, among others, air and water pollution, deforestation, desertification, and resource depletion, while efforts at amelioration and restoration include river and waterway cleanups, reforestation and desert control, restoration of fisheries, and creation of national nature reserves, along with energy conservation and development of renewables like solar and wind power. A comprehensive examination is provided of a country that has rapidly transformed from a serial environmental abuser into a leading force on environmental protection and restoration, including the international challenge of climate change.

This volume also provides portraits of prominent individuals in government and society in China, including the many newly created ENGOs that are confronted with the multifaceted challenges in the PRC of reversing years of environmental damage and creating—in the words of Chinese president Xi Jinping (2013–)—an "ecological civilization" and "beautiful China." With the largest economy on Earth and one-fifth of the world's population, the fate of China and indeed the entire planet depends on what is undoubtedly a herculean effort of undoing decades of woeful and unbridled ecocide.

Assisting in the preparation of this volume is Professor Robert Paarlberg, John F. Kennedy School of Government, Harvard University, who provided invaluable consultation on Chinese agriculture and related matters important to the Chinese environment, and Ms. Nicole Carty for her excellent editorial work.

Lawrence R. Sullivan
Nancy Liu-Sullivan
Long Beach, New York

Reader's Note

The romanization used in this dictionary for Chinese language terms is the Hanyu pinyin system, developed in the 1950s and currently used in the People's Republic of China (PRC). Chinese language terms generally unknown to Western readers are italicized, with some words different in English having the same Romanized Chinese spelling, for instance, *ge* and *ge* for the chemical elements cadmium and chromium, respectively. In Chinese and East Asian culture, generally, the family name comes first, preceding the given name.

Throughout the text, the conversion rate between the Chinese "people's currency" (renminbi or yuan) and the U.S. dollar is constant at 6:1. The data point PM 2.5 incorporated into the Air Quality Index (AQI) refers to tiny particles of two and one-half microns or less in width, which, in the air, pose serious health hazards ranked by parts per cubic meter ($\mu g/m^3$). The Water Quality Index (WQI) in China is divided into six grades: I–III = fit for human contact; IV = industrial and agricultural use only/unfit for human contact; and V–V+ = unsuitable for any use.

To facilitate the rapid and efficient location of information and make this book as useful a reference tool as possible, extensive cross-references have been provided in the dictionary section. Within individual entries, terms and names that have their own entries are in **boldface** type the first time they appear. Related terms that do not appear in the text are indicated by *See also* references. *See* references refer to other entries that deal with similar topics.

Acronyms and Abbreviations

ACCA 21	Administrative Center for China's Agenda 21
ADB	Asian Development Bank
AI	artificial intelligence
API	active pharmaceutical ingredients
AQI	Air Quality Index
ARG	antibiotic-resistant genes
BASIC	Brazil, South Africa, India, China
BOD	biochemical oxygen demand
BOF	basic oxygen furnace
BP	British Petroleum
BTCE	billion tons coal equivalent
BYOC	"Bring Your Own Chopsticks"
CAG	controlled environment agriculture
CAS	Chinese Academy of Sciences
CASS	Chinese Academy of Social Sciences
CB	cellulosic biofuel
CCP	Chinese Communist Party
CCTV	China Central Television
CEA	China Ecological Agriculture
CERN	China Ecological Research Network
CITES	Convention on International Trade in Endangered Species of Wild Fauna and Flora
CMA	China Meteorological Administration
CNG	compressed natural gas
CNNC	China National Nuclear Cooperation
CNPC	China National Petroleum Company
CO_2	carbon dioxide
COD	chemical oxygen demand

CPPCC	Chinese People's Political Consultative Conference
CRFS	controlled-release fertilizers
CSA	Community Supported Agriculture
CSEE	Chinese Society for Environmental Education
CSEJ	Chinese Society of Environmental Journalists
CSRME	China Society for Rock Mechanics and Engineering
DPRK	Democratic People's Republic of Korea
EAF	electric arc furnace
ECOIA	ecological impact assessment
ECS	East China Sea
EEI	Environmental Education Initiative
EFCA	Ecosystem Function Conservation Area
EIA	environmental impact assessment
ENGO	environmental nongovernmental organization
EPB	Environmental Protection Bureaus
EPL	Environmental Protection Law
EPR	Extended Producer Responsibility (system)
ERI	Energy Research Institute
EV	electric vehicle
FDIA	Farmers Drainage and Irrigation Association
FIP	fish improvement projects
FON	Friends of Nature
FTW	floating treatment wetlands
GDP	gross domestic product
GGDP	green gross domestic product
GEF	Global Environmental Facility
GFN	Global Footprint Network
GHA	global hectare (measurement unit)
GHG	greenhouse gas
GMO	genetically modified organisms
GONGO	government-organized nongovernmental organization
GPI	Genuine Progress Indicator

GSHP	Green Small Hydropower Project
GW	gigawatt
HLW	high-level waste
IBRD	International Bank for Reconstruction and Development
IEK	indigenous ecological knowledge
IIS	intelligent inspection system
IPCC	Intergovernmental Panel on Climate Change
IPEA	Institute of Public and Environmental Affairs
ITS	Intelligent Transportation Systems
IUCN	International Union for Conservation of Nature
km	kilometers
kWh	kilowatt hour
LED	light-emitting diode
LEED	Leadership in Energy and Design
LID	Low-Impact Development
LNG	liquified natural gas
LPI	Living Planet Index
MCA	Ministry of Civil Affairs
MEE	Ministry of Ecology and Environment
MEP	Ministry of Environmental Protection
MMT	million metric tons
MNR	Ministry of Natural Resources
MNRA	Marine Natural Resource Areas
MOC	Ministry of Commerce
MSW	municipal solid waste
MW	megawatt
MWR	Ministry of Water Resources
N_2O	nitrous oxide
NDRC	National Development and Reform Commission
NEPB	National Environmental Protection Bureau
NFCP	Natural Forest Conservation Program
NFPP	National Forest Protection Program

NGO	nongovernmental organization
NNSA	National Nuclear Safety Administration
NO_2	nitrogen dioxide
NO_x	nitrogen oxide
NPC	National People's Congress
NRDC	National Resources Defense Council
OBOR	One Road, One Belt Initiative
PBOC	People's Bank of China
PLA	People's Liberation Army
PM 2.5	fine particulate matter
PPM	parts per million
PPP	public–private partnership
PRC	People's Republic of China
PRD	Pearl River Delta
PV	photovoltaic
PX	paraxylene
RMB	people's currency (renminbi)
ROK	Republic of Korea
RSW	rural solid waste
SAC	Standardization Administration of China
SCEP	State Commission on Environmental Protection
SCS	South China Sea
SDG	Sustainable Development Goals Index
SEPA	State Environmental Protection Administration
SEPC	State Environmental Protection Commission
SEZ	special economic zone
SFA	State Forestry Administration
SFPI	Single-Factor Pollution Index
SFWA	State Forestry and Wetlands Administration
SLCP	Sloping-Lands Conversion Program
SMPA	Special Maritime Protection Area
SNWDP	South-to-North Water Diversion Project

SO_2	sulfur dioxide
SOA	State Oceanic Administration
SOE	state-owned enterprise
SPC	Sponge City Concept; State Planning Commission
SRFS	slow-release fertilizers
SSTC	State Science and Technology Commission
STE	sludge-to-energy
SWM	solid waste management
TAC	total allowable catch
TAR	Tibet Autonomous Region
TCM	traditional Chinese medicine
TDS	total dissolved solids
TGD	Three Gorges Dam
TRHA	Three Rivers Headwaters Area
TSP	total suspended particles
TVE	township and village enterprise
UHI	Urban Heat Index
UNCLOS	United Nations Convention on the Law of the Sea
UNDP	United Nations Development Programme
UNEP	United Nations Environment Programme
UNESCO	United Nations Educational, Scientific, and Cultural Organization
USI	Urban Sustainability Index
VOC	volatile organic chemical
WCED	World Commission on Environment and Development
WHO	World Health Organization
WQI	water quality index
WSI	water-saving irrigation
WTE	waste-to-energy
WTO	World Trade Organization
WWF	World Wildlife Fund (renamed World Wide Fund for Nature)
XUAR	Xinjiang-Uighur Autonomous Region

YEP	Yunnan Environmental Program
YRD	Yangzi River Delta
YREB	Yangzi River Economic Belt

Maps

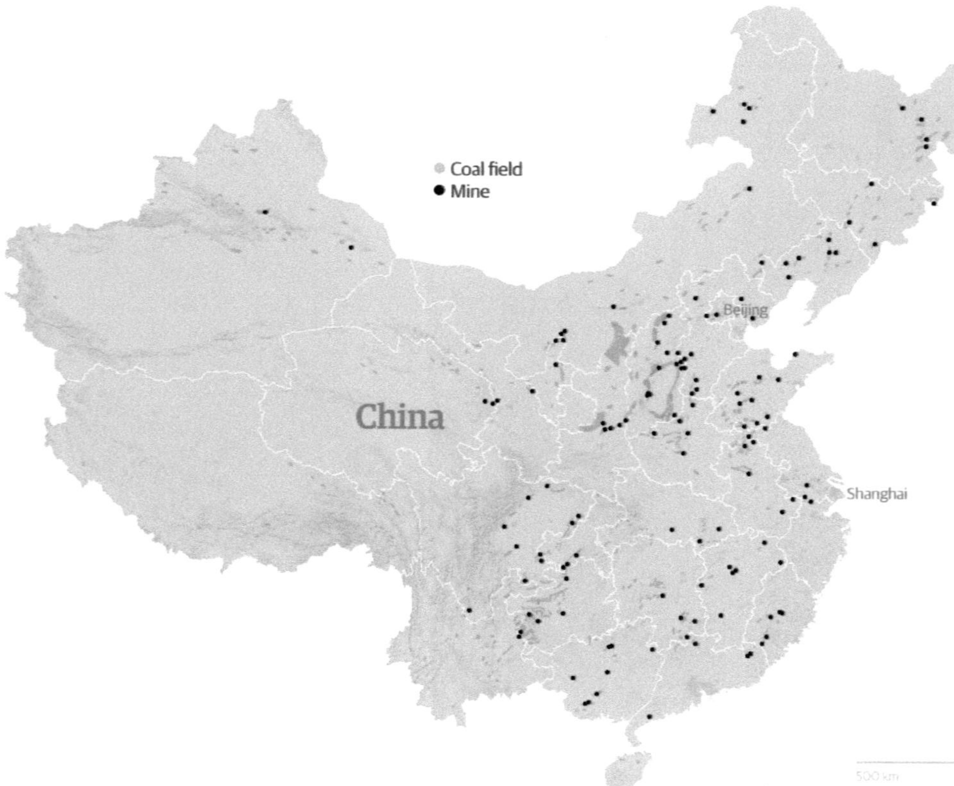

Distribution of Coal Fields and Mines

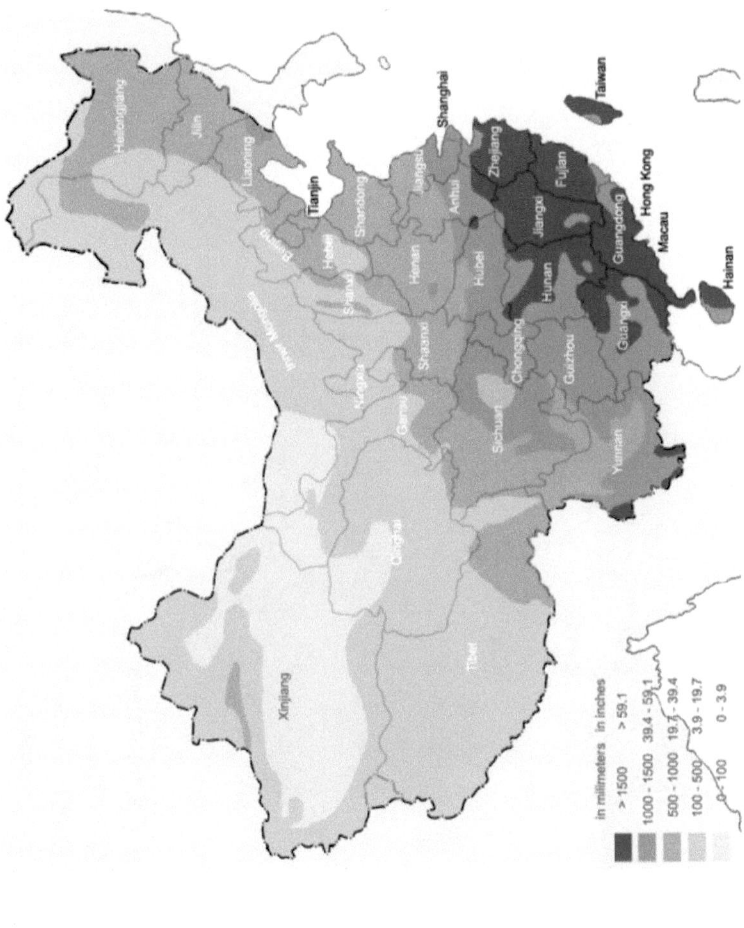

Average Annual Rainfall

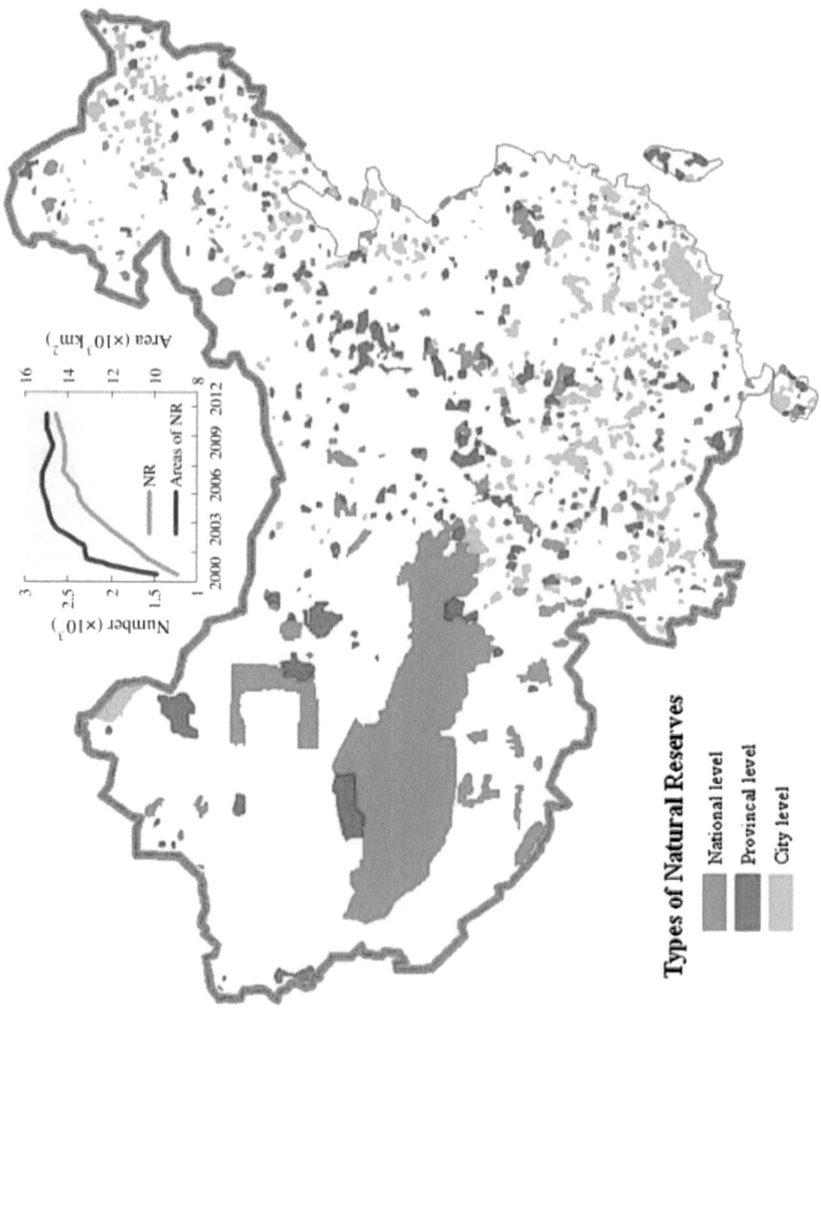

Types of Nature Reserves

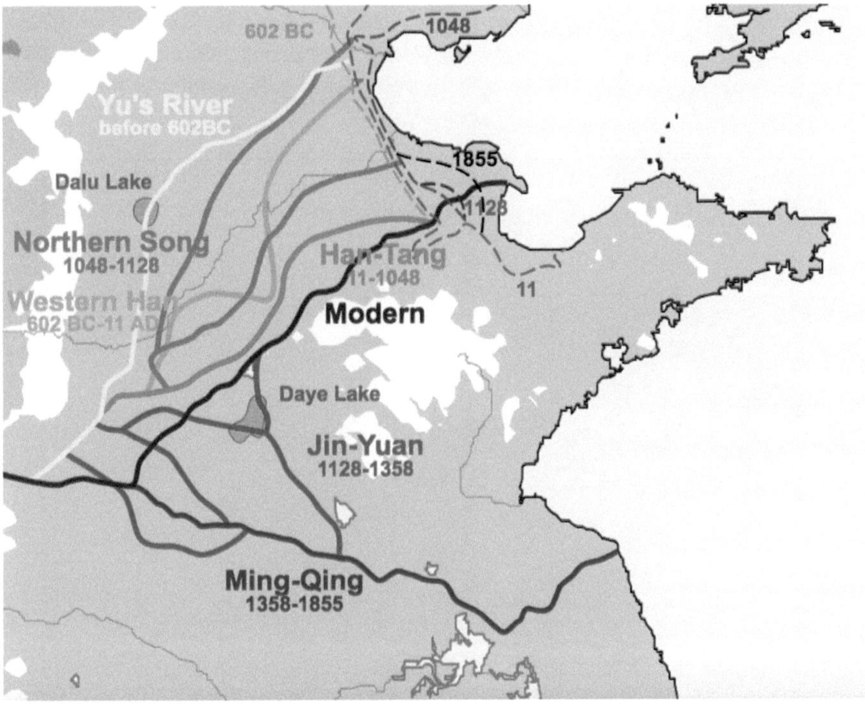

History of Course Changes Yellow River

Soil and Wind Erosion

Chronology

770–221 BCE Major deforestation begins to afflict the North China Plain.

200 BCE Persistent warfare induces desertification.

206 BCE–220 CE The Han Dynasty establishes dike construction and a maintenance regime to control periodic floods on the Yellow River brought about by severe deforestation along with major irrigation projects to support agriculture. The green tree belt is planted along the Great Wall. The Chinese population reaches 65 million.

605–610 The Grand Canal, linking the Yellow and Yangzi rivers between cities of Beijing and Hangzhou, Zhejiang province, is constructed.

618–907 The Tang Dynasty extends its empire into the far western regions of China and modern-day Sichuan province, with the total population reaching 90 million as colonization of southern China brings deforestation and soil erosion of hilly areas, notably subtropical zones.

960–1279 The Song Dynasty expands its production of basic industries, including copper and iron smelting, as a serious shortage of wood due to extensive deforestation leads to a shift to coal as the primary fuel source and the loss of forest cover increases the siltation of major rivers. The Chinese population reaches 120 million as land being cultivated totals 44 million hectares (109 million acres), 5.50 acres per capita.

1271–1368 The Yuan Dynasty oversees the rebuilding of the Grand Canal. A human-created hydraulic system of canals and dikes fundamentally alters the entire drainage system on the North China Plain. Massive amounts of timber and bamboo are used in the construction of the Grand Canal and the new capital in Beijing, with its elaborate palaces, exacerbating deforestation.

1368–1644 The Ming Dynasty extends its empire into the southwest as mining of silver and cinnabar leads to widespread pollution of rivers by mercury and other toxins. The Chinese population reaches 160 million.

1477–1485 Extensive flooding hits South China for nine consecutive years.

1642 Extreme man-made flooding of the Yellow River by siltation buildup results in the death of 300,000 people in the city of Kaifeng, Henan province, as the pace of deforestation accelerates in the Yellow and Xiang river valleys.

1644 Widespread and highly destructive drought leads to the collapse of the Ming Dynasty.

1644–1912 The Qing Dynasty, China's last imperial dynasty, comes to power.

1668 Emperor Kang Xi prohibits logging, mining, hunting, and fishing in the northeast (Manchuria). Widespread deforestation takes place in the Qinling Mountain range in southern Shaanxi by settlers.

1700–1720 Severe drought afflicts the North China Plain.

1798 *An Essay on the Principle of Population* by Thomas Malthus is published.

1850–1870 The Chinese population reaches 436 million on eve of the highly destructive Taiping Rebellion (1850–1864), while serious dike breaks along the Yellow River (1855) lead to massive flooding and a radical change in the river's course from south to north. Per capita land under cultivation drops to 2.71 acres, the smallest in Chinese history. Colonization of Inner Mongolia leads to the desertification of fragile grasslands.

1876–1879 A three-year drought results in 13 million deaths.

1886–1887 A massive Yellow River flood leads to 3 million deaths.

1896 Wilderness areas, especially in northeast, are opened to exploitation of timber and other natural resources by colonial powers, notably Russia.

1912 The Republic of China established with widespread plans for development, notably a large dam across the Three Gorges on the Yangzi River.

1915–1916 Severe flooding takes place in Central China. A state-run system of nature reserves is established in Tsarist Russia.

1917 The Bolshevik Revolution in Russia embraces science and technology.

1920s Foreign expeditions into China search for panda bears to capture or kill. Conservation measures are adopted in the Soviet Union during the liberal period of New Economic Policy (1921–1927).

1920 The first state-funded nature reserve for scientific research is established by Soviet leader Vladimir Lenin.

1921 The Chinese Communist Party (CCP) is formally established in Shanghai. Drought and famine hit North China.

1924 The All-Union Organization for the Preservation of Nature is established in the Soviet Union.

1928 Emergent Soviet leader Josef Stalin announces the "Great Break" with the Russian past involving rapid industrialization and the collectivization of agriculture, leading to severe environmental degradation.

1929 The first All-Russian Congress for Conservation is held in Moscow.

1931 Floods in Central China along the Yangzi and Huai rivers cause massive loss of life and physical destruction.

1938 Nationalist forces destroy Yellow River dikes to halt the Japanese military advance, causing severe flooding in Henan province and large-scale human fatalities.

1940 CCP chairman Mao Zedong declares that man "must conquer and change nature."

1945 Mao Zedong authors "How the Foolish Old Man Removed the Mountains." "High Stalinism" of Communist Party–led economic development is inaugurated in the Soviet Union.

1948 The Stalinist Plan for the Transformation of Nature is promulgated in the Soviet Union. DDT is invented.

PERIOD OF POLITICAL CONSOLIDATION AND CRASH INDUSTRIALIZATION, 1949–1957

1949 1 October: The People's Republic of China (PRC) is formally established.

1950 The population in China reaches 552 million, with 38 million people in the previously uninhabited northeast. The policy of "environmental hygiene" is enacted to improve medical care and provide clean water to cities. Measures on Protecting Rare Wildlife Animals is enacted.

1952–1953 Major flooding of the Yellow River occurs on the North China Plain. A plan to establish forest shelter belts in Northeast China is announced.

1953 The first national census is conducted as China's population reaches 600 million. The First Five-Year Economic Plan calls for recycling industrial wastewater but with no mention of the environment.

1955 July: The Plan for Permanently Controlling the Yellow River and Exploiting Its Water Resources is enacted. The first National Forest Conference is held, with the announced goal of reforesting 100 million hectares (161 million acres), as the planting of commercial rubber trees begins in the rainforests of Xishuangbanna, Yunnan province.

1956 The Plan for Natural Forest Nature Reserve Construction is enacted. The Regulation of the Protection of Mineral Resources bans the pollution of waterways by mining operations, while the Ministry of Forestry bans logging in selected natural forests. The Regulations on Factory Safety call for the proper disposal of waste materials by factories, with limitations imposed on the construction of new factories near residential areas and in water protection zones. The first national pollution standards are issued. **May:** The Minamata incident in Japan, involving mercury poisoning of water resources by industrial pollution, leads to 900 deaths.

1957 The National People's Congress (NPC) supports the establishment of forest reserves. The Provisional Program on Water and Soil Protection is enacted, largely aimed at disaster reduction. **April:** Construction of the Sanmenxia Dam begins on the middle reaches of the Yellow River.

PERIOD OF THE GREAT LEAP FORWARD, THE CULTURAL REVOLUTION, AND MASSIVE ENVIRONMENTAL DEGRADATION, 1958–1972

1958–1960 The Great Leap Forward leads to massive construction of dams, reservoirs, and ill-conceived irrigation projects, causing serious environmental degradation involving alkalization, salinization, and soil erosion, along with extensive deforestation. The giant panda bear and golden-hair monkey are designated as officially protected wildlife. The National Conference on Desert Reclamation is held in Beijing.

1959 A major insect infestation occurs as fallout from the Four Pests campaign. The Great Famine breaks out, extending into 1961. A major oil field is discovered in Daqing, Heilongjiang province.

1960 The Changbai Mountain Conservation Park is established in Jilin province.

1961 Drought afflicts North China, with major floods in South China, exacerbating the Great Famine, engendered by the Great Leap Forward.

1962 The State Council urges provinces in China to protect and reasonably use wildlife resources with the Directive Concerning the Active Protection and Rational Use of Wildlife and Natural Resources. The national pollution standards are revised as regulations from the pre–Great Leap era are revived. *Silent Spring* by Rachel Carson is published.

1963 The first giant panda bear reserve is established as proposals on nature conservation are presented to the NPC.

1964 The Third Front of industrial production for rearguard defense in the event of war is inaugurated. The "Taking Grain as the Key Link" campaign begins. Regulations on Forest Conservation are enacted as wildlife protection guidelines are issued. The Second national census puts China's population at 700 million. **February:** The campaign to "Learn from Dazhai" is inaugurated in agriculture.

1966 The Cultural Revolution begins, with environmental policy subordinated to political goals as political infighting leads to widespread neglect of the environment. Youth "sent down" to the countryside engage in environmentally destructive practices of land reclamation and extensive cuttings of forests.

1968 *The Population Bomb* by Paul R. Ehrlich published.

1971 The Leading Group for Environmental Protection is established informally under the State Council as the installation of air pollution monitoring systems throughout the country begins. A birth planning program is initiated by Premier Zhou Enlai, aimed at restricting population growth. The PRC is admitted to the United Nations.

1972 *The Limits to Growth* published by the Club of Rome postulates effect of pollution, population growth, and consumption of non-renewable natural resources on economic development. **Spring:** Major dry-up of the Yellow River indicates severe water stress on the North China Plain. The first environmental research institute is established in Heilongjiang province. **April:** A conference on dust stack removal is held in Shanghai. **June:** Premier Zhou Enlai, after receiving reports on widespread environmental pollution throughout China, authorizes a delegation to the United Nations Conference on the Human Environment in Stockholm, Sweden, headed by Qu Geiping, a leading advocate of strong environmental protection policies.

DEVELOPMENT OF INSTITUTIONAL AND LEGAL FRAMEWORK OF ENVIRONMENTAL PROTECTION, 1973–1993

1973 A conference on air emissions control is held in Shanghai. **June:** The First National Conference on Environmental Protection is convened, while the Rules on Protecting and Improving the Environment are issued by the State Council as the second national standards of pollution. Provisional Regulations on Nature Reserves are issued. The Leading Group on Environmental Protection set up under State Council as a temporary body. The "Three Wastes" campaign is inaugurated, targeting wastewater, solid waste, and waste gasses.

1974 Testing of noise levels begins in major cities and industrial plants. **January:** The Provisional Regulations on the Prevention of Pollution of Coastal Waters are enacted. **May:** A major pollution disaster hits Huai River in Henan province and other areas. The Leading Group on Environmental Protection, under the State Council, is formalized, as the Environmental Protection Office is established. The goal to control pollution within five years and eliminate it within 10 years is announced. Qu Geping is appointed as a delegate to United Nations Environment Programme (UNEP). The State Council authorizes a major media campaign to promote environmental protection. The Desert Conservation Park is set up in Minqin County, Gansu province, on the edge of the Badain Jaran Desert.

1975 The Wolong National Nature Reserve is established in Sichuan province as a protected area for giant panda bears. **May:** The 10-Year Plan for Environmental Protection is announced.

1976 The Chinese population reaches 950 million. Large numbers of giant panda bears die in the midst of bitter winter cold in Gansu province.

1978 The state constitution is amended to include a commitment to environmental protection. The inauguration of economic reforms leads to increasingly severe problems of air and water pollution and massive cutting of forests for the booming commercial timber industry. The National Environmental Protection Research Conference is held in Taiyuan, Shanxi province. Political decision-making authority and financial powers revert back to local governments as regional comprehensive pollution prevention is given priority. The first phase of the Three-North Shelter Forest Program ("Great Green Wall") of tree planting and reforestation is inaugurated, extending to 1985, aimed at controlling desertification, primarily in the northwest. China seeks aid from the World Bank for environmental protection projects.

1979 The Trial Environmental Protection Law is enacted. The National Conference on Forestry is held. The first Forestry Law is adopted. The creation of nature reserves accelerates. The One-Child Policy is enacted. The local Environmental Protection Offices are elevated to the level of bureaus. The Pollution Discharge Fee System is enacted. The Chinese Environmental Sciences Association is founded. China joins the United Nations Educational, Scientific, and Cultural Organization's (UNESCO) Man and the Biosphere Program. **March:** China declares 12 March as National Tree-Planting Day. The first conference on the environmental noise problem is held in Guangzhou.

1980 *World Conservation Strategy* jointly published by International Union for Conservation of Nature (IUCN), United Nations Environmental Program (UNEP), and World Wide Fund for Nature (WWF). **May:** China and World Wildlife Fund (WWF) reach an agreement on a joint project to protect giant panda bears. The Sixth Five-Year Economic Plan (1981–1985) includes environmental protection. **October:** The Society for Land Studies is established in Beijing.

1981 Sichuan province is hit by major floods linked to massive deforestation, leading to a national compulsory campaign of reforestation. **January:** China ratifies the 1973 Convention on International Trade in Endangered Species of Wild Fauna and Flora (CITES).

1982 The third national census puts the population of China at more than 1 billion. The Marine Environmental Protection Law is enacted. The state constitution is amended to expand the commitment to prevent and control pollution. Large-scale assessment of water resources in China is carried out by Ministry of Water Resources. A national monitoring network of air pollution is gradually installed as air quality standards are adopted.

1983 Environmental protection, along with family planning, is listed as a top national policy priority by the State Council. The Administrative Order on Wildlife Protection is issued. The Wildlife Conservation Organization is established. The Ministry of Urban and Rural Construction and Environmental Protection is created, consolidating existing environmental protection units into one ministry. The Chinese government bans use of DDT and BHC pesticides. Tough pollution controls are applied to newly constructed industrial facilities as local authorities issue a licensing system for certain industries. The Ministry of Forestry convenes the National Conference on Nature Reserves.

1984 A new Forestry Law bans clear-cut logging. The Water Pollution Prevention and Control Law is enacted. Sulfur dioxide is declared a major threat to the natural environment by an international coalition of 40 nations. **Janu-**

ary: The Second National Conference on Environmental Protection is held. **May:** The State Council announces the creation of the Environmental Protection Commission as an interorganizational body. **December:** The Environmental Protection Bureau (EPB) becomes the National Environmental Protection Bureau (NEPB), increasing authority over enforcement.

1985 The Grassland Law and Rangeland Law are enacted. China becomes a signatory to the London Convention on Prevention of Ocean Pollution by Dumping Waste and Other Matter. Mikhail Gorbachev assumes power in the Soviet Union. The first national survey of industrial pollution is carried out. The Temporary Article on the Management of Science and Historical Areas is enacted. **September:** The China–World Wildlife Fund Giant Panda Survey begins. **December:** China signs UN Convention Concerning the Protection of the World Cultural and Natural Heritage, opening up the country to establishment of World Heritage Sites.

1986 The Land Administration Law, Fisheries Law, and Mineral Resources Law are enacted. The second phase of Three-North Shelter Program ("Great Green Wall") begins, expected to extend to 1995. A national survey of water resources is conducted. Environmental protection bureaus are established in the majority of Chinese provinces. **April:** The Chernobyl nuclear disaster occurs in the Soviet Union.

1987 A national survey of water pollution is conducted. The National Climate Center is established. **March:** The concept of sustainable development is introduced by the United Nations World Commission on Environment and Development (WCED), known as the Brundtland Report, with contributions from Chinese experts. China enacts the Law on the Prevention and Control of Atmospheric Pollution. **May–June:** A massive forest fire breaks out in the Greater Khingan Mountains on the border of Heilongjiang and Inner Mongolia, lasting for weeks and destroying 700,000 hectares (1.28 million acres) of forest land, with 200 human fatalities.

1988 The Water Law and Standardization Law are enacted. **March:** The NEPB is granted independent regulatory authority. China World Water Week inaugurated to raise public awareness over water resources. **November:** The Wildlife Protection Law is enacted. **December**: The United Nations Intergovernmental Panel on Climate Change (IPCC) is created.

1989 The Inspector of Environmental Protection is created for conducting independent investigations into cases of pollution. A climate change research program is organized. The concept of "environmental diplomacy" is promulgated. **January:** A list of rare and endangered species in China is issued. **February:** A major toxic plume in the Huai River affects water supplies downstream in Anhui and Jiangsu provinces. **May:** At the Third National

Conference on Environmental Protection, major failings are acknowledged in the implementation of environmental laws and regulations. **December:** The new Environmental Protection Law, enacted as an "environmental responsibility system," becomes official government policy, with a major commitment to protecting public health.

1990 The Regulations on Control and Treatment of Pollutants is enacted. The Administrative Litigation Law is enacted, allowing lawsuits against government agencies. The fourth national census puts the Chinese population at 1.13 billion. International ban on ivory trade enacted. **December:** China increases its cooperation with international bodies to address global pollution.

1991 China ratifies the 1987 Montreal Protocol on Substances That Deplete the Ozone Layer. The Water and Soil Conservation Law is enacted. The Coastal Protective Forest Project begins. **March:** United Nations adopts 22 March as World Water Day. **June:** The Ministerial Conference of Developing Countries on Environment and Development is held in Beijing. **August:** Massive floods on the Yangzi and Huai rivers in Central China and the Liao and Songhua rivers in the Northeast afflict 230 million people in 24 provinces. **December:** China ratifies the 1989 Basel Convention on the Control of Transboundary Movement of Hazardous Wastes and Their Disposal.

1992 January–February: During a trip to Shenzhen Special Economic Zone (SEZ) in Southern China, Deng Xiaoping supports economic reforms leading to the creation of township and village enterprises (TVEs), many heavily polluting. **May–July:** The dry-up of the lower reaches of Yellow River from overuse leads to substantial reductions in both agriculture and industry. **June:** The United Nations Conference on Environment and Development, held in Rio de Janeiro, Brazil, issues the United Nations Framework Convention on Climate Change and Convention on Biodiversity, with the PRC expressing reluctance to accept the concept of sustainable development.

1993 Chinese textile exports are rejected by the United States for containing heavy metals and other pollutants. China bans the use of body parts of tigers and rhinoceros in traditional Chinese medicine (TCM). Major marine environmental protection regulations are enacted. **March:** The Natural Resources and Environmental Protection Committee formed in the NPC. **December:** China ratifies the United Nations Framework Convention on Climate Change.

THE EMERGENCE OF THE ENVIRONMENT MOVEMENT AND STATE PROMOTION OF SUSTAINABLE DEVELOPMENT, 1994–2019

1994 The requirement to maintain cropland leads to large-scale reclamation projects to offset losses of arable lands to commercial and industrial development. The Huai River Pollution Control Plan is announced, along with the Regulations on Nature Reserves. Loess Plateau Watershed Rehabilitation Project of ecological restoration inaugurated. **March:** Friends of Nature (FON) is registered by Liang Congjie as China's first environmental nongovernmental organization (ENGO). The PRC Agenda 21 document, based on the United Nations "Earth Summit I," is issued. **June:** The Action Plan for Biological Diversity Protection is issued. **July:** A major pollution crisis occurs along Huai River, as the waters turn black from industrial effluents and agricultural run-off, killing millions of fish and causing major illnesses among the local population. The fourth National Conference on the Environment is held.

1995 Chinese photographer Xi Zhinong documents activities of threatened golden snub-nosed monkey in Yunnan province. *Who Will Feed China?* by Lester Brown published. National Environmental Protection Agency (NEPA) joins People's Bank of China (PBOC) in denying credit to Chinese companies violating environmental regulations. The Prevention and Control of Environmental Pollution by Solid Waste Law is enacted. Electric Power Law establishes preference for clean and renewable fuels. Regions and cities ordered by State Council to draw up pollution control plans and institute a pollution-discharge permit system.

1996 State Council issues mandate for all industrial enterprises in China to abide by national emission standards or confront closure as air quality standards expanded. Hundreds of polluting factories along the Huai River shut down. American and European ships carrying solid waste to China turned away. Noise Pollution Law enacted. Platform for Action on National Environmental Publicity and Education issued along with 21st Century Ocean Agenda of China. National Irrigation Agricultural Water-Saving Program inaugurated with 300 counties selected for pilot water-saving projects. **July:** The Fourth National Conference on Environmental Protection promotes the concept of sustainable development as President Jiang Zemin emphasizes environmental protection as a "core policy" of the PRC.

1997 The Yellow River dries up for 236 days, reaching 400 miles inland as President Jiang Zemin announces policy of improvement and beautification of rivers in the country's northwest. The Energy Conservation Law is enacted. China adopts the Ninth Five-Year Plan for Environmental Protection and Distant Targets for 2010. **March:** National Conferences on Environmental Protection are set to be held annually through 2007. **July:** Sixty-seven thousand small, heavily polluting factories and facilities are closed. **October:** The Revised Criminal Law is enacted, containing a provision dealing with the "sabotage of the protection of environment and resources."

1998 Large-scale flooding on the Yangzi River traced to massive deforestation leads to a ban on logging as the Natural Forest Conservation Program (NFCP) is enacted. Beijing prohibits the burning of leaded gasoline. The State Environmental Protection Commission (SEPC) is abolished, as the Leading Group on Environmental Protection is raised to the ministerial level. Government cadres are directed to avoid leadership of nongovernmental organizations (NGOs). Zhu Rongji is appointed premier and moves to make environmental protection an important national priority as the preservation of national biodiversity and coral reefs in the South China Sea intensifies. **January:** The Zero-Hour Operation is announced to clean up the Huai River. **March:** The State Environmental Protection Administration (SEPA) is established in the midst of a major reorganization of central government ministries. **May:** China ascribes to the Kyoto Protocol on global climate change.

1999 The Center for Legal Assistance to Pollution Victims is established by Wang Cangfa as "ruling the country by law" is incorporated into the state constitution. China launches the "Open to the West" campaign to reduce the wealth gap between poor western and wealthy eastern regions of the country, threatening environmental damage in the former. A ban is imposed on Styrofoam lunch boxes following a visit by Premier Zhu Rongji to the Three Gorges Dam reservoir, where power generation stations were blocked by a massive accumulation of the boxes and other debris. The New Wildlife Protection Law is enacted, while harvesting of coral reefs is outlawed. **March:** President Jiang Zemin proclaims companies failing to comply with environmental protection laws and regulations will face closure.

2000 China announces a major campaign to prevent and control desertification. The most severe drought in 50 years causes a major decline in grain production, extending into 2001. The fifth national census puts the Chinese population at 1.2 billion. Conservation Action Plan adopted along with nationwide ban on logging in protected areas as depletion of forest cover accelerates through 2015. Major clean-up of lakes and construction of sewage facilities inaugurated. **January:** China adopts first set of vehicle fuel and emissions standards. **March:** China commits to spending 1.2 percent of gross domestic product (GDP) to environmental protection and restoration.

2001 China is accepted as a member of World Trade Organization (WTO). An "ecological migration" policy is adopted for Inner Mongolia. The World Bank issues a report entitled "China: Air, Land, and Water." The 10th Five-

Year Plan for Environmental Protection (2001–2005) is adopted. **July:** Major floods occur along the Huai River as waters filled with garbage, yellow foam, and dead fish inundate Anhui province.

2002 China ratifies the 1997 Kyoto Protocol to the United Nations Framework Convention on Climate Change. The Water Law is amended. The South-to-North Water Diversion Project is approved. The Sand Control and Management Law is enacted. President Jiang Zemin endorses sustainable development for "bettering the natural environment." **March:** CCP constitution commits to "establishment of ecological civilization." **June:** Cleaner Production Promotion Law adopted. **August:** The World Summit on Sustainable Development is held in Johannesburg, South Africa.

2003 National Wetland Survey conducted as Huai River experiences major flooding. China accounts for seven out of ten most polluted cities in the world. SARS virus spreads to south China. Confrontation between villagers in Qiugang, Anhui province and local officials over dumping of chemical wastes into local rivers. **March:** Hu Jintao appointed president and Wen Jiabao premier. **July:** Major rock slide in area of Three Gorges Dam. **September:** Environmental Impact Assessment Law implemented. **December:** Gas explosion near city of Chongqing kills 198 people.

2004 China embraces the concept of sustainable development with an emphasis on controlling emissions of air pollutants, protecting water quality, promoting "green" development, and establishing international cooperation on the environment. The green gross domestic product (GGDP) is adopted as a measure of the environmental costs of economic growth. The PRC ratifies the Stockholm Convention on Persistent Organic Pollutants. **22 April:** As many as 100,000 Chinese students take part in environmental activities on international Earth Day.

2005 China is cited as the world's most polluted country as pollution-related social protests number 51,000 nationwide. The minister of environmental protection, Pan Yu, authorizes a "storm of environmental assessment." China signs the Cartagena Protocol on Biosafety dealing with genetically modified foods. The 11th Five-Year Economic Plan (2006–2011) calls for building low-carbon-intensity industries. The National Flood Management Strategy is issued by the Ministry of Water Resources (MWR). **November:** A major chemical spill of benzene into the Songhua River occurs in northeastern Jilin province. **December:** Environmental protection is given top priority in highly polluted regions of China with concomitant restraints imposed on economic development.

2006 The Three Gorges Dam on the Yangzi River is completed. Citizens in major Chinese cities launch the "Bring Your Own Chopsticks" (BYOC) campaign in an effort to cut down on the use of disposable chopsticks in local restaurants. The Yangzi River dolphin (*baiji*) is declared extinct. An online public database tracking air pollution in major urban centers is inaugurated by environmental activist Ma Jun. The Wetland Conservation Program is enacted. **January:** The Renewable Energy Law is enacted. **April:** Beijing is hit with largest sandstorm since 2001. **August:** The Guidelines on Establishing the Green Finance System are issued. **October:** Sections of the Yangzi River turn red from industrial effluents and algae plumes. **November:** The First National Assessment Report on Climate Change is conducted. The Agricultural Food Safety Law is enacted.

2007 China overtakes the United States as the world's largest emitter of greenhouse gases (GHG) and issues the first national plan on climate change. An international study indicates the prospective devastating effects of climate change for coastal cities in East China. The Medium- and Long-Term Plan for Renewable Energy is issued. Environmental activist Wu Lihong is sentenced to prison on charges of corruption. Environmental courts are established by the Supreme People's Court. Annual reports on GGDP of the PRC are put on hold. The State Council announces that environmental conditions will be used in determining promotions and bonuses for provincial and local CCP and government officials. The Measures on Open Environmental Information are enacted. **May:** A blue algae outbreak occurs in Lake Tai, Jiangsu province. Social protests in Xiamen, Fujian province, bring a halt to plans for a paraxylene (PX) plant in the city. **June:** China gains Forest Stewardship Council (FSC) international certification for responsible management of forests. **October:** At the 17th CCP National Congress, President Hu Jintao makes his first major policy speech endorsing environmental protection. **December:** A United Nations Climate Conference held in Bali, Indonesia, considers a successor to the Kyoto Protocol.

2008 China buys huge cache of raw ivory from African nations as Chinese investors in the wake of the global financial crisis rush to buy large quantities of ivory and rhinoceros' horn as a hedge. **March:** SEPA is elevated to the level of Ministry of Environmental Protection (MEP). The Measures for the Disclosure of Environmental Information are enacted. **May:** A massive earthquake in Sichuan province kills 70,000 people. **August:** The Summer Olympics are held in Beijing, where authorities shut down factories and restrict transportation, ensuring relatively clear skies.

2009 Designated "Year One of the Garbage Era," government authorities and ENGOs begin to address the major issue of trash collection and garbage disposal. President Hu Jintao announces a plan to increase forest cover in

China by 40 million hectares (98 million acres) by 2020, as China conducts the National Forest Survey. **May:** Premier Wen Jiabo suspends construction of a hydropower station on the Nu River in Yunnan province. **December:** At the United Nations Climate Change Conference in Copenhagen, Denmark, a Chinese delegation temporary walks out over the exclusion of the PRC from major events as the conference fails to reach a consensus on a commitment to targeted goals of pollution amelioration.

2010 The first survey of national pollution sources is carried out, with the study identifying 459 "cancer villages" in China. The fifth national census puts the population in China at 1.3 billion. Regional Joint Prevention and Control of Air Pollution plan announced by State Council. China becomes largest automobile market in the world, surpassing the United States. **June:** The Ministry of Commerce (MOC) calls for restrictions on the production of disposable chopsticks to reduce deforestation. **July:** A pipeline explosion in the northeast city of Dalian produces a major oil spill into the Yellow Sea, threatening wildlife and fisheries. Tougher standards are issued on the release of pharmaceutical pollutants into waterways.

2011 The UNEP issues a report on sustainability and equity. **March:** The 12th Five-Year Economic Plan (2011–2015) sets major environmental goals involving carbon intensity targets and pilot cap-and-trade programs. **April:** Several tons of chromium-contaminated carcinogen chemicals are dumped into reservoirs in Yunnan province. **May:** Mongol herdsmen are killed while protesting the pollution of traditional grazing grounds. **August:** The Chinese government announces a plan to protect remaining farmland from pollution and soil degradation. International pressure is brought against foreign shoemakers in China for the use and disposal of toxic materials. Social protests break out in Dalian, Liaoning province, against the proposed construction of a PX plant. **November:** The State Council issues the white paper entitled *China's Policies and Actions for Addressing Climate Change*. The China Council of International Cooperation on Environment and Development, involving Chinese and international experts, recommends an expenditure of RMB 5.7 trillion ($863 billion) to transition away from polluting and energy-intensive industries. At the United Nations Climate Change Conference in Durban, South Africa, China pressures developed countries to do more to confront the problem. **December:** Protesters in Haimen, Guangdong province, block roadways in opposition to the proposed construction of a coal-fired power plant. The Second National Assessment Report on Climate Change is conducted by the Chinese government.

2012 A campaign against toxic pollutants is expanded to foreign textile firms operating in China, one of which is Levi Jeans. China issues the National Report on Sustainable Development and enacts the National Ambient Air

Quality Standard. The Civil Procedure Law is amended to allow environmental public interest lawsuits. **November:** Xi Jinping is appointed general secretary of the CCP.

2013 The Asian Development Bank (ADB) finds that 99 percent of China's largest cities do not meet the minimum air quality standards set by World Health Organization (WHO). **January:** An "Airpocalyspe" in Beijing and other northern cities produces air pollution levels 40 times greater than the safe level established by the WHO. The existence of "cancer villages" in China is officially recognized by MEP. China lifts the ban on the production of Styrofoam lunch boxes, which had been widely ignored. **March:** Xi Jinping is selected as president of the PRC, with a commitment to give greater priority to environmental protection. Newly appointed premier Li Keqiang calls for more transparency in government and increased public supervision to improve environmental compliance. Ten thousand dead pig carcasses flow down the Huangpu River into Shanghai. United Nations establishes 3 March as World Wildlife Day. **May:** The southern city of Guangzhou is hit with major flooding as Beijing, Shanghai, and Guangzhou are declared "barely suitable for living." The Guangzhou government reveals that 40 percent of rice samples sold in the city contain cadmium from contaminated soils. **July:** Workshop held on plans for carbon-capture projects in China with the international Global Carbon Capture and Storage Institute. **September:** The Ten Specific Measures of Air Pollution Prevention and Control Action Plan is issued. **October:** China Coal Cap Project announced calling for reduction of annual coal consumption to 3.5 billion tons by 2020.

2014 The PRC participates in the international Census of Marine Life. Second national wetlands survey conducted while Chinese government extends logging ban to forests in the northeast. Central canal of the South-to-North Water Diversion Project completed. **January:** Central Committee Document Number One calls for "reform" in agriculture with emphasis on promoting food security and sustainable agriculture. **March:** Premier Li Keqiang declares a "war on pollution" as the 2030 Agenda for Sustainable Development is announced. **April:** A new Environmental Protection Law is enacted. **December:** Interim Regulations on Administration of Carbon Emissions Trading issued.

2015 China issues a Master Plan for the reform of government institutions and systems to build "ecological civilization," notably an Action Plan on Water Pollution as Paris Climate Accords reached including PRC. Solid Waste Pollution and Control Law enacted. **November:** The Deep Reform Plan for Ecological Civilization is announced. **December:** A mountain of discarded construction debris and trash collapses in the southern city of Shenzhen, killing 69 people.

2016 The Action Plan on Soil Quality is issued. Highly polluting steel mills, coal mines, and aluminum smelters are shut down with ecological "red lines" established to restrict environmentally destructive development. MEP sets up a Central Environmental Inspection System to monitor the local enforcement of environmental laws. The new Wildlife Protection Law is enacted. **October:** "EU-China Cooperation on Emissions Trading in China: Achievements and Lessons" published. **November:** A United Nations environmental report calls for a further cut of 25 percent in GHG emissions from predicted 2030 levels to limit global warming. The Paris Climate Accords go into force.

2017 Global atmospheric carbon levels reach 445 parts per million (PPM), the highest recorded in human history. Evidence of microplastics consisting of three polymers have been found in the atmosphere in Dongguan city, Guangdong province.**June:** The United States announces its withdrawal from Paris Climate Accords. China adopts the National Climate Change Program, with the goal of reducing carbon intensity per unit of gross domestic product (GDP) by 40 to 45 percent. **July:** A proposal to ban the import of foreign waste into China for recycling is announced. **October:** At 19th CCP National Congress, General Secretary Xi Jinping commits to environmental protection through the "Beautiful China" initiative. Tens of thousands of heavily polluting factories shut down in a major government push for stronger environmental protection. **December:** China inaugurates a major carbon emissions trading plan for domestic companies. China imposes total ban on domestic ivory carving factories and retailers.

2018 January: China officially bans the import of foreign waste and scrap for recycling, with an estimated market value of $18 billion annually. A revised Environmental Protection Law is enacted, including an environmental tax on offending companies. The reclamation of coastal lands for business-related activity is banned. China imposes ban on all ivory imports. **Spring:** A study estimates that flooding in China during a period spanning from 2016 to 2035 will impose economic costs of as much as $400 billion. Former Premier Wen Jiabao says water shortage threatens very survival of Chinese nation. **March:** A major reorganization of government structure is implemented, one facet of which is the formation of the Ministry of Ecology and Environment (MEE) and the Ministry of Natural Resources (MNR). Poverty reduction and pollution control are listed as top national priorities by Premier Li Keqiang. Beijing hit with massive sandstorm driving air quality index to "very unhealthy" levels. **June:** The International Carbon Dioxide Conference is held in China. **July:** A chemical plant explosion in Sichuan province, southeast of the provincial capital Chengdu, kills 19. **August:** The Symposium on Environmental Governance in China is held, stressing law-based measures for environmental protection. Glacial lake bursts occur in Xinjiang province from rising average water temperatures of two to three degrees. **September:** The PRC participates in the United Nations Conference on Marine Biological Diversity. **October:** A giant landslide of rock and debris into the Jinsha River, on the upper reaches of the Yangzi River in Tibet, creates a barrier lake. The northern city of Zhangjiakou, Hebei province, a major site of the upcoming 2022 Winter Olympics, is rocked by the explosion of a vehicle carrying methanol outside a chemical plant, resulting in 22 deaths. **November:** Following international outcry, China reverses decision to allow import of rhino horn and tiger bone.

2019 January: Carbon dioxide in earth's atmosphere measured at 409.92 parts per million (ppm), up 2.87 ppm over previous year. **February:** Hainan province adopts total ban on all plastics. **March:** An explosion at a chemical plant producing highly flammable anisole in Yancheng City, Jiangsu province, kills 78 people and contaminates local waterways. Major forest fires break out in drought-afflicted regions of Shanxi and Sichuan provinces, killing civilians and local firefighters in the latter. A dead whale found in the Philippines has its stomach full of plastics. State Council authorizes "zero waste" pilot program with ten cities selected as national models. **April:** International Horticultural Exhibition convenes in Beijing. China launches two earth observation satellites for monitoring air pollution with high spectral resolution. **May:** United Nations announces one million global species threatened with extinction largely human-induced while China is selected as co-chair for a UN Conference on Climate Change slated for September. **June:** Global Plastics Summit convened. At G-20 meeting in Osaka, Japan, China agrees to "scale up efforts to tackle climate and biodiversity crisis." **5 June:** World Environment Day organized under auspices of the United Nations celebrated in Hangzhou, Zhejiang province. **July:** Southern China struck by torrential rains and massive floods, the most severe in half a century, destroying homes, flooding farmland, and causing large-scale evacuations. **August:** Record high temperatures in the Artic, Greenland, and Alaska lead to massive glacier and ice sheet melts (10 billion tons in one day in Greenland) while wildfires in the Amazon and Siberia destroy millions of acres of forest land and increase CO_2 releases into the atmosphere with major outbreaks of algae blooms occurring off Denmark and Mexico. United Nations warns climate change could threaten world's food supply as July 2019 is earth's hottest month on record.

Introduction

The history of the environment in the People's Republic of China (PRC) is a history of two, diametrically opposed eras, the first of enormous environmental destruction and degradation beginning in 1953 and extending into the 1990s, and the second of a national commitment to environmental protection and restoration from the mid-1990s to the present (2019). Inheriting a country already suffering serious and widespread environmental degradation from hundreds of years of extensive agriculture into marginal, semiarid lands and overexploitation of water and other limited natural resources, the new Chinese government established in 1949 adopted a system of central economic planning from the Soviet Union, itself a serial environmental abuser, embarking on a strategy of rapid industrialization and agricultural collectivization with little or no regard for the natural environment. Emphasizing such heavy-polluting and resource-guzzling economic sectors as steel and coal production, and imposing agricultural collectivization on a small-scale peasant economy, deleterious environmental effects ensued, including air and water pollution, and soil erosion and contamination, along with highly inefficient and destructive usage of the country's natural resources.

DECLARING "WAR ON NATURE"

Exacerbated by radical campaigns of breakneck development provoked by Chinese Communist Party (CCP) chairman Mao Zedong (1949–1976), the country pursued economically irrational policies in both the industrial and agricultural sectors, including the notorious "backyard steel furnaces" (80,000 in number) and agricultural practices like close planting and deep plowing, inherited from quack Soviet biologist Trofim Lysenko, which resulted in nationwide environmental devastation, especially to the country's already denuded forest land. Equally ill-informed were such policy initiatives as "Learn from Dazhai" and "Taking Grain as the Key Link" in agriculture, and the Third Front of industrial relocation to the country's interior as the Maoist era of environmental deterioration continued throughout the ensuing political battles during the Cultural Revolution (1966–1976), when the country's urban youth were "sent down" to the countryside, where adhering to Mao's proclamation of a "war on nature" wreaked havoc on local environments with wild-eyed notions of "increasing production" by reclaiming lands from pristine lakes and replacing old-growth forests with water-guzzling rubber trees. With Mao advocating the principle that "with many people, strength is great" and opposing any form of family planning, the Chinese population underwent explosive growth, expanding from 550 million people in 1950, to 950 million at the time of the chairman's death in 1976, creating serious problems of overpopulation, particularly in the eastern regions of the country, with major deleterious effects on the environment.

The passing of the political mantle in the PRC to Deng Xiaoping (1978–1992) as the country's paramount leader produced enormous changes in economic policy as industry and especially agriculture were liberalized and the previously autarkic country was "opened up" to foreign trade and investment. "Getting rich is glorious" was, however, no better than Maoist "making revolution" for the nation's environment as the explosion of small-scale township and village enterprises (TVEs) and the expansion of large-scale state-owned enterprises (SOEs), particularly in the chemical industry, along with shifts from organic to chemical fertilizers and pesticides by Chinese farmers, wreaked havoc on the nation's air, water, and soil quality. And while family planning was instituted in the form of the One-Child Policy in 1979, population growth continued to 1.4 billion in 2018, with continued negative effects on both the urban and rural environments. Transfer of power from the "old revolutionary" Deng Xiaoping, who, like his predecessor, was generally oblivious to environmental matters, to the engineer Jiang Zemin (1989–2002) led to increased attention to the deteriorating environment, most notably in the smog-infested cities, but with little in the form of major policy initiatives.

HU JINTAO AND "SCIENTIFIC DEVELOPMENT"

The emergence of the leadership of President Hu Jintao and Premier Wen Jiabao (2002–2012), both trained in science and engineering, and the proclamation of "scientific development" as the core of government policy led to major initiatives on the environment. While environmental matters had drawn the attention of former Premier Zhou Enlai, who sent the first PRC delegation to the United Nations Conference on the Human Environment in Stockholm, Sweden, in 1972, and the subsequent passage of Environmental Protection Laws in 1979 and 1989, the implementation of stringent environmental protection and restoration, especially at the local level, was lacking, with the Ministry of Environmental Protection (MEP) exercising weak enforcement powers. Even as environmentally questionable policies were still pursued, for example, the Three Gorges Dam, the Hu-Wen administration, invoking environmental protection and restoration as a "vital interest" to the nation and coining the term "ecological civilization," engaged a series of major initiatives, including the reforestation of seriously denuded mountains and hillsides, and the development of such "green" technologies as solar and wind power, while citing the importance of climate change for the nation and the entire planet. While resistance to stringent and dramatic environmental protection and restoration remained intense, chiefly at the lower levels of the frequently unwielding and fragmented Chinese bureaucracy, where economic growth remained supreme, environmental mitigation became a policy priority at the central level and in many key provincial and urban areas, for instance, heavily polluted Hebei province and such prominent but dirty cities as Beijing, Shanghai, and Shijiazhuang.

XI JINPING AND THE PROMISE OF A "BEAUTIFUL CHINA"

The year 2013 was a seminal moment for both the environment in China and government policy on environmental protection and restoration of a heavily polluted and degraded natural world. The year began with the outbreak in January of pervasive and long-lasting smog, known as the "Airpocalypse," threatening public health in major cities, notably the capital, Beijing, followed by major floods in the spring, along with official recognition of so-called "cancer villages" in highly polluted rural areas. Into this environmental maelstrom came Xi Jinping, selected as PRC president in March, with Li Keqiang as premier, both strong advocates of pursuing environmental protection on the national and international stages.

With the clarion call for a "war on pollution" declared by Premier Li in March 2014, major commitments were made to environmentally friendly policies, including sustainable development and the expansion of renewable energy, with a concomitant emphasis on breaking the country's addiction to highly polluting coal and environmentally destructive mining in pristine and minority-populated regions, especially in Tibet and Xinjiang. Concerned with the highly fragmented and overlapping jurisdictions plaguing the persistently ineffective environmental bureaucracy headed by MEP, major institutional changes were carried out, including the creation in 2018, of two superministries—the Ministry of Ecology and Environment (MEE) and the Ministry of Natural Resources (MNR)—with enhanced oversight and enforcement powers.

A series of action plans on air pollution and soil quality were also inaugurated, along with the establishment of an environmental inspection system for overseeing enforcement of the frequently ignored environmental statues, while major shutdowns were pursued of highly polluting industrial facilities, particularly in the chemical and paper pulp industries, which for decades had contaminated the country's rivers and lakes. After years of environmental offenders getting away with paying excessively low pollution fees rather than installing expensive pollution-control equipment, as mandated by law, higher taxes were imposed on pollution violators, who were also subjected to public-interest lawsuits by increasingly aggressive environmental protection lawyers and organizations. Along with legally required environmental impact assessments (2002) and the establishment of approximately 3,000 nationwide environmental enforcement bureaus, China has instituted a more integrated and organizationally functional regime devoted to creating, in President Xi's words, a "beautiful China" with cleaner air and water, and more productive soils, while tackling serious problems of advancing deserts and declining wildlife.

Advancing the national and international agenda of environmental protection and restoration is the emergence in the PRC of an active and increasingly vibrant civil society centered on the appearance of several thousand environmental nongovernmental organizations (ENGOs), which provide crucial scientific and policy expertise on the complex policy-making and policy-implementation process on environmental matters. With an increasingly investigative media and journalism looking into environmental problems and major ecological disasters like the poisoning of the Songhua River (2005) and the explosion of a chemical storage facility in Tianjin (2015), greater transparency of environmental data and government policy-making are being pursued in an increasingly open society, where debate and discussion have replaced state dicta and ideological orthodoxies that ignored and abused mother nature.

PROGRESS AND PERSISTENT PROBLEMS OF THE CHINESE ENVIRONMENT

In a country under the iron-clad political control of the CCP without viable opposition and a heavily state-controlled media and judiciary, central policies on environmental protection and restoration are broadly ignored, especially in poorer, backward, and largely inland regions, where economic development and the environment are still considered diametrical opposites. With 2020 as the target year for achieving real progress on multiple environmental fronts, the Xi–Li administration has witnessed some areas of real progress, principally in marginal improvements in urban air quality and cleaner waters, but with enormously problematic areas, especially involving the continued heavy reliance on coal burning in antiquated power plants, constituting 60 percent of national electric power production. In a country with huge coal reserves (estimated at 114 billion tons) but relatively sparse oil and natural gas reserves, renewable energy is a prime target for rapid development, which will undoubtedly require years of enhanced investments and technological advancements.

No less problematic is the serious contamination of the country's surface waters (rivers, lakes, and coastal areas) and groundwaters, upon which agriculture and many cities, notably Beijing, increasingly rely. With environmental damage and habitat destruction occurring in previously pristine and untouched areas, for example, the Qinghai–Tibetan Plateau and Xinjiang province, the Chinese government is confronted with serious dilemmas, with the two areas providing rich mineral and energy resources but with environmentally important features, like glaciers, now melting at accelerating rates, and such prized wildlife as the Tibetan antelope (*chiru*) threatened with extinction.

CHINA AND THE INTERNATIONAL ENVIRONMENTAL FRONT

For more than two decades, the PRC was an outlier in the development of global environmental policy, denouncing proposals for sustainable development as a "conspiracy" of developed countries, largely in the West, to maintain domination of the international economy at the expense of poorer countries like Brazil, India, and China. Following the PRC accession to the United Nations in 1971, and under the leadership of then-premier Zhou Enlai, Chinese delegations were sent to such international gatherings as the periodic United Nations conferences on the environment, posing as a defender of poorer nations but also embracing the need for international action on major environmental problems, for instance, ocean pollution, which transcended national borders. Signing a myriad of international agreements and joining international bodies focusing on specific environmental issues, for example, the preservation of coral reefs, China emerged in the 2010s as a major leading force on international environmental action, especially climate change, where the withdrawal of the United States from the Paris Climate Accords (2015) opened the door to PRC leadership in dealing with this compelling international environmental problem. If the climate, including rising global temperatures, is to be stabilized as a key requirement of maintaining global environmental health, the leader in this effort may well be the PRC.

A

ACID RAIN (*SUANYU*). A major problem afflicting both **air quality** and **water quality** in the People's Republic of China (PRC), acid rain is estimated to impact as much as 30 percent of the country's land mass, including more than 250 **cities**, many in the south and southwest, particularly in **Sichuan province**. The principal components of acid rain are sulfur dioxide (SO_2 [*eryanghua liu*]) and nitrogen oxide (NO_x [*yanghua dan*]) from anthropogenic activities, mainly the burning of **coal** and **oil** in factories and power plants, and **transportation**, including **motor vehicles** and ships. Released high into the atmosphere, SO_2 and NO_x bond with water, oxygen, and other **chemicals** and then fall to the ground as sulfuric acids and nitrate acids (*liusuan, xiaosuan*). Widespread areas of the country experience rainwater with pH readings below 5.0, making the PRC the third-largest country in the world suffering from acidic precipitation.

Half of the rainfall in the most affected regions is acidic wreaking havoc on croplands, **forests**, and vegetation while acid rain originating in China has also fallen on neighboring countries including North Korea, South Korea, and Japan. The combined deleterious effects in China on **soil quality** and water quality have reduced productivity in **agriculture** and freshwater **aquaculture** along with excessive and long-term corrosion of building exteriors and national historical monuments including ancient temples and statues, most notably the giant Leshan Buddha statue on the **Min River** in Sichuan province. Acid rain-associated **air pollution** in China has also contributed to growing **health** problems in the PRC including higher rates of lung and heart disease and childhood asthma, especially in the most severely-affected areas.

Acid rain was recognized as a major environmental problem in the late 1970s, when the PRC became the largest single polluter of SO_2 in the world. The problem was rooted in the policies of massive and rapid industrialization pursued during the period of central economic planning (1953–1978), based on the model of economic growth with little or no regard for the environment borrowed from the **Soviet Union**. Exacerbated by the **"war on nature"** promoted by **Chinese Communist Party (CCP)** chairman **Mao Zedong** (1949–1976), which intensified in the early years of economic reform introduced by **Deng Xiaoping** in 1978–1979, China engaged in a single-minded pursuit of economic growth generally unconcerned with damaging environmental effects, notably acid rain, which expanded from a few isolated pockets in South China in the mid-1980s to one-third of the entire country by the mid-1990s. Spewing as much as 25 million tons of SO_2 into the atmosphere in 2006, twice the level that is considered environmentally safe, and representing a 27 percent increase from 2000, with average annual increases of 9 percent, factories, particularly in the metallurgy (*yejin*), petrochemical (*shiyou huaxue*), and **energy** sectors, contributed an estimated 85 percent of China's total SO_2 emissions, with a small but growing percentage from transportation, especially automobiles. Nitrous oxide (N_2O) emissions also grew at accelerated rates, particularly from the excessive use of chemical **fertilizers** in agriculture, where acidification affected an estimated 2.7 million hectares (*gongqing*/6.5 million acres) of cropland.

Seven provinces in the south and southwest experienced the greatest acidification, chiefly Sichuan, where acid rain with pH values below 4.5 have reduced soil fertility and caused paddy rice to turn yellow, as precipitation in some areas was rated 100 percent acidic. Experiencing higher humidity and excessive burning of sulfur-rich coal, these regions are most heavily afflicted by acid rain, while the north, with more alkaline soil dust, along with a positive ion exchange rate (*zheng lizi jiaohuan lü*), which increase the level of absorption, is better able to deal with elevated levels of acids, despite its concentration of heavy **industry**. Agriculture nationwide suffers an average 4.3 percent production decline from acidic precipitation, which, combined with other environmental damage, generates a total economic cost, according to the Chinese government, of $13 billion, while the **World Bank** offers a much higher estimate of $35 billion. Among the major cities severely afflicted are **Guangzhou**, **Chongqing**, and Chengdu (Sichuan), in addition to cities long considered some of the nicest places to live in China, for example, Xiamen (Fujian) and Dalian (Liaoning).

The harmful effects of acid rain on the environment are multifaceted and difficult to reverse. For soil quality, acidification disrupts soil chemistry primarily by reducing levels of such crucial nutrients as potassium (*jia*), sodium (*na*), calcium (*gai*), and magnesium (*mei*), which are key in supporting **plant species**, while reactions with soil minerals like mercury (*gong*) and aluminum (*lü*) create harmful components that are absorbed into plants. Increased levels of acid rain are also considered a major factor in the growing frequency of **landslides** afflicting China, now the world leader, as rock layers are weakened and made spongy by microbes fertilized by acid rains, which throughout time can break down rock structures. Similar effects have afflicted forests, with leaves turning brown and bark suffering damage, producing significant defoliation and disabling the capacity for photosynthesis (*guanghe zuoyong*), as well as denying **habitat** to **wildlife**.

For China's many **rivers** and **lakes**, acid rain is equally destructive, creating heightened levels of algae plumes (*zaolei yumao*) and depriving the waters of flora and fauna (*zhiwuqun he dongwuqun*), on which **fisheries** and human livelihood depend. As for air quality in major urban areas, an increase in fine particulate matter (PM 2.5) produced by acid rain has contributed to an increasing frequency of the so-called "Airpocalypse," where grey, swampy air blankets entire cities for long periods, particularly during the winter months, making breathing open air especially harmful to human health, causing an estimated 18 percent of annual fatalities in the country.

Responding to this widespread and well-documented deterioration of the country's environment made possible by enhanced acidic precipitation monitoring sites in almost 700 cities and counties, along with popular **social protests** demanding action, the Chinese government has initiated increasingly aggressive policies aimed at remediation. Beginning most significantly with the 10th Five-Year Economic Plan (2001–2005), major reductions in SO_2 emissions (*eryanghua paifang liang*) were targeted for the early 2000s, with an even larger reduction of 15 percent slated for 2020. Considerable international assistance with monitoring the problem from countries like Norway and strong state action against major SO_2 emissions by relatively small but numerous coal- and oil-fired operations have yielded significant decreases in national emissions of SO_2 (although not NO_x), even as previously unaffected regions, including the northeast and such major cities as **Beijing**, now grabble with the problem.

Backed by a commitment to spend $175 billion on remediation that began in 2006, major policy initiatives include requirements that large coal-fired plants install scrubbers (*jingqi qi*) and other desulfurization technologies with links to a central monitoring and national registry system, imposing a 10 percent increase in costs on commercial enterprises. Equally important is the ongoing shift by power plants to low-sulfur fuels, including natural gas (*tianran qi*), from such domestic sources as the huge underground field in Ordos, Inner Mongolia, along with increased natural gas imports. Efforts are also being pursued to close down high sulfur–content coal mines and establish greater centralized control over the highly fragmented coal supply and distribution system.

Through the establishment of acid rain and SO_2 control zones, cost-effective approaches for mitigating acidic deposition are being pursued, along with a major tree-planting campaign aimed at the reforestation of 20 million hectares (49 million acres) by 2020. While significant reductions in acid rain have been recorded, especially for the south, the frequency of acid rain episodes rose in 2017, as local governments highly dependent on profitable coal-fired plants suffer from weak enforcement, cutting corners to avoid major economic disruption, particularly in poorer areas, where a culture of monitoring and verification has yet to take hold. While the environmental bureaucracy pushes for major reductions in coal- and oil-fired facilities, other bureaucratic agencies pursuing such projects as the West-to-East Electricity Transfer Plan are calling for even more coal-fired plants in poor and highly polluted regions like the southwestern province of Guizhou, ensuring acid rain as a persistent if somewhat less severe problem for the Chinese environment. Factors contributing to continuing acid rain and other forms of air pollution include numerous and highly dispersed users of coal, scarcity of investment in environmental controls, weak institutional capacity for managing air pollution by more systematic monitoring and regulation of sulfur emissions, and a still-underdeveloped permit system.

ACTIVISTS AND ACTIVISM. Responding to rapid environmental deterioration and reflecting the growing importance of **civil society** in the country's political landscape, activists and activism have become a major feature of the environmental movement in the People's Republic of China (PRC). Beginning with a few individuals in the early 1990s, environmental activists emerged throughout the country in both wealthy urban areas and economic backwaters, composed of a rich diversity of individuals from university professors and prominent lawyers to average workers and peasant farmers, all of whom were no longer willing to sit idly by as China suffers through growing problems of **air pollution** and **water pollution**, degradation of agricultural lands, and destruction of irreplaceable natural **habitat** and **forests**, along with equally irreplaceable **wildlife**. Riding the wave of a virtual explosion in environmental nongovernmental organizations (ENGOs) percolating up from the country's grassroots, which numbered more than 3,500 in 2017, many environmental activists emerged as national and internationally prominent figures, shaping the policy contours of China's rapidly evolving struggle to preserve an environment that seemed on the verge of destruction and perpetual decay.

Leading the move to individuals and organizations engaging environmental issues outside the realm of state authority was **Liang Congjie**, who, in 1993, joined with a few intellectual and professional colleagues to form **Friends of Nature (FON)** as China's first environmental **nongovernmental organization (NGO)**. Promoting such politically "safe" activities as tree planting campaigns and the preservation of the **Tibetan antelope (*chiru*)**, Liang set the tone for Chinese environmental activism, pursuing a pragmatic agenda of focusing on specific environmental problems, while avoiding the kind of grand, systematic political ideology and activities that the Chinese government does not tolerate. Maintaining close ties with relevant central government agencies and limited by law to organizing in specific locales, FON sparked the emergence of similar environmental activists and groups composed heavily of **journalists**, along with filmmakers, photojournalists, and even businessmen, many lacking a background in science but with a knack for maximizing publicity via China's expanding community of social **media** consumers willing to draw public attention to ongoing environmental problems. Led by **Ma Jun** and his **Institute of Public and Environmental Affairs (IPEA)**, China's largely volunteer corps of environmental activists operate according to a shared principle that the most effective tool for spurring government action, where too often indifference and outright corruption prevent any move toward environmental remediation, is to concentrate on publicity campaigns revealing to the general populace crucial data on environmental conditions like local **air quality** and **water quality**, along with egregious cases of environmental transgressions by factories and corporations, domestic and international. Spurred on by individuals whose livelihood and/or community has suffered from some form of environmental degradation or major disaster, notably the growing problem of **cancer villages**, individuals and groups roused to take action, including bringing public interest **lawsuits**, generally understand the risks involved. Despite overall government tolerance and even support for such grassroot efforts, in the case of unsanctioned assembly and **social protests**, state harassment or even prosecutions can result, as has occurred in several prominent cases, for example, computer-technician-turned-environmental-activist Tan Kai, founder of Green Watch, from Hangzhou, Zhejiang province.

While individuals with prominent backgrounds and all-important political and personal "connections" (*guanxi*) operating in the name of middle-class interests in major urban areas are generally, although not always, immune from state interference, the same cannot be said for those choosing to engage environmental issues in relatively poor and often **minority**-dominated areas of the country, which oftentimes have suffered the most egregious damage, along with government inaction. Benefiting from more experienced and professional international organizations like **Greenpeace** and their own experience abroad, Chinese environmental activists remain beset by inadequately developed networks and poor means of communication, as well as undeveloped organization, which is exacerbated by national and local state authorities unwilling to tolerate the emergence of a full-blown environmental movement on a national scale. Among rural victims of environmental degradation and pollution, a kind of "resigned activism" has emerged, consisting of low-key and indirect opposition to official corruption and malfeasance in tolerating violations of environmental laws and regulations.

AGENDA 21 (*21 SHIJI YICHENG*). A nonbinding action plan adopted at the United Nations Conference on Environment and Development ("Earth Summit I") held in Rio de Janeiro, Brazil, in 1992, for promoting **sustainable development** in the 21st century, Agenda 21 (20 chapters, 78 program areas) was adopted by 178 governments, one of which was that of the People's Republic of China (PRC), which officially endorsed the guidelines in

1994. Promoted by the International Union for the Conservation of Nature and World Commission on Environment and Development, China's response involved input from 300 experts, notably a working group composed of several state ministries, with drafts prepared by the State Planning Commission (SPC) and the State Science and Technology Commission (SSTC). Approved by the State Council in 1994, the *White Paper on China's Population, Environment, and Development in the 21st Century* made the PRC the first developing country to endorse the plan to implement national strategies to halt environmental degradation and promote resource use on a sustainable basis.

The adoption of Agenda 21 led to the creation of new administrative bodies by the central government, for example, a Leading Group on Agenda 21, along with an Administrative Center for China's Agenda 21 (ACCA 21), with guidelines for governments at every level in the PRC to develop concrete plans at the regional and local levels for achieving the environmental goals that were laid out for the years 2000 and 2010. Incorporated into the five-year economic plans beginning with the Ninth Five-Year Economic Plan (1996–2000), 62 projects were selected for major funding, with priority given to the **energy** sector involving newly developed **coal** combustion technologies. Along with ancillary documents, one of which was 10 Point Tasks for Environment and Development, China hosted international conferences on Agenda 21 (1994, 1996), while maintaining coordination with the international community through the ACCA 21 and involvement in international efforts at training relevant personnel and the development and use of new technologies for environmental amelioration.

While adopting the basic principle of "polluter pays" (*wuranzhe fu qian*) for addressing major environmental problems, China's commitment to environmental protection still competes with the important goals of economic development and **poverty** alleviation. With President **Xi Jinping** reiterating the PRC's commitment to Agenda 21 at the United Nations Sustainable Development Summit (September 2015), China also endorsed the 2030 Agenda for Sustainable Development in 2016. Among the problems of implementing Agenda 21 goals is the opposition of local **Chinese Communist Party (CCP)** and government officials whose careers have generally depended largely on economic performance. Also problematic is the fractured nature of **policy implementation** by state enforcement authorities, especially at the local level.

AGRICULTURE (*NONGYE*). A nation of 1.4 billion people in 2017, with a system of agriculture that is largely self-reliant, especially in basic grains (*liangshi*), the transformation of the physical environment of land and **water resources** by agriculture in China throughout the centuries has been extensive. True for the premodern era from ancient times (3rd to 4th centuries

BCE) to the 20th century, massive **habitat** destruction, including clearing of **forests** and **reclamation** of land from **coastal** areas, **rivers**, and **lakes**, with extensive hillside terracing, even in remote areas, combined with gigantic **water-control projects**, was promoted by the Chinese dynastic state to facilitate the development of a farming culture. With two-thirds of China's arable land *(gengdi)* consisting of relatively infertile rocky and loamy soils (alpine, allitic, regosol, and luvisol) lacking in phosphorous potassium (*lin jia*) and nitrates *(xiaosuanyan)*, heavy pressure was placed on the overall environment up to the modern era.

Maximizing nutritional adequacy was the top priority in Chinese agriculture, as a great deal of **biodiversity** was maintained, with fields, gardens, managed semiwild lands, and forests being part of an integrated system of intensive and generally highly productive cultivated lands. Following the establishment of the People's Republic of China (PRC) in 1949, China's lands would be gradually transformed into an industrial agriculture, with traditional organic **fertilizers** and pest control replaced, especially after 1978, by inorganic **chemical** fertilizers and modern **pesticides**, resulting in extensive **air pollution** and **water pollution**, and degradation of **soil quality**.

Throughout the dynastic era (221 BCE–1911), development of China's unique garden-type of farming, with its relatively high productivity, was driven by both the food requirements for the preparation and prosecution of frequent warfare and support for a constantly increasing population, particularly during the 18th and 19th centuries, when the numbers exploded from approximately 150 to 430 million people, creating severe pressure on fragile environments, chiefly in the northwest, where sedentary agriculture consisting primarily of wheat (*xiaomai*) and barley (*damai*) gradually encroached on the steppe **grasslands**. The lifeline of the Chinese state, cereal production kept pace through extensive reclamation of land from forests, along with river and lake beds, including sections of the **Yellow River**, for massive water-control projects primarily for **irrigation**. The last included such elaborate projects as the Dujiangyan system on the **Min River** in **Sichuan province**, built during the 3rd century BCE and still in use today, and the massive Qiantang Seawall in Hangzhou Bay constructed to protect coastal agricultural lowlands. As forest lands and **wetlands** disappeared, so too did **wildlife**, as China's once-rich variety of animal species, including elephants (*daxiang*) and rhinoceros (*xiniu*), and **birdlife**, were driven to near-extinction as vast areas of the landmass became virtually denuded of wild animals in the face of the inextricable expansion of farmland.

With the coming to power of the **Chinese Communist Party (CCP)** and the pronouncement by Chairman **Mao Zedong** that "man must conquer nature" (*ren ding sheng tian*), destructive practices intensified, including the reclamation of lands in such evermore fragile environments as **Dian Lake (Yunnan province**) and the **Qinghai–Tibetan Plateau**, and grasslands in

Inner Mongolia. Combined with single-minded policies like "taking grain as the key link" (*yi liang wei gang*), which converted lands unsuitable for cereals to cropland with predictably deleterious results, Chinese agriculture, as in past years, no longer absorbs its carbon output. Poorly planned and unnecessary water-control projects, notably 98,000 reservoirs, and other equally destructive agricultural practices pursued during such hyperintensive campaigns as the utopian **Great Leap Forward (1958–1960)** brought severe blows to China's rural areas, including worsening air pollution from ill-conceived ideas like **backyard furnaces**.

The death of Mao Zedong in 1976, and the adoption of the Agricultural Responsibility System by **Deng Xiaoping** in the early 1980s, brought greater rationality to the agricultural sector in China. Yet, with 60 percent of the lands infertile, the promotion of rapid agricultural growth encouraged by the new regime led to heavy application of state-produced chemical fertilizers and pesticides, which along with the use of untreated wastewater for irrigation brought further widespread environmental damage. Of greatest import is the deposition in an estimated 20 percent of the soils of heavy metals, including arsenic (*shen*), cadmium (*ge*), lead (*qian*), nickel (*nie*), mercury (*gong*), and chromium (*ge*), which set off a phototoxic reaction in plants that reduces agricultural yields. Along with residue from chemical fertilizers and pesticides affecting 90 percent of China's agricultural land, an estimated 20 percent, 3.3 million hectares (8 million acres), has been left unsuitable for agricultural production. A product of excessive **mining**, especially surface open-pit **coal** mines in poorer regions like Guizhou province and inadequately regulated smelting and petrochemical industries, arable land in China has actually declined, with approximately 8 million hectares (20 million acres), according to a 2013 land survey, lost since 1997, to industrialization, along with urbanization and expansion of forest land.

Consuming 35 percent of the world's supply of nitrogen fertilizer, Chinese farmers on average use twice as much as needed, while livestock breeding and **aquaculture** have left millions of tons of waste on farmland and in the country's waterways. Cleanup costs in the PRC are estimated at RMB 1 trillion ($159 billion), and cleanup could take years to be fully implemented, as current efforts consist largely of pilot projects in highly impacted areas of the country. Soil salinization (*yanhua*), alkalization (*jianhua*), and waterlogging (*lao*) have accompanied the continued drop in soil fertility as the pursuit of higher and higher yields has led to less fallow time and poor soil maintenance, particularly in triple-cropping areas in the south. Environmental damage, especially to the country's land and waters, has been exacerbated by the continued policy of "self-reliance" (*zili gengsheng*), with minimal grain imports, which if expanded from grain-surplus nations like Australia and the United States could reduce pressure on China's marginal lands.

Dry-zone crops grown without irrigation by tapping available moisture in the country's extensive semi-arid and arid regions include winter wheat, corn, beans, grapes, vegetables (such as tomatoes, eggplant, broccoli, potatoes, zucchini, and jalapeno peppers) along with herbs (including basil and lavender). For areas primarily in the south and southeast with substantial precipitation, wet-zone crops are grown including most importantly paddy rice constituting 23 percent of all cultivated land mostly in the south. Other crops include asparagus, berries, persimmons (native to China), sugar beets, cotton, and tobacco. With more than 100 million hectares (247 million acres) of saline and alkaline soils in China, new rice strains known as "sea rice" (*hai mifan*) have been developed with plans to expand production to six million hectares (14.8 million acres) of previously uncultivatable lands along the **coasts** and in areas with substantial salinization, such as **Xinjiang** province, by 2020. Also developed since the early 1990s are the many varieties of juncao wild grasses used as a substrate for growing edible and medicinal mushrooms and forage for livestock along with serving as a green barrier against **desertification**. Growing up to eight meters (26 feet) high and harvested three to six times per year, the deep-rooted and low-water-consuming juncao also protects against **soil erosion** especially along the banks of the **Yellow River**.

One of the largest consumers in the world of meat, especially pork (28 percent of global demand), with attendant deleterious environmental effects of industrial-level production, China has embraced the alternative meat market by manufacturing and importing since the 1990s **plant**-based meat substitutes with dramatically less ecological footprint, such as mushrooms and soy protein isolates. Drawing on a long tradition among Chinese Buddhists and vegetarians in mimicking meat dishes and supported by a well-educated and increasingly **health**-conscious middle class, China is slated by government mandate to reduce meat consumption by 50 percent.

AIR POLLUTION (*KONGQI WURAN*). Accounting for seven of the 10 **cities** with the most air pollution in the world in 2003, the People's Republic of China (PRC) has in the last 15 years made substantial strides in improving urban **air quality**. By 2017, only two cities, Baoding and Xingtai, both in **Hebei province**, ranked in the world's top 10 (numbers 9 and 10, respectively), while for 62 cities, air pollution levels dropped by an average of 30 percent between 2013 and 2016. Concentrations of particulate matter (*tebie shi*) less than 2.5 micrometers (PM 2.5) are, however, still a major urban problem, with virtually no major Chinese city reaching levels recommended by the World Health Organization (WHO) at PM 2.5 of 10 $\mu s/m^3$, as the constant presence of toxic gray clouds rob all but 1 percent of the country's

560 million urban dwellers, particularly in North China, of truly safe air quality, most notably during periods of extreme air toxicity, generally in the winter, dubbed "Airpocalypse."

Estimates of major **health** problems and premature deaths from high levels of air pollution and poor indoor air quality from antiquated **coal**-fired heating systems, especially in rural areas, number between 350,000 and 400,000 annually (an average of 4,400 a day in 2015), largely from heart and lung disease. Cited as the country's biggest health threat in 2012, by the president of the China Medical Association, air pollution from carbon emissions grew an average of 10 percent a year from 2000 to 2017, surpassing the United States, with the PRC contributing 25 percent of global carbon dioxide (CO_2) emissions.

Ninety percent of Chinese cities (74 of 82) failed to meet government pollution standards of PM 2.5 equivalence in 2014, with predictions that total cleanup will require 15 years of continued remediation. Major culprits include large numbers of coal-fired power plants, **steel** plants, and other highly polluting industrial facilities, often in close proximity to residential areas in cities like Xingtai, the country's most polluted urban area in 2017; the growing number of privately owned **motor vehicles**, especially automobiles; the burning of agricultural biomass; and the fact that for years during the period of central economic planning (1953–1978) the Chinese government provided free coal for boilers to residences and offices located north of the **Huai River** between the **Yellow River** and **Yangzi River**, known as the "Huai River Policy." Exacerbated by increasingly frequent dust storms stemming from **desertification** and greater **energy** intensity in rapidly expanding urban areas, cities like **Beijing**, Harbin, **Tianjin**, **Shijiazhuang**, **Shenyang**, and Taiyuan still periodically record high levels of PM 2.5, sometimes as high as 1,000 $\mu s/m^3$. Air pollutants also cause damage to **plant species** and **agriculture**, with sulfur dioxide reducing the rate of photosynthesis in soybeans, rice, potatoes, and winter wheat, and, combined with hydrogen fluoride, increasing stomatal resistance, potassium leakage, superoxide dismutase, and peroxidase activity, while reducing chlorophyll, causing an overall decline in production yields.

With traces of Chinese smog containing elevated levels of mercury (*gong*) and lead (*qian*) reaching western areas of the United States and Canada, the Chinese government adopted two action plans, the Air Pollution Prevention and Control Action Plans (2013–2017 and 2017–2020), with goals for reducing PM 2.5 by 10 percent for the first period and as much as 25 percent for an expanded number of cities in the second plan. Under the so-called "Blue Skies" initiative, highly polluting **paper** pulp mills, cement plants (*shuini chang*), and **chemical** plants and smelters (*yelian chang*) were either closed down, retrofitted with scrubbers (*xidiqi*), or converted from coal to natural gas (*tianranqi*), while local households and small-scale enterprises were banned from burning coal altogether. That many of the larger facilities were state-owned enterprises (SOEs) made this costly and highly controversial policy more achievable with reliance on an expanding and increasingly accurate and comprehensive monitoring system.

Launched in 2013, China employs a four-color air pollution alert system consisting of blue ("heavy pollution" for the next 24 hours), the lowest level; yellow ("hazardous" for the next 24 hours); orange ("heavy pollution and hazardous" for three consecutive days); and red (heavy pollution for four consecutive days). For residents confronting potentially serious air pollution, real-time data (*shishi shuju*) on air quality (once kept secret) is available from such cell phone and web browser apps as the China Air Quality Index and Air Visual. There are also apps for specific cities, for example, the Shanghai Air Quality Index. For northern cities like Baoding and Shijiazhuang, two of the most heavily polluted urban areas in China, reductions were 30 percent or higher, with expectations of extended life expectancy throughout time, although some complain these goals were achieved simply by moving the more polluting industrial facilities into less sensitive and politically weak areas in remote areas of Shandong province and elsewhere. Similar results were achieved in areas of the **Yangzi River** and **Pearl River** deltas, particularly in Guangdong province, where dramatic improvements in the environment led to eliminating the province from the second action plan. Entitled "Winning the Blue Sky" and offered as a model of environmental amelioration for the entire nation, Guangdong was replaced as a major target area by the highly polluted, coal-rich Fenwei plain in Shanxi and Shaanxi provinces, encompassing such cities as Xi'an, where PM 2.5 levels continue to increase.

Ozone pollution, although not a major problem in China, was also addressed, while work continues on developing, by 2019, regulations for issuing pollution discharge licenses to manage stationary pollution sources and coordinate efforts at controlling multiple pollutants both in air and **water**. Future problems include, most notably, **transportation**, as the expansion of private automobiles and reliance on diesel-powered trucks to carry freight create new sources of air contamination, requiring even stricter regulations, along with imposing controls on the emission of methane gas (*jiawan qi*) from the accumulation of open **garbage**, which in some urban areas forms virtual mountains. Efforts, some bordering on the farcical, to achieve cleaner air especially in urban areas include bicycles equipped with jerry-rigged air purification devices, smog-cleaning towers (including the world's largest in Xi'an, Shaanxi province), anti-smog water cannons and flying drones, polymer coatings in parking lots to absorb car exhaust, and the growing of algae in huge greenhouses to suck up carbon dioxide, which is then used in the production of **fertilizers**. Cans of "fresh air" have also been given away by Chinese billionaire Chen Guangbiao.

See also INSTITUTE OF PUBLIC AND ENVIRONMENTAL AFFAIRS (IPEA/*GONGGONG HE HUANJING SHIWU YANJIUSUO*); MA JUN (1968–).

AIR QUALITY STANDARDS (*KONGQI ZHILIANG BIAOZHUN*). First established in 1982, and expanded in 1996 and 2012, air quality standards for the People's Republic of China (PRC) are divided into two basic classes: class I, applying to special areas like **nature reserves and national parks**, and class II, applying to all other areas slated for completion in 2016. Regulated by the **Ministry of Environmental Protection (MEP)**, now the Ministry of Ecology and Environment, initial standards set in 1982 established limits on total suspended particles (TSP), including sulfur dioxide (SO_2), nitrogen oxide (NO_x), lead (Pb), and benz(a)pyrene (BaP), with standards strengthened in 1996, and revised in 2000, with less stringent limits on certain pollutants. Further revisions were instituted in 2012, including first-time limits on fine particulate matter (PM 2.5), which were introduced in **Beijing** and **Tianjin**, along with **Hebei province** and the **Yangzi River** and **Pearl River** deltas, that same year, with an extension to prefecture-level **cities** in 2015, and nationwide in 2016. A new Air Quality Index was also introduced, including PM 2.5 and **ozone** levels, with goals of reducing the former by 15 to 25 percent in key urban areas, in accord with the Air Pollution Prevention and Control Action Plan (2013).

The five Chinese cities with the best air quality in terms of P.M. 2.5 concentrations averaging daily scores between 100 to 50 and below, are in rank order: Sanya (Hainan), the so-called "East Asia Hawaii"; Guilin, (Guizhou), located in karst landforms region; Kunming (**Yunnan**), the so-called "spring city"; Lhasa (**Tibet**), located on a highland plateau, surrounded by high **mountain** peaks; and Xiamen (Fujian), coastal city and winner of United Nations Habitat Award in 2004. Other cities with good air quality include Ankang (Shaanxi), Beihai (Guangxi), Haikou (Hainan), and Zhanjiang and Zhuhai (Guangdong).

ALIBABA. The largest e-commerce company in China and possibly the world, with annual revenues of more than $15 billion, and led by Jack Ma, Alibaba has been involved in Chinese environmental issues, most prominently by establishing the Alibaba Foundation as a **nongovernmental organization (NGO)** in 2012. With a commitment of 0.3 percent of Alibaba's annual revenues, the foundation promotes **environmental awareness** by the Chinese public, especially involving the pollution of major **rivers** like the **Yangzi River** and Qiangtang River, the latter flowing near Alibaba headquarters in Hangzhou, Zhejiang province. Joining with other major Chinese environmental NGOs, including **Friends of Nature (FON)**, the **Institute of Public**

and Environmental Affairs (IPEA), and Green Hunan, the foundation is also part of an alliance engaged in assisting the Chinese government in overseeing the role of state-appointed "river chiefs" (*heliu qiuzhang*) for individual sections of the Yangzi, with responsibilities including resource protection, pollution prevention and control, and ecological restoration.

Spurred to action by rising pollution levels, particularly in the nearby Qiangtang River, Jack Ma became increasingly devoted to using his enormous personal wealth from Alibaba to address China's increasingly severe environmental problems, including the destruction of ocean sharks (*haiyang shayu*) through the Chinese consumption of shark fin soup (*yuchi tang*), which is also opposed by international Chinese basketball star Yao Ming. Alibaba has also joined in the effort to preserve China's **birdlife** by refusing to post on its website the sale of such threatened wild birds as the yellow-breasted bunting (*huangxiong caiqi*). Despite these commitments, the company has been criticized for some of its commercial activities, for example, the highly successful Single's Day (10 November), when the sale and delivery of 1 billion packages throughout the country creates an enormous waste of packaging material.

ALTERNATIVE ENERGY. *See* BIOENERGY (*SHENGWU NENGYUAN*); RENEWABLE ENERGY (*ZAISHENG NENGYUAN*).

AO XIAOQIANG (1964–). Trained as an engineer at the South China University of Technology and a graduate of Peking University with a master of business administration (MBA), Ao Xiaoqiang is chairman of Beijing SDL Technology and Trade Co., a major supplier of environmental monitoring equipment to such major industries as electric power generation, **steel**, and **chemicals**. One of China's first "environmental billionaires," Ao founded the company in 1998. It also provides **water quality** monitoring equipment essential in solving the immense problems of **water pollution** in the People's Republic of China (PRC). Ao also serves on the board of directors of the China Association of Environmental Protection Industry.

AQUACULTURE (*SHUICHAN YANGZHI*). With the largest output of freshwater and marine **fisheries** in the world, constituting 60 percent of total global output, aquaculture has had a substantial and generally negative impact on the environment in the People's Republic of China (PRC), as the industry is small-scale and largely unregulated. Fish farms in freshwater ponds (*danshui chitang*), **lakes**, reservoirs (*shuiku*), and rice paddies (*daotian*), combined with operations in mud flats (*nitan*) and other areas in the various seas (**Bohai**, **East China**, **South China**, and **Yellow**) off the country's shores, yield an annual output of more than 50 million tons, 76 percent of the country's entire fish and aquatics production. Major pollutants include fish antibiotics, organic wastes, and chemical releases, most notably nitrogen (*dan*) and phosphorus (*lin*), which end up in the atmosphere and sediment and when spewed into water lead to eutrophication (*shuti fu yingyanghua*) sparking highly-toxic algae plumes (*zaolei yumao*) that require major waste management that has yet to be fully implemented.

Concentrated in the middle and lower reaches of the **Yangzi** and **Pearl** rivers, including **Tai Lake** and off the coasts, especially those of Guangdong and Hainan provinces, aquaculture relies on enclosures like cages, net pens, and closed ponds, generating substantial discharges of unused nutrients into aquatic ecosystems and ultimately the sea, with deleterious effects, notably the release of such neural toxins as BMAA (beta-methylamino-L-alanine) into fish and other aquatic species, for example, dolphins. Major aquatic products include freshwater fish, chiefly carp, shrimp, mollusks (snails, mussels, and octopus), and crustaceans (crab, crayfish, and lobsters), which has dramatically improved **food security** with increased protein intake in the PRC but at the expense of environmental degradation due to the release of NO_x (nitrogen oxide), a powerful greenhouse gas (GHG), into the atmosphere, exacerbating **climate change**.

Relying on data from surveys of national pollution sources begun in 2010, major remediation efforts include the establishment of a licensing system for governing and managing aquaculture, along with more dramatic actions, like clearing out the most heavily polluting and often-unlicensed fish farms, especially from inland lakes and the country's more than 98,000 reservoirs. As many as 300,000 cages and 160,000 hectares (395,000 acres) of aquaculture sites were destroyed in 2018 alone, often with little warning by bulldozers, as official policy is to convert to more efficient and large-scale fish farms with a concomitant increase in deep-water production in the seas. Increases in feed efficiency and nutrient recycling through integrated multitropic aquaculture in which wastes from one aquatic species becomes an input into another are also being pursued, along with increases in production of nitrogen- and phosphorus-consuming seaweed.

Unreported and unregulated fishing remains a major problem, contributing to substantial polluting effects on algae and phytoplankton used for fish meal in such major bodies as Tai Lake, the third-largest freshwater lake in China. Extensive pumping of groundwater for ponds and other man-made fish farms has also contributed to extensive land subsidence (*dimian chenjiang*) and salinization (*yanhua*). Among the major **nongovernmental organizations (NGOs)** committed to promoting a sustainable fishing industry, China Blue is developing a sustainability database, along with facilitating a program for aquaculture zonal management, especially for China's coastal fisheries.

Examples of environmentally friendly aquaculture include the building of solar power units over crayfish ponds, which both generates **renewable energy** and improves crayfish production by reducing water temperatures as is being pursued in Jinhu county, Jiangsu province, a major aquaculture center.

ARBORICULTURE. *See* FORESTS (*SENLIN*); NATURE RESERVES AND NATIONAL PARKS (*ZIRAN BAOHU QU, GUOJIA FENGJING MINGSHENGQU*).

ATMOSPHERIC SCIENCES (*DAQI KEXUE*). Dating back to the early 20th century, atmospheric sciences and meteorology have been important research disciplines in the People's Republic of China (PRC), with major contributions to determining the country's approach to **climate change**. Among major institutions conducting research and training personnel are the School of Atmospheric Sciences, Nanjing University, along with programs at the University of Science and Technology, Hefei, Anhui province, the Institute of Atmospheric Sciences, Fudan University, **Shanghai**, the College of Oceanic and Atmospheric Sciences, Qingdao, Shandong province, and Beijing Atmospheric Environmental Research Institute. Weather forecasting and other meteorological services like climate prediction and **drought** and **flood** monitoring are provided by the China Meteorological Administration (CMA), with 681 weather stations located throughout the country, along with several research institutes, notably the Chinese Academy of Meteorological Sciences. Under CMA is also the Environmental Meteorological Center of China, established in 2014, which concentrates on monitoring and forecasting the atmospheric environment, along with measuring the **health** effects of atmospheric conditions, a process that quantum computers will greatly enhance with their capacity to manage massive amounts of data simultaneously.

AUTOMOBILES. *See* MOTOR VEHICLES.

B

BACKYARD FURNACES (*XIAO GAOLU*). A centerpiece of the utopian goals promoted primarily by **Chinese Communist Party (CCP)** chairman **Mao Zedong** to achieve dramatic increases in Chinese **steel** production during the disastrous **Great Leap Forward (1958–1960)**, backyard furnaces had a devastating effect on the Chinese environment, chiefly in bringing about massive **deforestation**, which throughout the years would contribute to serious problems of **soil erosion**, **landslides**, and increasingly frequent dust storms. A product of the Maoist philosophy of engaging in a **"war on nature"** and reflecting the lack of **coal** supplies in China's rural areas for fuel to fire up the furnaces, **forest** wood was stripped from entire **mountains**, including old growth like the rare parasol (*wutong*) tree and bamboo groves (*zhulin*), the prime habitat of China's giant **panda bear** population.

Estimates are that in a period of a few months during the height of the steel-making campaign, when 500,000 of the simple smelters made of sand, stone, fire clay, or brick were built in even the most backward villages, one-tenth of China's entire forest reserves were felled (30,000 to 40,000 km^2) with some provinces, for example, Hunan and Hubei, losing as much as one-third of their total forest cover. Much of the output of backyard furnaces consisted of useless iron ingots and slag smelted out of broken pots, bed frames, and other metal objects contributed by the local population, which proved useless for modern rolling mills. Working in 24-hour shifts at a breakneck pace driven by a militant and utopian urgency while leaving crops to rot in the fields, China confronted the Great Famine (1959–1961), when **birds** and other **wildlife** were driven to near- or complete extinction at the hands of a population fighting to stave off starvation, ultimately costing 30 million lives, largely in the countryside.

See also GRASSLANDS (*CAOYUAN*); WATER-CONTROL PROJECTS (*SHUILI GONGCHENG*).

BEACHES (*HAITAN*). With a coastline of 14,500 kilometers (9,000 miles) and 6,500 offshore islands, the People's Republic of China (PRC) is replete with attractive beaches, from subtropics in the south, especially on Hainan island, to equally appealing sites in the country's central and northern zones. Offering pristine waters and colorful sands, and other natural land forms, attracting tourists and naturalists alike, China's beaches confront increasingly severe environmental degradation, as many have become dumping grounds of detritus from the country's rapid economic growth and mass consumption, especially hard-to-clean-up **plastics**, which constitute approximately 60 percent of beach rubbish.

Emblematic of this seemingly inexorable deterioration of beach **ecosystems** is Silver Beach (*Yintan*), Beihai, Guangxi province, which long known as the finest beach in China for its white quartz sand and soft waves, has been inundated by tons of rubbish, notably polystyrene and bottles casually discarded by sunbathers or washed onshore. Notorious for large crowds exhibiting the worst abuses of the country's pervasive **"waste culture"** Chinese beaches have also been seriously afflicted by unbridled development projects and illegal **reclamation** often involving collusion between aggressive property developers and corrupt local officials turning once-pristine sands and sea waters into trash dumps and cesspools.

Major amelioration efforts have involved the creation of local beach-cleanup environmental groups, for instance, Shanghai Rendu Ocean Development Center, along with dramatic reductions in economic growth plans to stem the tide of beach destruction, along with bans on plastics, as was announced on beach-rich Hainan. Improvements in beach cleaning and management systems are also being pursued, in addition to joint projects involving beach cleanups and **environmental education** organized by such volunteer groups as China Ocean NPO Forum, especially on World Ocean Day (8 June).

Major beaches, many resort sites, in China, include the following: Yalong Bay (Hainan), with white sands and turquoise waters, known as the "Oriental Hawaii"; Golden Sand Beach (Qingdao, Shandong), with soft yellow sand; Gulangyu Island off Xiamen (Fujian), known as the "Garden of the Sea"; Beidaihe (**Hebei**), a major **bird**-watching site; Wuzhizhou Island (Hainan), known for its attractive **coral reefs**; Golden Pebble Beach (Dalian, Liaoning), known as the "Garden of Stone," with its pink rocks and golden stones; and Putuoshan (Zhejiang), a center of Buddhist temples and religious sites. Others include Fujiazhuang (Dalian, Liaoning); Golden Beach (Zhuhai, Guangdong); Qinhuang Island (Hebei); Rizhao (Shandong); Shimei Bay (Hainan); Songlanshan (Ningbo, Zhejiang); and Tianya Haijiao (Hainan).

BEES AND BUTTERFLIES (*MIFENG/HUDIE*). The largest producer of honey in the world, with 6 million domestic bee colonies and 1,317 species of butterflies, the People's Republic of China (PRC) is confronting a decline in both populations, along with other insect varieties, a problem afflicting much of the globe since 2006. Bees are especially crucial to the pollination of such crops as almonds, apples, cherries, melons, and tomatoes with prime pollination areas including more than 100,000 acres of canola flowers, Luoping county, **Yunnan province**, and in the "world's pear capital," Hanyuan county, **Sichuan province**. Genetically diverse native varieties of wild bees have suffered declines mainly due to the overuse of **fertilizers** and **pesticides**, along with the deleterious impact of **deforestation**, open-pit **mining**, large-scale infrastructure projects, and **climate change**. While the rich diversity of native Chinese bees has kept losses in the PRC relatively small, at 10 percent, compared to much larger losses in other parts of the world, some areas, like Hanyuan, have suffered a near-complete eradication, forcing farmers to either rent bees from commercial keepers or engage in labor-intensive hand pollination.

Importation of Western bee species has helped fill the gap, although this more uniform variety avoids pollinating many native Chinese plants. Remediation efforts of bee breeding grounds include planting strips of wild flowers and retaining natural vegetation on croplands, while the conversion to **genetically modified** crops, although limited in China, requires less pesticides. Major Chinese research institutions involved in tracking bee populations in such regions as the **Qinghai–Tibetan Plateau**, which sports the world's most diverse species of bumblebees, include the **Xishuangbanna** Tropical Botanical Garden and the Institute of Apicultural Research, **Beijing**.

Butterflies are found throughout much of China from the hesperioidea (skipper), lycaenidae (gossamer-winged), nymphalidae (brush-footed), papilionidae (swallow tails), and peridae (whites and yellows) families. Subtropical (*ya redai*) and tropical (*redai*) regions are home to rich concentrations, especially in Yunnan province, where, in Jinping Butterfly Valley, large-scale migrations of as many as 130 million butterflies, 400 different varieties, occur periodically during the spring. Documentation of the country's butterfly population was carried out by European and Chinese entomologists in the 19th and early 20th centuries, with joint Chinese–Soviet expeditions during the 1950s, especially into Central Asia. More contemporaneous expeditions involving the Entomological Society of China led to the discovery of Jinping Valley in 2009, and the sighting of new varieties, notably giant butterflies in Guangdong province with an eight-inch wing span. With recent declines in butterfly numbers in places like Jinping Valley, the Hani Terraced-Field Administrative Center was established, the only body in the world dedicated to the protection of butterflies.

BEIJING. The national capital of China in the imperial era from 1271 onward and throughout the People's Republic of China (PRC), *Beijing* ("northern capital") is a city where the natural environment has had a profound influence on the history and lifestyle of its population. Located in the arid region of North China, **water** scarcity has been a major factor in the city's development, while as a center of **industry** and with a growing population (20 million in 2017), problems of **air pollution** have bedeviled the urban area since the 1950s. Surrounded on three sides by **mountains**, Beijing Municipality (*Beijing shi*) is 10,000 km^2 (3,860 mi^2) in size and includes both urban and rural areas, where economic improvements, especially since the inauguration of reforms in 1978–1979, have led to the emergence of a well-off middle class with increased consumption habits that have intensified important environmental issues.

Most significant is the ongoing problem of air pollution, where natural surroundings and weather patterns, combined with industrialization and enhanced consumerism, have led to a significant deterioration in **air quality**, which was substantially reduced but not eliminated in the 2010s. Beginning in the 1950s and 1960s, a substantial base of highly polluting industries was developed in Beijing consisting of electronics (*dianzi chanpin*), textiles (*fangzhipin*), **chemicals**, and **steel** factories, which, along with **coal** burning for residential and commercial buildings, spewed billowing smoke throughout the city and its immediate environs. While the Chinese government reversed course with a decision in 1983, to gradually eliminate the city's industrial base, relocating such industrial behemoths as the huge Capital Iron and Steel Corporation complex out of the municipality, air quality continued to deteriorate, as 35 to 60 percent of air pollutants into Beijing came from sources outside the city, especially from steel and power plants in neighboring **Hebei province**, which annually burns 300 million tons of coal.

Combined with the effects of the dramatic growth in private **motor vehicles**, particularly automobiles, which numbered 5.5 million in 2015, periodic "haze" (*wumai*) composed of sulfur dioxide (SO_2), nitrogen oxide (NO_x), carbon dioxide (CO_2), and PM 2.5 concentrates has broken out, on occasion reaching 40 times the recommended safety level. In the peak year of 2013, Beijing's air was declared "unsafe" for 60 percent of the days in the first half of the year, with PM 2.5 levels reaching record levels in January 2013, described as an "Airpocalypse," making the city, according to its mayor, virtually "unlivable." Along with periodic sandstorms, especially in the spring, that blow out of the nearby Gobi and Tianmo deserts with sand dunes (*shaqiu*) advancing inexorably on the city, daily life in the capital was severely disrupted, with frequent closure of schools, city highways, and airports, while public-sector workers were ordered to remain at home. More important was the deleterious impact on public **health**, most notably among children, whose exposure to unsafe air has spawned increased behavior disorders and slower mental development, even as schools have built artificial bubbles to avoid the outdoors, while air purifiers are relied on in residences, along with elaborate face masks when venturing outdoors.

Spurred on by periods of air quality remediation brought on by widespread factory shutdowns and other controls, for example, during the run-up to the 2008 Summer Olympics and by the demand of government leaders, one of whom was Premier **Li Keqiang**, in 2014, to bring down pollution levels, concerted efforts at improving air quality in the city were inaugurated, with a goal of reducing air pollution by 25 percent, at a cost of $12 billion. Among the more important measures were plans to cut coal consumption by 2.5 million tons by 2017, in part by converting local power plants and residential heating systems to natural gas (*tianranqi*), while shutting down antiquated coal-burning factories, including the last coal-fired power plant in 2017. Plans have also been announced to limit the number of registered private automobiles, as auto emissions contribute 70 percent of the city's air pollution, with additional plans to replace the thousands of heavy-polluting trucks and taxis with vehicles that meet tougher **fuel emission standards**, now set at Euro 4 levels, the only city in China to implement these rules, along with 3,800 natural gas-powered buses and construction of the city's first bicycle-only road.

Relying on 27 monitoring centers scattered throughout the city, the Beijing **Environmental Protection Bureau (EPB)** provides residents with accurate data that is supplemented by a Twitter feed from the U.S. Embassy, which also gathers information on the city's air quality. From one of the 20 most air polluted cities in the world, Beijing underwent a 53 percent reduction in air contamination by 2018, although variable weather patterns in winds and temperature and humidity remain major factors in overall air quality. The evident incapacity of the municipal government to stem the increasing number of illegal and unlicensed taxicabs that continue to operate despite official restrictions and undoubtedly contribute to days with extremely unhealthy pollution levels reflect the problem of making such improvements in air quality permanent.

Equally problematic for Beijing is the deterioration in **water quality** from both industrial and residential sources of pollution. As measured by standard chemical oxygen demand (COD), five of the largest waterways in the municipality suffer from serious pollution as much as 95 times higher than what is considered safe, including the North Canal (*Beilun*), which supplies 70 percent of the city's drinking water. Ancient water sources like the Yongding River, which was originally built during the Ming Dynasty (1368–1644), are now abandoned as a water source because of excessive pollution and extensive and prolonged dry-up. While several **sewage** plants to treat the city's enormous volume of wastewater and sludge (1.2 billion tons annually) were built in the run-up to the 2008 Summer Olympics, 30 percent of the city remains unconnected to any sewage system, discharging wastewater directly into nearby rivers. Of the total number of 140 **rivers** within the municipality, 52 are now cleared, with the rest slated for cleanup by the end of 2018.

Drawing water supplies from the Baiyangdian network of **lakes** near the city of Baoding (Hebei province), Beijing has faced a severe shortage of water supply in the nearby Miyun Reservoir, which has been severely depleted to one-third of its capacity by **drought**, with some nearby rivers completely drying up.

With nearby springs destroyed by **deforestation** and the Guanting Reservoir, built in 1954, too polluted to furnish drinking water, the Beijing municipal water authority has sought new supply from an emergency project to divert water from Hebei province and faraway Shanxi province. Fees for water usage in the municipality have also been raised from low levels to reduce consumption, which, at 300 m^3 per capita, is still far below the international benchmark of 1,000 m^3 per capita for an area suffering acute water shortage. In 2013, demand for water in the city reached 3.6 billion m^3, while only 2.1 billion m^3 was available from nearby sources. And while the completion of the central canal of the **South-to-North Water Diversion Project (SNWDP)** in 2014 has increased supplies into the city of 1.1 billion m^3, demand continues to grow and may soon once again outstrip supply as groundwater (*dixiashui*) sources from 100,000 wells, which supply two-thirds of the city's needs, continue to be depleted as annual consumption, at 4,000 million tons, outstrips yearly precipitation.

Among proposed administrative solutions to the air and water pollution that confronts not only Beijing, but also nearby **Tianjin** and Baoding, is a proposal aired by President **Xi Jinping** to knit the three areas into one massive economic zone that will require local leaders to better coordinate their policies on pollution, which remains a regional and not purely a local problem. Composed of 130 million people and popularly known as the Jing-Jin-Ji Regional Authority, it would hopefully prevent local leaders from pursuing competing economic policies that have enhanced pollution and other problems, thus achieving **sustainable development**. Efforts to continue pollution remediation have also been spurred by the selection of Beijing as the site of the 2022 Winter Olympics.

With an estimated 25,000 tons of trash and waste produced per day, municipal solid waste (MSW) in Beijing is collected by a fleet of 10,000 trucks supplemented by as many as 130,000 pickers, with migrant workers, including many children, accounting for 30 percent of the total. As much as 40 percent is buried in local landfills or simply deposited in 1,000 open dumpsites, with the rest burned in incinerators, generally opposed by the public, or subjected to biochemical treatment. Committed to eliminating landfills (*lese tian maichang*) and dumping by 2035, Beijing authorities are pushing for greater reliance on **recycling**, precollection sorting of rubbish by city residents, and reducing wasteful consumption habits that generate solid waste, most notably damaging **plastics**, which can end up being dumped into rivers and ultimately the ocean. A vivid description of Beijing's waste problems can be found in the film *Beijing Besieged by Waste*, by Wang Jiuliang.

BIODIVERSITY ***(SHENGWU DUOYANGXING)*****.** Virtually ignored in the People's Republic of China (PRC) during the period of central economic planning (1953–1978), the preservation of China's rich biodiversity of animal and **plant species** became a major priority from the late 1970s onward, consistent with the overall attention to environmental issues. In China, as throughout the world, biodiversity is defined as the variability among living organisms from all sources, including terrestrial, marine, and other aquatic **ecosystems** within species, between species, and of ecosystems. One of 12 mega-biodiversity countries in the world, China, with its topographical isolation provided by high **mountain** ranges and large deserts, harbors almost 10 percent of all plant varieties and 14 percent of animal species on Earth. This includes 35,000 types of vascular plants, second largest in the world, of which 17,300 are endemic to China; 2,340 terrestrial vertebrates, of which 667 are endemic; and more than 10,000 fungi species (14 percent of global total). China is also home to the largest number of **bird** species (avifauna) in the world (1,314) and the most gymnosperm varieties, plants unprotected by ovary or fruit, with 15 to 20 percent of plant varieties listed as **endangered**, along with 40,000 species of related organisms.

Destruction of biodiversity in China from 1953–1978 was widespread with the **"war on nature"** propounded by **Chinese Communist Party (CCP)** chairman **Mao Zedong** and embodied in such campaigns as the **Great Leap Forward (1956–1960)**, combined with rapid population growth from 610 million in 1955 to 993 million in 1980, wreaking havoc on the country's natural **habitat** and attendant animal and plant life. With 90 percent of **grasslands** and 40 percent of **wetlands** degraded by overgrazing, overextraction of natural resources, and massive land **reclamation**, while **forests** suffered equal devastation from exponential growth in **timber** production, at 30 to 60 million metric tons a year, drastic reductions of bird populations ensued, all of which was exacerbated by campaigns like the **"Four Pests"**—in which 210 million sparrows (*maque*) were killed. Prized wildlife like the giant **panda bear** and the **Tibetan antelope (*chiru*)** were driven to near-extinction by loss of habitat and excessive **hunting and poaching**, as one-half of vertebrate species in the country disappeared altogether. While some legal protection and remediation was attempted with the adoption of a Plan for Natural Forest Nature Reserve Construction (1956) and Regulations on Forest Conservation (1964), these limited state efforts had little effect on stemming the tsunami of biodiversity destruction rendered by the Maoist imperative of transforming the Chinese landscape into a vast system of collective production.

Spurred to action on the environment by attendance at such international gatherings as the United Nations Conference on the Human Environment (1972), along with periodic **floods** on the **Yellow River** and **Yangzi River** in 1981, preserving biodiversity became a major goal of Chinese government policy with ancillary support provided by the emergence of the many environmental nongovernmental organizations (ENGOs) from the early 1990s onward, now numbering more than 3,500 nationwide (2017). Domestically, a series of laws were enacted aimed at protecting habitat and **wildlife**, including the Forest Law (1984), the Grasslands Law (1985), and the Law on the Protection of Wildlife (1988), along with major programs of biodiversity remediation. Among the last are the Three-North Shelter Forest Program (*Sanbei Fanghulin*), also known as the "Great Green Wall" (*Weida de Lü Qiang*), in 1978, which was aimed at combatting **desertification** and promoting **soil** and water conservation; the Action Plan for Biological Diversity Protection (1994), which laid out rules and regulations for ecosystem protection; and the Green for Grain Program (1999), which was aimed at curbing flooding and mitigating **soil erosion** by limiting farming on marginal lands. **Nature reserves**, which had been initially authorized in 1956, with the Dingshanhu Preserve in Guangdong province as the country's first, were substantially expanded from 34 (1978) to 500 (1991) and 2,750 (2017) at both the national and local levels, with the largest in northern **Tibet**. Reinforcing this commitment was the PRC's involvement in such international bodies as the Convention on Biological Diversity, where China is one of the few contracting parties to complete the convention's action plans, along with the Convention on International Trade in Endangered Species of Wild Fauna and Flora (CITES), which the PRC joined in 1980. International conferences have also been hosted in the PRC, for example, the International Conference on Biodiversity Conservation in the **Qinghai–Tibetan Plateau** (2016), while joint programs are pursued with United Nations Educational, Scientific, and Cultural Organization (UNESCO) World Heritage Sites and the World Wildlife Fund (WWF), especially in the latter's preparation of *Ecological Footprint Reports on China* (2008, 2010, and 2012) and *Living Planet Report China* (2015).

Among the major Chinese government bodies charged with overseeing policies addressing biodiversity in China are the Ministry of Agriculture; the Ministry of Construction; the **Ministry of Environmental Protection (MEP)**, now the Ministry of Ecology and Environment; and the **Ministry of Natural Resources (MNR)**—along with the State Forestry Administration, the State Oceanic Administration, and the Inter-Ministerial Joint Meeting for Protection of Biodiversity Resources, all of which have had input into such major policy guidelines as the China National Biodiversity Conservation Strategy and Action Plan (2011–2013). Academic institutions like the Nanjing Institute of Environmental Sciences and the Chinese Academy of Sciences (CAS) are also major policy players, along with **nongovernmental organizations (NGOs)**, including the Wildlife Conservation Society and Panther Protection of Wildlife, as popular awareness of biodiversity is enhanced by such public events as celebration of the International Day of Biodiversity (22 May) and birding festivals held in many urban areas.

Major achievements of these policy initiatives include the stabilization and growth of bird populations from the late 1970s to the 2010s, although 10 percent of birds remain threatened with extinction; restoration of reclaimed lands; and bans on further reclamation of China's **coastal** areas, along with such international projects as protection of the Nanpu coastal wetlands on the **Bohai Sea**, with removal of seawalls to allow tidal infusion. Projects for saving endangered plants and increasing plant varieties from 13,000 to 21,000 have also been announced by CAS (2013), along with similar projects to save endangered wildlife through the establishment of breeding centers. Special programs have also been enacted for protecting individual species, including the giant panda, red ibis, bears, and tigers, while antipoaching efforts continue on behalf of the Tibetan antelope. Support for these efforts by the country's top political leadership was also evident at the 19th National Congress of the CCP (October 2017), when general secretary and state president **Xi Jinping** reiterated the government's commitment to a "Beautiful China," while also calling for support of environmental protection from commercial enterprises and public participation as China seeks to balance economic growth with ecological protection.

Persistent problems include poaching of endangered animals, with 233 vertebrate species still threatened with extinction and numbers of wild animals continuing to decline; plundering of plants, one-fourth of which are still threatened, particularly as ingredients for **traditional Chinese medicine (TCM)**, even within nature reserves, along with the erosion of wild crops, notably wild rice; lack of thorough documentation on endangered species; undertrained personnel at nature reserves; introduction of **invasive** plant and animal species, for instance, bullfrogs (*niuwa*), from outside the country; destructive activities by tourists in fragile environs; poor education in biodiversity, with **ecology** as an academic discipline only begun in the last 20 years; overcentralization of **policy-making** by the national government on biodiversity, constituting a "one-size-fits-all" approach, with inadequate involvement of local communities and dilution of authority for conservation measures among multiple levels of bureaucracies; and continuation of major **hydropower** projects, with a destructive impact on wildlife habitat, fish spawning grounds, and wetlands. China has also been urged to include biodiversity protection in the country's One Belt, One Road Initiative (OBOR) of international infrastructure construction and economic development. Host for the UN Convention on Biodiversity (2020), the PRC also maintains the China Biosphere Reserve Network with numerous UNESCO Man and Natural Biosphere sites.

See also BOTANY (*ZHIWUXUE*).

BIOENERGY (*SHENGWU NENGYUAN*). A form of **renewable energy** derived from organic material like corn, cassava, and grass, bioenergy can be used to produce heat, electricity, **transportation** fuels, and a variety of industrial products. Considered a viable alternative source of **energy** in the People's Republic of China (PRC) to fossil fuels, including plentiful but highly polluting **coal** and relatively scarce **oil** and natural gas (*tianran qi*), requiring imports from the Middle East and the United States, various forms of bioenergy, including, most prominently, methane gas (*jiawanqi*) and biofuels (*shengwu rangliao*), are produced in 4,000 production facilities. The former is used for heating, mainly in rural areas, by as many as 20 million households, while the latter consists of low-carbon ethanol (*yichun*) produced from corn (*yumi*) and agricultural waste, and, absent a nationwide mandate, is available in 11 Chinese provinces, where, mixed with gasoline, it is a means of reducing **air pollution** from the country's rapidly growing fleet of **motor vehicles**, chiefly automobiles. The third-largest producer of ethanol in the world, at 3 billion liters annually, along with 1.4 billion liters of biodiesel for trucks and buses, China's goal is to have every gasoline-burning motor vehicle contain at least 10 percent ethanol by 2020, relying primarily on the country's huge surpluses in corn and corn stalks (*yumi jiegan*), estimated at 200 million tons.

Plans also exist to produce cellulosic biofuels (CBs) from grasses, trees, and agricultural wastes, the last burned by Chinese farmers after harvest, contributing to rural air pollution, although the difficult two-stage production process for CBs makes the goal harder to achieve. China has also devoted considerable public and private research and development resources to creating biofuel and biodiesel from lipid, oil-rich microalgae, which can be grown on marginal lands in salt water and has several advantages over ethanol, including lower carbon emissions. Numerous strains of algae are available in the country with the necessary solar and carbon dioxide resources, the latter from **industry**, and nitrogen and phosphate from wastewater, for use in photo bioreactors to generate the biomass. Major problems include high energy costs associated with algae cultivation and harvesting. Engineering solutions, such as nitrate starvation, would make for commercially viable levels of production along with co-production of food, industrial enzymes, medicines, and cosmetics.

Along with increased production of **electric vehicles (EVs)**, primarily automobiles by several Chinese and international car companies, China ultimately wants to ban the sale of gasoline-burning vehicles nationwide to dramatically reduce the serious air pollution plaguing its major **cities**. With bioenergy also used in electricity generation, amounting to five gigawatts (GW) in 2010, China's goal is to produce 30 GW from bioenergy by 2020, which, along with other clean sources of electrical power, primarily **hydropower**, is considered essential for reducing the country's heavy dependence on highly polluting coal in power plants and industrial facilities. In 2006, China hosted the Asia Biofuels Conference and Expo in **Beijing** as China has emerged as one of the top five global producers of algae biomass with gasification turning algae with high lipid content into crude oil while also bringing about significant reductions in carbon dioxide emissions.

BIOTECHNOLOGY (*SHENGWU JISHU*). Introduced into the People's Republic of China (PRC) in the 1970s, biotechnology has become a major weapon in combatting pollution in the country's environment and had a major impact on **agriculture** and the use of natural resources. Buoyed by dramatic increases in funding from such sources as the National High-Technology Research and Development Program (Program 863), Chinese scientists have worked on developing a multiplicity of biotechnological mechanisms for addressing a range of pollution issues plaguing the country, from **air pollution** to the degradation of **soil quality** to the enormous discharge of industrial and organic wastes (annually 8.3 million tons and 4 billion tons, respectively) and the eutrophication (*shuiti fu yingyanghua*) of **lakes** and other bodies of water from toxic runoff, which can lead to the death of animal and **plant species** from lack of oxygen (*yang*). Most development of biotechnical remediation has been at the experimental and pilot program stage, with toxin and organic **chemicals** subject to biotreatment involving microbial degradation, which can metabolize various destructive compounds, minimizing their impact on the environment.

Various biotechnological processes and technologies are under development in China, both for monitoring pollutants and applying remediation, cleanup, and **recycling** and reuse. These include bioflocculation, in which biodegradable biopolymers from living organisms are used to remove particles, suspended solids, and dyes (*ranliao*) from wastewater; biosorption, consisting of biomass that binds and concentrates heavy metal ions, including from aqueous solutions; biofiltration for remediation of air pollution; and microbial electrogenesis for monitoring soil degradation and reducing greenhouse gas (GHG) emissions. Major technologies and methods applying these processes include bioreactors (*shengwu fanyingqi*), in which living material is employed to capture and biologically degrade pollutants by processing wastewater, capturing harmful chemicals and microbiologically oxidizing contents in the air; constructed **wetlands**, a low-cost ecological system that combines chemical and biological processes in an engineered and managed system used to treat domestic wastewater, agricultural waste, mine drainage water, and contaminated **river** water; expanded granular sledge beds for the treatment of industrial and municipal **sewage** and waste; and microbiological fuel cells used for waste recovery.

Industries benefiting from developments in biotechnology include **genetically modified organisms (GMO)** in agriculture, especially cotton (*mian*) and development of new and improved crop seed varieties that experienced serious deterioration in the early 2000s. Also benefiting are marine and freshwater **aquaculture** through genetic breeding of **fish**, disease control, and aquaculture seed production. Among the major biotechnology operations in China are the Beijing Genomics Institute and the China National GenBank (CNGB), largest in the world, which preserves and manages a genome data base and biological samples and conducts research on enhancing plant photosynthesis. Research institutes, many within the Chinese Academy of Sciences (CAS), include the Chengdu Institute of Biology, Institute of Biophysics, Institute of Genetics and Developmental Biology, Institute of Hydrobiology, Institute of Life Sciences and Biology, Institute of Microbiology, Northeast Institute of Plateau Biology, Qingdao Institute of Biological Energy, Shanghai Institute for Biological Sciences, and Tianjin Institute of Industrial Biotechnology, along with several state key laboratories.

In 2011, Chinese leaders stated that biotechnology was a major priority for science and technology spending, while the number of research and development personnel in biotechnology number 200,000 nationwide. Priority is also given to the field of biotechonomy, also known as bioeconomy, namely the development of economic activity derived from the scientific understanding of mechanisms and processes at the genetic and molecular levels, and applied to creating and improving industrial processes.

BIRDLIFE (*NIAOLEI*). Totaling 1,373 avifauna species out of a global total of between 9,000 and 10,000, the People's Republic of China (PRC) has a rich variety of birdlife that reflects the country's widely contrasting environment. Included are 52 species endemic to China and 84 species of birds of prey and owls, the largest in the world, along with 707 species classified as protected and 314 as **endangered**, many globally. Bird environments range from desert drylands in the northwest, to humid and heavily forested **mountains** in the central and southern regions, to tropical rainforests (*redai yulin*), mainly in **Xishuangbanna, Yunnan province**, with many regions, especially Yunnan, suffering severe loss of birdlife.

Protected under the Wildlife Conservation Protection Law (1989/amended 2017), birdlife in China has suffered from human-induced environmental degradation, primarily **deforestation**, **desertification**, and loss of **wetlands** and riverine mudflats, along with political campaigns targeting birdlife, most notably the attack on the **"Four Pests"**—singling out sparrows (*maque*) for destruction. Inaugurated by **Chinese Communist Party (CCP)** chairman

Mao Zedong, especially during the environmentally destructive **Great Leap Forward (1958–1960)**, sparrows were killed by the millions based on a mistaken belief that the birds, technically Eurasian tree sparrows, consumed enormous amounts of seed grain, while ignoring their crucial role as pest predators (*haichong bushizhe*). Near-disappearance of the sparrow at the hands of marauding groups, notably the young armed with slingshots and clanking gongs, driving the birds to exhaustion, led to an explosion of locusts, an ecological imbalance that contributed to the Great Famine (1959–1961), with the estimated death of 30 million people.

Threats to Chinese birdlife in more recent years have come from the loss and fragmentation of **habitat** to overdevelopment, **invasive species**, overharvesting and illegal **hunting and poaching**, and environmental pollution, along with the growing impact of **climate change**. Included are large migratory flocks numbering an estimated 50 million, with 240 separate species traversing China along the East Asian–Australasian Flyway. Relying on habitat and food in wetlands, **mangroves**, tidal mudflats (*chao nitan*), and **lakes** in China, especially **Poyang Lake** (Jiangxi province), along with 14 posting sites on the tidal mudflats of the **Yellow** and **Bohai** seas during their long migrations between Siberia in **Russia** and Australia, 10 percent of these flock including snow cranes (*xuehe*) and the rare oriental white stork (*dongfang bai guan*) have faced extinction. Prominent native species threatened include the once-plentiful yellow-breasted bunting (*huangxiong caiqi*), also known as "rice birds," whose numbers have dropped 90 percent since 1980, as the bird is considered a delicacy in southern Chinese cooking for reputedly promoting sexual vitality. Others are threatened by their use in **traditional Chinese medicine (TCM)**, along with hunting and poaching, and such large-scale development projects as **hydropower**. These include the Baer's Pochard diving duck (*qianya*); common barn owl (*gucang maotouying*); Chinese-crested tern (*fengtou yan'ou*) and the river tern (*he yanou*) with 50 and 5, respectively, left in the wild; brown-eared pheasant (*zongse yeji*); blue-crowned laughing thrush (*languan xiao huamei*); green peacock (*lü kongque*); Japanese-crested ibis (*Riben zhulu*); red-crowned crane (*danding he*); rufous-headed robin (*hong toufa zhigeng niao*); spoon-billed sandpiper (*shao zui yu*); swan (*tian'e*); Tibetan-eared pheasant (*Xizang zhiji*); white-eared night heron (*bai er canglu*); black-faced spoonbill (*heilian pilu*); and white-shouldered ibis (*bai jian yibisi*).

Among protective policies pursued by the Chinese government in recent years is the requirement of a license to capture wild birds, which is apparently widely violated, along with the expansion of green spaces amenable to birdlife in urban areas. The emergence of bird-watching societies in China, for example, Birding Beijing, promoted by Birdlife International, also aid in the protection of the bird population, with appeals for expanded protected conservation areas and **nature reserves**, as birds are still under assault from commercial and other predatory interests, especially in urban areas. Classified into two orders, passerine (perching) and nonpasserine (nonperching), major bird families found in China include the following: albatross (*xintianweng*); babbler; bulbul (*yeying*); bunting (*caiqi*); cormorant (*luci*); crow/raven (*wuya*); cuckoo (*bugu*); dove/pigeon (*gezi*); duck (*ya*); eagle/hawk (*ying*); egret (*bailu*); falcon (*gu*); finch (*que*); flamingo (*huolieniao*); flycatcher (*zhuzhu*); goose (*e*); grebe (*piti*); grouse (*song ji*); gull (*ou*); hen (*muji*); heron (*lu*); ibis (*yibisi*); jay (*song ya*); kingfisher (*cui niao*); lark (*yunque*); laughing thrush (*xiaozhe huamei*); loon (*lan*); magpie (*que*); osprey (*yu ying*); ousel (*helei xiao niao*); parrot (*yingwu*); partridge (*zhugu*); peacock (*kongque*); pelican (*ti*); pheasant (*yeji*), which originates in China; plover (*hang*); quail (*anchun*); sandpiper (*yu*); shrike (*bolao*); starling (*ouliang*); stork (*guan*); swallow (*tun*); tern (*yan'ou*); thrush (*huamei*); tit (*shanque*); tree creeper (*shu paxingzhe*); wagtail (*jiling*); warbler (*ying*); water snipe (*shui yu*); and woodpecker (*zhuomuniao*). Collection of bird-flying data is essential to tracking ongoing changes in several important **ecosystems**.

BOHAI SEA (*BOHAI WAN*). Also known as the Bohai Gulf and forming the innermost gulf of the **Yellow Sea**, the Bohai Sea is enclosed by land on three sides and is 59 nautical miles wide, covering 77,000 km^2 (29,722 mi^2), with shallow waters averaging 18 meters (59 feet). Connected to the Yellow Sea through the Bohai Straits and bordering **Hebei**, Liaoning, and Shandong provinces, along with **Tianjin** municipality, the Bohai Sea sits astride one of the most economically developed regions in the People's Republic of China (PRC) surrounded by dense population, heavy **industry**, and intensive **agriculture**. A major shipping lane and dotted with offshore **oil**-drilling rigs, the sea serves as the drainage for the sediment-laden Hai River and 15 other major waterways, with intense pollution runoff from the entire North China Plain of 410,000 km^2 (158,100 mi^2). This includes annual inflow of 5.7 billion tons of toxic wastewater and 2 billion tons of solid waste (2007), along with **sewage** discharges, 81 percent untreated. **Water quality** monitoring indicates 26 to 41 percent of the Bohai Sea consistently fails to meet required environmental protection standards, especially in **coastal** waters near such major **cities** and ports as Huludao in northeastern Liaoning province.

Afflicted by frequent red tides (*hongchao*) of phytoplankton (*fuyouzhiwu*), eutrophication (*shuiti fu yingyang hua*), and **habitat** destruction from major land **reclamation** projects, the once-rich **fisheries** in the Bohai Sea, traditionally known as the "Emperor's Fishery" (*Huangdi de yuye*), have suffered near-total destruction with major infestation of shellfish by toxic per fluorinated alky substances (PFAS) from local factories. Prized prawn stocks and clam beds have been decimated by years of overfishing and **water pollution** from events like major oil spills, the last occurring in June 2011. A primary target of Chinese government amelioration and cleanups, current goals include making 73 percent of coastal waters fit for human contact by 2020 through "significant reduction" in industrial wastewater, as previous efforts have been largely ineffective, with pollution levels continuing to increase, while the State Oceanic Administration (SOA) warned that the sea is so polluted it could "die" as early as the late 2010s.

BOTANY (*ZHIWUXUE*). The second-largest collection of vascular plants (characterized by conducting tissue) in the world, numbering 35,000 varieties, botany is a vital scientific discipline in the People's Republic of China (PRC), with as many as 4,000 to 5,000 **plant species** in the country confronting the possibility of extinction. Beginning in the 19th century, Western botanists came to China to collect flora (*zhiwuqun*), while their Chinese counterparts began conducting graduate work in the field in the 1910s and 1920s. With the establishment of the Institute of Botany in 1928, under the Chinese Academy of Sciences (CAS), research on plant **ecology**, vegetation, and **forest** geography was carried out in major regions of the country. Influenced by such American-trained researchers as Hu Hanxu, botany in China focused on taxonomic (*fenlei*) studies concerned with the collection and classification of specimens, along with developing Chinese-language equivalents of Latin-based botanical terms.

Following the establishment of the PRC in 1949, more than a dozen large-scale vegetation and ecology expeditions were organized, with detailed vegetation maps produced tracking species ranging from alpine permafrost (*gaoshan yongjiu dongtu*) plants and Himalayan shrubs (*guanmu*), to aquatic plants in marshes and **wetlands**, to desert and sandy land xeric plants—for example, fern bushes (*juelei zhiwu*), cacti (*xianrenzhang*), sage brush (*shanai shu*), and tamarix (*chengliu*), the last employed in anti-**desertfication** efforts. In the 1950s, CAS set up more than 60 research stations nationwide, establishing the organizational basis for the China Ecosystem Research Network and the China Biosphere Reserve Network in the 1980s. **Nature reserves** numbering 2,750 have been created in China for in situ conservation and protection of **biodiversity**, including the more than 600 **endangered** plant species, in line with provisions of the United Nations Treaty on the Convention on Biological Diversity, signed by the PRC in 1992.

Given the enormous ecoenvironmental degradation of the past decades, especially during the **"war on nature"** carried out during the 1950s and 1960s, with significant **deforestation** going back several hundred years, research on the extent of damage to vegetation and forests throughout the country has been conducted, along with assessment of the various programs aimed at restoration and increasing overall plant diversity. Halting the large-scale loss of wetlands and forests has also been pursued through such programs as the Hunan Forest Restoration and Development Project. Impact on vegetative cover from **agriculture**, particularly the replacement of natural flora by crops like tobacco (*yancao*), corn (*yumi*), and rice (*baifen*), and by the tilling of **mountain** slopes, has also been examined, along with the deleterious effects on plants of urbanization, **mining**, overgrazing, and **climate change**.

Fragile areas targeted for remediation and restoration include **grasslands** and rangelands in Inner Mongolia and **Xinjiang** provinces, especially lands suffering **desertification**, along with tropical rainforests in **Xishuangbanna** in **Yunnan province**, the home to one-half of the vascular plants in the country. Major botanical gardens are located in **Beijing**, Hangzhou (Zhejiang), **Lanzhou** (Gansu), Nanjing (Jiangsu), **Shanghai**, and Wuhan (Hubei), the last with more than 24,000 species, including rare and endangered plant varieties. China also maintains a major seed bank in Kunming, the only one in the country, which would allow for the reintroduction of plants into the wild if confronted with extinction. Overall national goals involving vegetation and plant life are reviewed in *China's Strategy for Plant Conversation: Progress of Implementation* (2012), while the PRC has also hosted such prominent global gatherings as the International Botanical Congress (2017). Major research institutes dealing with botany include, in CAS, the Institute of Botany (Beijing), Kunming Institute of Botany, South China Botanical Garden, Wuhan Botanical Garden, and Xishuangbanna Tropical Botanical Garden, along with the Key State Laboratory of Plant Diversity.

BROWNFIELDS (*ZONGDI*). With an estimated 5,000 sites in major Chinese **cities**, brownfields have emerged as a serious environmental challenge as policy makers at the central, provincial, and local levels struggle to create a comprehensive regulatory and legal framework for remediation of heavily polluted **soil** and groundwater (*dixiashui*). Defined as land previously used for industrial or commercial purposes with various pollutants that can remain tainted for years, brownfields became a problem requiring identification and remediation in the 2010s, following a major shift in industrial facilities out of urban areas beginning in 2000, both to improve **air quality** and prepare cities for major international events, for example, the 2008 Summer Olympics in **Beijing** and the 2010 World Expo in **Shanghai**.

During the period of central economic planning and rapid industrialization (1953–1978) modeled after the **Soviet Union**, production facilities, including such heavily polluting industries as **steel**, coke (*jiaotan*), **pesticides**, and **chemicals**, were built on the periphery of major cities, which, by 2000, were enveloped by urban expansion and growth. With little or no previous attention to dealing with abandoned industrial sites, China lacked both the regulatory framework and the technological infrastructure for brownfield remediation, for instance, specialized facilities to treat polluted soils ex situ and properly dispose of hazardous wastes. With inadequate transparency given to brownfields and half-baked efforts at remediation, housing and community parks were often built on **soils** with contamination as much as 10 meters (33 feet) deep, along with contaminated groundwater, subjecting residents to harmful toxins with deleterious effects on **health** and the overall quality of life.

Lacking national policy guidelines and a coherent regulatory and technical framework, cities were often on their own to develop strategies of both monitoring and acquiring the funding for brownfield remediation in an effort to accelerate urban renewal. Following studies by the **World Bank** and relying on the experiences of multinationals from countries like the United States and the United Kingdom that had confronted their own brownfield problems, China gradually moved in the direction of an integrated approach involving updates to the Environmental Impact Assessment Law (2002) and assistance and loans from the World Bank ($150 million in 2016) for addressing serious cases of brownfield contamination, for example, in Zhuzhou city (Hunan province), one of the eight major industrial bases with nonferrous metallurgy (*feitie yejin*) and chemicals developed in the 1950s. Rudimentary "dig and dump" methods, for instance, incineration, have still been employed, while cases of construction workers suffering from exposure to poisons (*du*) and other contaminants have occurred, even as sale of abandoned lands to municipal governments is required to ensure proper monitoring of polluted sites. Too often, close alliances between urban governments and local land developers prevent the kind of comprehensive identification and remediation of brownfield sites prior to major construction projects, which dot China's urban landscape, as the country still lacks a comprehensive brownfield law, as found in many developed countries, like the United States.

C

CANCER VILLAGES (*AIZHENG CUN*). A phenomenon largely unique to the People's Republic of China (PRC), "cancer villages" numbered more than 450 in 2017, and are located in every province except for Qinghai and **Tibet** in the far west. Defined as rural communities with death rates from cancer considerably higher than the national average, which having undergone an 80 percent increase in the last 30 years makes the disease the leading cause of death in the country, most of the afflicted villages are located near **chemical**, **pharmaceutical**, and power plants. The facilities spew wastewater into local **rivers**, **lakes**, and ponds, and are considered a major factor in the high levels of the disease among young and older residents alike. First reported by environmental **activists** and investigative **journalists** in 1998, existence of the villages, generally concentrated in Eastern China, was not officially recognized by the Chinese government until 2012, after years of denial and open intimidation of whistle-blowers by factory owners and local state officials, the latter often dependent on the polluting but profitable industries for revenue.

Even as residents of the affected villages petitioned the government all the way up to central authorities in **Beijing** for remediation and filed lawsuits, state action was slow in coming, requiring substantial pressure from environmental nongovernmental organizations (ENGOs) and publicity, including coverage by China Central Television (CCTV), as victims were generally poor, living in remote areas and lacking political influence. With the government often withholding from the public test results of local water supplies and **soil quality** samples, with detection of the disease elusive for years, "proof" of a causal link between pollution and cancer has been hard to ascertain, leaving victims to confront mounting medical bills and loss of livelihood. Water supplies for crops are severely contaminated and agricultural soils subject to substantial degradation from such heavy-polluting facilities as leather (*pige*) **tanneries** and **paper** mills, many recently shifted from eastern urban areas to inland sites with less oversight and regulation.

Amendments to China's basic Environmental Impact Assessment Law now provide for harsher punishment of polluters, while RMB 660 billion ($100 billion) has been slated by the central government for cleaning up China's tainted water resources, estimated at 70 percent, with one-half of the supply rated unfit for human contact (Grade IV and higher). Prohibitions on the production of the most lethal chemicals have been issued by the **Ministry of Environmental Protection (MEP)**, along with shutdowns of older, more offending factories in such areas as the **Huai River** Delta, where thousands of small factories dump waste directly into the river. Yet, even as plants are closed or move elsewhere, they often leave behind landfills and waste dumps in industrial **brownfields**, often containing chemical carcinogens (*zhi'ai wu*), ensuring continued exposure of China's rural population to a harmful and potentially life-threatening environment.

CAP-AND-TRADE (*SHANGXIAN YU JIAOYI*). A market-based approach to controlling pollution by providing economic incentives to cut emissions, in contrast to top-down, command-and-control environmental regulations, cap-and-trade was adopted by the central government of the People's Republic of China (PRC) in December 2017. With China burning more **coal**, primarily for power generation, than the rest of the world, combined with provincial governments in China having failed in their attempts to impose their own cap-and-trade policies, the plan is designed to create a giant nationwide market for enterprises, state-owned and private, in the power-generation sector to trade credits for the right to emit planet-warming greenhouse gases (GDG/*wenshi qiti*), while giving firms a financial incentive to operate more cleanly. With the burning of fossil fuels tripling since 2000, which on a per capita basis approaches developed countries like the United States, the new trading system begins with firms in the power **industry**, including the six big firms led by Shenhua-Guodian, the largest power company in the world—together emitting approximately 26,000 tons of carbon annually, a figure equal to 10,000 tons of coal.

In the system, enterprises that cut emissions can sell permits to more heavily polluting firms at a healthy price but only if the latter are not provided excessive credits, a practice that undermined previous provincial policies, leaving heavily polluting firms with little or no incentive to cut their emissions and restricting the sale of permits to a mere $400 million over the previous four years. When fully implemented by 2020, the plan is slated to cover 3.5 billion metric tons of emissions from 1,700 stationary sources in China, roughly 34 percent of total national emissions, with a goal of extending the program to other highly polluting sectors, for example, cement (*shuini*), aluminum (*lü*), and the country's antiquated and highly polluting heating sector. With China slated to hit peak carbon emissions in 2030, the Asian

Development Bank calculates that the country needs to capture 10 million tons of carbon dioxide (CO_2) by 2020, 40 million tons by 2030, and 2.4 billion tons by 2050.

CHEMICAL INDUSTRY. A high-priority sector of the **economy** of the People's Republic of China (PRC), the chemical industry has approximately 21,000 plants nationwide, discharging contaminated effluents into **rivers** and **lakes**, with periodic spills and explosions causing substantial damage to the environment. The largest producer in the world of chemical **fertilizers**, along with **pesticides**, paints (*youqi*), and **plastics**, China's chemical and petrochemical plants turn out huge quantities of such highly toxic component inputs as aniline (*ben'an*), anisole (*benjiami*), benzene (*ben*), cadmium (*ge*), nitrobenzene (*xiaojiben*), paraxylene (*duierjiaben*), and xylene (*erjiaben*), with factories often operating at maximum capacity to meet growing demand, which, in 2016, constituted one-third of the global total.

Among the major spills and explosions in recent years have been the massive spill of benzene into the **Songhua River** after several explosions at petrochemical plants in Jilin City, Jilin province, leading to the cutoff of the water supply to the city of Harbin (November–December, 2005); an explosion at Yingte chemical plant in **Chongqing** (November 2005); discharge of cadmium into the Bei River near **Guangzhou**, Guangdong province, from a zinc (*xin*) smelter (December 2005); the discharge of cadmium into the Xiong River in **Henan province** (January 2006); toxic chemical releases on an almost-daily basis into the Songhua River (September 2006); a truck accident that led to the release of **coal** ash (*meihui*) into the Dasha River, Shanxi province (June 206); the spill of nine tons of aniline into the Zhang River in the coal-**mining** area of Shanxi province (January 2013); 16,000 dead pigs in Huangpu River, running through **Shanghai**, followed by 100 tons of poisoned fish in Fuhe River in central Hubei after a chemical spill that included arsenic (*shen*) (March 2013); the explosion of ammonium nitrate (*xiaosuan an*) at a privately run container-storage facility in the port of **Tianjin**, resulting in 173 deaths, with 797 injuries (August 2015); an explosion involving methanol used in the **pharmaceutical** industry in Chengdu, **Sichuan province** (July 2018); and a major explosion at a chemical facility that produced highly flammable anisole in Yancheng City, Jiangsu Province, which killed 47 people and contaminated local waterways.

Chemical spills have also occurred as the result of shipping accidents in the **Yangzi River**, along with the frequent discharge of wastewater from the country's **paper industry**, the second largest in the world. Among the problems and failings contributing to these toxic events are the lack of enhanced safety equipment at plants producing benzene; the cover-up of major incidents by local government officials, along with weak regulatory regime; the absence of early pollution warning systems and emergency response mechanisms; and the violation of regulations governing the location of chemical storage facilities near residential areas with apparent collusion of local governments, as occurred in the devastating explosion in Tianjin in August 2015. Contributions of the chemical industry to environmental remediation include production of coal-based methanol, a clean source of fuels, of which the PRC is the world leader.

CHEN JIE. An award-winning photojournalist, Chen Jie worked at *Beijing News* (2005–2014), where among his many portrayals of everyday life in the People's Republic of China (PRC) were photographs of pollution sites and environmental damage that provoked government and public reaction. The most dramatic were his revealing photos of highly contaminated and totally unregulated evaporation pits (subsequently banned) of toxic **chemicals** located on the edges of the Tengger Desert in Inner Mongolia, which led President **Xi Jinping** to launch an investigation that resulted in the sacking of provincial officials accused of covering up the illegal dumping. Chen has also taken photos of China's distressed miners and the damaged site of the massive chemical explosion and fire in an illegal storage facility in **Tianjin** in August 2015, which resulted in 173 fatalities and damage to 8,000 residences. Winner of the China Environmental Press Award in 2015, Chen has won notoriety for topics other than the environment, most notably his photos revealing difficult conditions among school children in a cliff village in **Sichuan province**, which also led to government action.

CHINESE COMMUNIST PARTY (CCP/*ZHONGGUO GONGCHANDANG*). The single ruling political party since the establishment of the People's Republic of China (PRC) in 1949, the Chinese Communist Party (CCP), governed by top leaders, dominates domestic and national **policymaking**, with **policy implementation** in such areas as environmental protection and conservation carried out by government institutions. Operating on the basic principle of democratic centralism formulated by Vladimir Lenin, the first leader of the **Soviet Union**, political authority and power is concentrated in the top leadership of the CCP with subordinate organs at the provincial, municipal, district, county, and local levels, mandated to carry out the reigning party line.

During the period of central economic planning (1953–1978) and the early years of the economic reforms (1980s–1990s), the CCP pursued a policy of rapid economic growth and industrialization with little or no attention to environmental issues, as a purely utilitarian view of nature pervaded party ranks. During radical phases inaugurated by Chairman **Mao Zedong**, most notably the **Great Leap Forward (1958–1960)** and the **Cultural Revolu-**

tion (1966–1976), an outright **"war on nature"** was carried out, with party cadres (*ganbu*) obligated to follow central directives, many enormously damaging to the environment.

While the environment was spared such direct assaults following the passing of Mao (1976) and purge of his radical followers, labeled the "Gang of Four," the new leadership, headed by **Deng Xiaoping** (1978–1992), generally pursued a single-minded policy of rapid economic growth, which resulted in considerable environmental damage to **forests**, **rivers**, **lakes**, and **air quality**, as party cadre were evaluated for promotion on the sole basis of economic progress with virtually no consideration of the environment. With the pursuit of more balanced policies of economic growth and environmental protection inaugurated during the reign of CCP leader and president **Hu Jintao** (2003–2013), party cadres are evaluated on both criteria, although economic progress still often trumps the environment, especially among local party officials in the country's poorer and more backward regions.

While the creation of environmental policies is largely reserved for government institutions headed by the **Ministry of Environmental Protection (MEP)**—now the Ministry of Ecology and Environment—and affiliated organs, particularly the **environmental protection bureaus (EPBs)** at subnational levels, party cadres, through CCP branches in EPBs, local government organs, and even private enterprises, domestic and international, will often intervene in environmental disputes, for example, in the courts. Under the current CCP leader, general secretary and PRC president **Xi Jinping** (2013–), environmental protection and conservation have become a top national priority, as indicated by his speeches at such events as the 19th National Congress of the CCP in October 2017.

CHONGQING. The largest inland city in the People's Republic of China (PRC) built on steep hills with a total population in the municipality (*shi*) of 32 million people (2017), Chongqing was one of the dirtiest urban areas in the country and in the world dubbed "fog city" with serious **air** and **water pollution**. Landlocked and located at the confluence of the **Yangzi** and Jialing rivers in the **Sichuan** basin surrounded by **mountains**, Chongqing became a major industrial base in China beginning during the Second Sino–Japanese War (1937–1945), when, serving as the wartime capital of the Nationalist government (1927–1949), the city (then known as Chungking) underwent a massive infusion of armaments and other vital industries. Targeted after 1949, as a center of industrialization, including **Third Front** facilities, by the central government of the People's Republic of China (PRC) during the period of central economic planning (1953–1978) and a beneficiary of the economic reform policies (1978–1979), Chongqing underwent rapid urban growth involving such heavily polluting industries as cement (*shuini*), **paper** mills, and **motor vehicle** (automobile and motorcycle) production, with the population growth averaging 1,300 new residents per day.

By the 1990s and early 2000s, the city confronted a multiplicity of environmental problems, including acrid daily smog and frequent **acid rain**, as sulfur dioxide (SO_2) emissions grew from 15.5 million tons in the 1990s to 1.4 billion tons in the 2000s. No less a problem was municipal solid waste (MSW) and **garbage**, which piled up in unsanitary dumps scattered throughout the city, along with industrial wastewater amounting to 25 billion tons and **sewage** that largely ended up being dumped untreated into the local rivers, which often turned red from toxic pollutants. While local **forests** in nearby mountains suffered serious deterioration from dirty air and pollution-related insect infestations, with one-third of crops in the rural areas of the municipality damaged by persistent acid rain, the urban population experienced major **health** problems, most notably a spike in cancer rates of 45 for every 100,000 people versus 27 for every 100,000 in nearby rural areas.

Major efforts at ameliorating environmental conditions in the municipality began in the late 1990s and early 2000s, with an important role played by the **World Bank**, which, under the Chongqing Urban Environment Project, provided $200 million in loans to facilitate wastewater collection and sewage treatment facilities. Altogether, RMB 19 billion ($2.8 billion) has been slated for the construction of water treatment infrastructure in the municipality, which, by 2004, consisted of 23 sewage disposal plants, including service for rural areas that for decades lacked treatment facilities. By 2008, 90 percent of sewage discharge was subject to treatment, as dramatic improvements in **water quality** have resulted in 100 percent rated grade III (suitable for domestic use), with both chemical oxygen demand (COD) and biochemical oxygen demand (BOD) declining significantly, even as the absolute amount of sewage produced in the municipality has increased.

With both the Yangzi and Jialing rivers experiencing dramatic slowdowns in water flow stemming from the construction of the **Three Gorges Dam** 600 kilometers (372 miles) to the south, Chongqing confronted the threat of the formation of a giant cesspool, which the cleanup has effectively prevented. **Air quality** has also undergone major improvements, with many of the most heavily polluting factories either shut down or subject to dramatic production reductions and some of the most serious offenders moved outside the city to remote rural areas, while the construction of a monorail and the introduction of nonpolluting public buses have reduced the spewing of toxins by motor vehicles into the city's notoriously humid air. Massive landfills (*lese tian maichang*) like the giant Changshengqiao mega pit have been created to handle the approximately 3,500 tons of MSW produced daily; however, the problem of toxic seepage from these sites into the local groundwater (*dixia shui*) remains a constant issue.

Public–private partnerships for handling pollution include the Chongqing Resources and Environment Exchange, which serves as a platform for local enterprises to trade pollution rights, although the success of such efforts at the provincial level has proved elusive. The exchange also includes an equity fund with investments devoted to environmental protection. Private firms like Chongqing Kangda Environmental Protection have also emerged, specializing in environmental infrastructure construction, while heavy-polluting facilities confront substantial municipal taxation. Beginning in 2006, a major effort at industrial **recycling** was also inaugurated, as Chongqing was selected as one of 10 **cities** in the country to initiate a pilot program. While sewage slurry has replaced **coal** as the fuel of choice by some local factories, for example, the city's giant cement facility, Chongqing has focused on attracting new, less polluting industries, like laptop computer and software production, as the city is now touted as a model of low-carbon growth, even as the continued expansion in local automobile ownership, at a rate of 10,000 per month, poses serious challenges to environmental amelioration.

Chongqing is also home to numerous environmental nongovernmental organizations (ENGOs), for instance, the Green Volunteer League of Chongqing, which brought a **lawsuit** against major food delivery services for their default distribution of disposable **chopsticks**, and the Chongqing Liangjiang Voluntary Service Center, which monitors and investigates sources of pollution. Plans for a proposed **hydropower** station on the Yangzi, known as Xiaonanhai, with promises of substantial outside investment but subject to criticism by environmental **journalists**, were canceled in 2015, due to threats of the project to the Yangzi River Native Fish Reserve, the only free-flowing stretch of the river that can support hundreds of different fish, many of them species endemic to China.

Also developed are two major pilot projects, one involving the diversion of waters from the Jialinga River, into city buildings replacing air conditioning with cooler river water in the summer and heating systems with warmer water in the winter, and the other a shale gas production site at Fuling, the largest in the country, yielding 10 billion m^3 annually of clean **energy** with concomitant reductions in carbon emissions of 12 million tons. Major "greening" sites include Chongqing Garden Expo Park with various gardens covering 3,300 hectares (8,151 acres) and the nearby Longshuixia Fissure Gorge with dramatic karst formations, steep cliffs, and stream waterfalls.

CHOPSTICKS (*KUAIZI*). In a country where food consumption borders on the religious, wooden disposable chopsticks have exploded in demand, with annual production of 57 to 80 billion pairs, with commensurate damage to the Chinese environment, especially **forests**. Made from cottonwood (*sanye yangshu*), birch (*huamu*), and spruce (*yunshan*) trees, and the more costly bamboo (*zhuzi*), with 4,000 pairs per tree, estimates are that disposable chopsticks consume 20 million trees annually in the People's Republic of China (PRC) for an average of 100 acres cut per day. At less than two cents a pair, with no need like reusable counterparts for costly sterilization, disposable chopsticks are a favorite of low-cost fast-food restaurants in China, with 18 million pairs exported annually to other Asian countries, primarily Japan and South Korea. In addition to the destructive impact on forests, the manufacture of disposable chopsticks also involves toxic **chemicals** and preservatives, including sulfur dioxide (*eryang hualiu*), hydrogen peroxide (*guo yanghua qing*), and paraffin (*shila*), which leach into the environment when dumped into landfills, along with the chopsticks' **plastic** wrapping.

Chinese government action aimed at stemming the flood of disposable chopsticks began in the 2000s, with a 5 percent tax per pair levied in 2006, which apparently reduced production rates, along with warnings issued by the Ministry of Commerce (MOC) and other government agencies of stricter regulation in 2010. **Social protests**, including those by prominent Chinese celebrities, have also been brought against free distribution by restaurants in **cities** like **Beijing**, leading to voluntary elimination, while **lawsuits** have been filed by environmental groups against default distribution by such bourgeoning food delivery services as Ele.me and Meituan-Dianping. Campaigns, most prominently "Bring Your Own Chopsticks" (BYOC), have also been promoted, along with public exhibits by Chinese artists the likes of Xu Xinhai, who erected artificial trees composed of disposable chopsticks, aimed at raising public awareness that apparently remains limited, as estimates put consumption at 130 million pairs a day and a whopping 45 billion annually.

Resistance to change has also come from the disposable chopstick industry, which employs as many as 100,000 workers in 399 plants and opposes any major crackdown on the ubiquitous utensils. With China as the world's largest importer of wood and commercial **timber**, continued mass production of disposable chopsticks and other major wood-based products, for example, furniture, a major Chinese export, the international environment remains under threat, especially in the forest-rich regions of Southeast Asia in Indonesia and Malaysia.

CITIES (*CHENGSHI*). With 685 cities of which 65 have populations of more than one million and a total of 813 million people living in urban areas including six "mega cities" of more than 10 million people each the likes of **Beijing**, **Chongqing**, and **Shanghai**, the People's Republic of China (PRC) confronts enormous environmental problems associated with urban life including **air** and **water pollution**, municipal **sewage** and **garbage**, and **noise pollution**. With less than 20 percent of the total population living in cities in the 1970s, urban areas in the PRC have grown to 60 percent of the total population in 2017, adding 200 million people, with projections of growth to

1 billion city residents by 2035, constituting 70 to 75 percent of the country's total population. Producing 75 percent of the country's gross domestic product (GDP), China's cities also consume 80 percent of all **energy**, while contributing 85 percent of the nation's emission of carbon dioxide (CO_2) from both **industry** and the rapidly expanding number of **motor vehicles**, which, in 2012, amounted to 9,723 million metric tons (MMT) of CO_2 from 288 cities.

Located primarily in the country's north and northeast, and dominated by such highly polluting heavy industries as **steel**, cement (*shuini*), and petrochemicals (*shiyou huaxue changpin*), with large numbers of motor vehicles, especially automobiles, major cities in China suffer high levels of CO_2 emissions. Included are densely populated Shanghai and the steel-producing city of Tangshan, **Hebei province**, which together produce 200 MMT of CO_2 annually, while 19 cities, three of which are Beijing, Chongqing, and **Tianjin**, generate more than 100 MMT (50 tons per capita), with the city of **Shijiazhuang** (Hebei) recording the highest reading of PM 2.5 pollutants.

With Premier **Li Keqiang** declaring a "war on pollution" (*dui wuran de zhanzheng*) in 2014, and responding to a fairly well-off and well-educated public increasingly concerned with the deleterious effects on **health** from dirty air, the Chinese government, at the central and provincial/municipal levels, inaugurated major campaigns aimed at improving **air quality**. Current plans call for providing clean air 80 percent of the time for the country's largest cities by 2020, with major measures aimed at the amelioration of air pollution levels. These include reductions in the burning of **coal** (10 percent in the north and south, 5 percent in Shanghai) by residential households, businesses, and **industry** in 28 cities; the construction of satellite cities around mega-cities, for instance, the proposed Xiong'an New Area, near Beijing, allowing for better urban planning and a shift of polluting sources away from population centers; the adoption of "smart city" (*zhihui chengshi*) technologies, employing artificial intelligence (AI/*rengong zhineng*) for the better management of urban traffic and improved environmental monitoring with pilot projects in 200 cities; reform of the **taxation** system by eliminating the incentive to convert low-carbon arable land to more carbon-intensive urban use (although not including expanded urban sprawl through land-creating projects); and shifting China's urban **economy** into such low-polluting service sectors as computer software, AI, and information technology. Low-carbon pilot programs have also been introduced in five provinces (Guangdong, Liaoning, **Hubei**, Shaanxi, and **Yunnan**) along with eight cities (Baoding, Chongqing, Guiyang, Hangzhou, Nanchang, Shenzhen, Tianjin, and Xiamen) inaugurated in 2009 with a target date of 2030 for substantial reduction of green-house gas (GHG) emissions. By 2016, air quality had improved substantially by an average of 33 percent in 74 cities, including significant reductions in PM 2.5, although other factors were at play, for example, benign weather conditions, while some cities with heavy-polluting industries, as in Heilongjiang province, recorded actual decreases in air quality.

Problems of urban water pollution were also addressed with planned cuts of 10 percent in chemical oxygen demand (COD), along with reductions in ammonia and nitrogen content, with cleanup campaigns aimed at important waterways like the North Canal in Beijing and the Qiangtang River in Hangzhou, Zhejiang province. Cities have also had to deal with a new source of urban pollution in the form of millions of discarded bicycles on city streets and parks from the unregulated bicycle-sharing boom. Cities suffering severe economic downturn from the loss of natural resources, especially by coal **mining** and other exhaustible minerals, are officially designated as "resource-depleted cities" (*ziyuan kujie de chengshi*), numbering 44 in 2010. Provided with financial assistance by the central government to cover the cost of finalizing mine closures and other necessary measures, a shift to more up-to-date "greener" industries has also been pursued but with limited success.

Also promoted since the late 1990s is the green roof (*lüse wuding*) program pursued in major cities the likes of Beijing, **Guangzhou**, Shanghai, Shenzhen, and Wuhan with local requirements that all new public buildings come with government-subsidized rooftop gardens. Plans also exist for creation in Nanjing and other cities of "vertical forests" with the balconies of tall buildings serving as platforms for carbon dioxide-absorbing and oxygen-producing trees and shrubs along with urban "green walls" in which plant modules filter and cool urban air. Hangzhou, Zhejiang province is China's "greenest" city centered on the 6 km^2 West Lake and nearby XiXi **wetlands** with trees and vegetation throughout the city and facilitated by low-carbon travel including large numbers of **electric vehicles** (**EVs**) and a booming bicycle rental service.

CIVIL SOCIETY (*GUONEI SHEHUI*). Defined as the aggregate of **non-governmental organizations (NGOs)** and other institutions that manifest the interests and will of ordinary citizens, the emergence of a nascent civil society in China has been intimately intertwined with the growth of popular interest in environmental protection and general **environmental awareness**. Often referred to as the "third sector of society" separate from government and business, civil society in the realm of environmental affairs was made possible by several legal and legislative actions of the Chinese government, most notably the Resolution on Registration and Management of Social Organizations (1989), along with ancillary regulations allowing public access to government documents and databases, making possible the formation of groups independent of direct control of the **Chinese Communist Party**

(CCP) and Chinese state authority, which, since the establishment of the People's Republic of China (PRC) in 1949, had maintained a monopoly on associations and organizations.

Since the registration of **Friends of Nature (FON)** in 1994, more than 3,500 environmental nongovernmental organizations (ENGOs) have been established in accordance with Chinese law, requiring registration via the Ministry of Civil Affairs (MCA) and with nationwide operations prohibited, limiting ENGO work to the local level. Cooperative action with other domestic and foreign ENGOs is tolerated, along with government-organized nongovernmental organizations (GONGOS), all dealing with the environment. Civil society in the environmental sphere has also been facilitated by more aggressive and often independent **journalists** and **media**, for example, filmmakers and such organizations as **Legal Assistance to Pollution Victims**, which have engaged in public-interest **lawsuits** also now tolerated by state authority, as the environment, unlike other more sensitive political issues, for instance, human rights, is recognized as a legitimate area of public participation.

CLIMATE CHANGE (*QIHOU BIANHUA*). The largest emitter of greenhouse gases (GHGs/*wenshi jiti*) in the world since 2007, contributing 25 percent of GHGs into the global environment, the People's Republic of China (PRC) has emerged as a global leader in policy and **environmental diplomacy** for limiting the effects of extreme weather events including significant temperature rise, dramatic shifts and increases in precipitation, and more frequent typhoons and tornadoes, the latter a general rarity in China. Declaring climate change as "posing a huge challenge to the survival and development of the human race," Chinese leaders are devoted to creating a low-carbon "ecological civilization" (*shengtai wenming*) for addressing the multiple consequences of climate change, including rising temperatures, especially on the **Qinghai–Tibetan Plateau**, dramatic changes in weather patterns, **sea-level rise**, and growing incidences of **floods** in the rain-soaked south and **drought** in the arid north. With an estimated one-third of China's land mass subject to the deleterious effects of climate change, major problems include increasing food insecurity leading to estimated deaths of 308 people per million (330,000 annually), with parts of the country becoming virtually uninhabitable.

China's emergence as a leading nation among others on climate change diplomacy followed the withdrawal of the United States from the 2015 Paris Climate Accords in 2017, when Chinese president **Xi Jinping** urged the other 193 participating nations to remain in the agreement, sponsored by the United Nations. Confronting imminent threats from sea-level rise with the possibility of mass migration from severely afflicted areas along the **coasts**, such as **Guangzhou**, the PRC considers addressing climate change an international priority with the country serving as a model of environmental amelioration. Policies include the adoption of a major carbon transfer plan for private industry (2017) and substantial national investments in **renewable energy** projects including **wind** and **solar power**, **hydropower** and **nuclear power**. With large **cities** along the coasts, for example, **Guangzhou**, confronting imminent threats from sea-level rise and 93 percent of the Chinese people accepting the reality of climate change, the PRC considers addressing climate change an international priority, with the country serving as a model of environmental amelioration, including a major carbon transfer plan for private industry, adopted in 2017, and substantial national investments in **renewable energy** projects, including **wind power**, **solar power**, **hydropower**, and **nuclear power**.

That China has recently pursued such an aggressive posture on climate change stands in stark contrast to the negative and highly critical stance taken earlier by the PRC at the United Nations Climate Change Conference in Copenhagen in December 2009. Declaring climate change and low-carbon (*ditan*) targets then proposed by the United States and other largely developed nations as a way to control and humiliate China, the PRC joined other underdeveloped nations, like India, in preventing any meaningful binding commitments to targeted temperature limitations and specific measures for reducing carbon emissions. Despite having approved the Kyoto Protocol in 2002, extending the United Nations Framework Convention on Climate Change, China continued to push policies of rapid economic growth destructive to the environment, including the construction of 500 **coal**-fired plants in the four years prior to the Denmark meeting, as the PRC opposed any internationally imposed cap on emissions. Yet, with the **economy** slowing down and such environmental problems as **air pollution** escalating in major cities and growing problems of **water** scarcity, along with attendant **health** problems, costing the country an estimated $100 billion annually, Chinese leaders began to shift position, seeing the need for greater **energy** efficiency and looking to the production of advanced low-carbon, "green" technologies as the next stage of economic progress.

Creating a National Coordinating Committee on Climate Change in the powerful National Development and Reform Commission (NDRC) in 2007, China reversed position, adopting an "open attitude" toward human causes to **atmospheric** changes from 2010 to 2011, according to the Chinese environmental minister, **Xie Zhenhua**, as the many Chinese books denouncing climate change, notably *The Global Struggle behind the Low Carbon Hoax*, disappeared from view, in addition to their climate change–denying authors. With climate change policies integrated into the nation's five-year economic plans and the first National Assessment of Global Climate Change indicating the country was already experiencing the effects of rising temperatures and dramatic shifts in precipitation, especially on the Qinghai–Tibetan Plateau,

China considers environmental protection as part and parcel of economic growth, with extensive short-term and medium-term goals for limiting emissions and cleaning up the country's rich but highly polluted **water resources**. Once the target of criticism for inaction by the developed world, including the United States, it is now China that excoriates the U.S. government for shirking its international responsibilities on climate change. Whether China will take the dramatic steps necessary to significantly reduce the country's contributions to worldwide GHGs is open to question. Peak extremes of global climate change is predicted for 2050 with the development of quantum computers necessary for handling the massive amount of data necessary to model the complexity of climate change.

COAL (*MEITAN*). Producing and consuming more coal than the rest of the world combined with more than four billion tons burned annually (10 tons a second), the People's Republic of China (PRC) still subsidizes coal production as the fuel of choice, especially for the electric power **industry** of which there are 2,770 plants in the country that have wreaked enormous environmental damage. Generally high in sulfur (*liu*) content, Chinese-mined coal provides more than 60 percent of the fuel for the country's electrical power generation along with substantial percentages for residential and business heating and indoor and outdoor cooking particularly in rural areas contributing to serious environmental problems most notably severe **air pollution**. Among major pollutants produced by coal production and burning in China (with percent of total emissions in parentheses) are mercury (84), sulfur dioxide (93), nitric oxide (70), and carbon dioxide (80) as every unit of **energy** produced by coal generates 80 percent more CO_2 than natural gas and 20 percent more than **oil**. Coal also accounts for 43 percent of all industrial wastewater effluents with coal **mines** releasing large amounts of methane (*jiawan*) into the atmosphere and contributing 25 percent of China's total green-house gas (GHG) emissions.

With only 20 percent of coal washed prior to burning, release of particulate matter (P.M. 2.5) into the atmosphere constituted 69 percent of total particulate emissions in the 1980s, especially in highly-industrialized urban areas in the north and northeast. **Cities** like **Shijiazhuang** and Tangshan (**Hebei province**), as well as Jilin City (Jilin province), have experienced daily particulate levels reaching 403 micrograms per m^3 in some areas, with total national emissions peaking at 25 million tons in 1988. Additional environmental damage from coal has included land subsidence and the release of wastewater and waste gas, especially from coal **mining**, of which there are approximately 4,000 deep- and open-pit mines, many in the arid northwest, along with heavy use of **water resources**, with as much as 20 percent of the country's available water supply consumed by coal processing.

During the period of central economic planning (1953–1978), mass industrialization produced an enormous demand for coal, as alternative fuel sources, for example, natural gas and **renewable energy**, were generally unavailable. In the course of 20 years, 600 coal-fired plants were built in China, with 620 such relatively inefficient ("subcritical") plants operating in 2014, while burning coal for heating and cooking was prevalent throughout the country, particularly in the countryside. Coal consumption during the push for industrialization doubled in the 1970s and continued to grow during the next three decades, with production and consumption peaking in 2013–2014, at approximately 3.8 billion metric tons annually, an increase from 2.14 billion tons in 2005. Demand for coal and profits in the industry remained strong, especially following the introduction of economic reforms in 1978–1979, as electric power (*dianli*) usage grew exponentially with only marginal effects from renewable energy sources like **hydropower**, **nuclear power**, **wind power**, and **solar power**.

Current plans call for limiting coal-fired generation of electricity to 1,100 gigawatts (GW) by 2020, as production is concentrated in the north and the far western provinces of Inner Mongolia, Shanxi, Shaanxi, and **Xinjiang** and in the southwest, especially Guizhou and **Sichuan** provinces. Severe environmental problems have afflicted these regions, including Hulun Buir (Inner Mongolia) and Linfen (Shanxi), the latter one of the 10 most contaminated places on earth, along with 2,000 slag heaps dotting the Shanxi landscape with acidic soot covering the ancient Buddhist grottos outside the city of Datong.

A net importer of coal since 2009, the central Chinese government has moved aggressively to rein in the industry. They have ordered shutdowns, especially of small-scale plants and factories with serious emission problems; expanded the washing of coal in the production process; and banned coal burning in heavily affected areas like the city of Taiyuan (Shanxi), the veritable capital of coal country in China, with offending parties subject to arrest and prosecution. So-called "zombie" (*jiangshi*) mines engaging in flagrant violation of environmental and safety standards have been shuttered, while local **Chinese Communist Party (CCP)** officials who tolerate regulatory transgressions have been summarily dismissed. New, more large-scale and technologically advanced mines in China must meet a bevy of "green" requirements, for instance, the planting of trees on mining sites, with existing mines required to undergo major environmental upgrades. Specialized storage sites for mine tailings are also being established as a way to prevent the contamination of local land and groundwater supplies, while large coal-fired plants are being equipped with electrostatic precipitators as filtration devices for removing fine particles from flowing gases. Cities where coal mining shutdowns have had a profound effect on the local **economy**, such as Xuzhou in Jiangsu province, have required enormous government assistance to develop new industries and restore landscapes scarred by decades of mining. Targeted reductions in emissions of methane, a major greenhouse gas, from coal mining have yet to be reached as of 2019.

Government assistance has also been provided to such new conversion technologies as catalytic hydromenthanation for creating natural gas (*tianranqi*) from coal, along with the production of coal-based gasoline, **chemicals**, and **fertilizers**, although these processes are extremely expensive and involve their own environmental problems, mainly heavy demand for water. China has also pursued clean-coal technologies since the 1990s, including the circulating fluidized-bed combustor (CFBC), integrated gasification combined cycle (IGCC), and magneto-hydrodynamic steamed combined cycle (MHD-CC), which have yet to move beyond the experimental power station stage, or, in the case of MHD-CC, have been dropped altogether. Most convenient is the development of highly efficient ultra supercritical (USC) technology for coal-burning power generation, with a test facility in the city of Taizhou, Zhejiang province. Coal liquefaction to synthetic gas (*hechengqi*) and then oil is a technology China has researched since the 1980s, with an industrial-scale experiment in Ordos, Inner Mongolia, begun in 2011, that has made enormous demands on local water resources and involved the substantial release of carbon dioxide into the atmosphere. Newer power plants built near lower sulfur-content open-pit coal mines, especially in the western region, employ passive cooling towers less reliant on water. More promising is the reining in of coal consumption through such policies as "coal-free zones" established in cities like **Beijing**, the largest coal-burning city in the world in the 1980s, and **Tianjin**, and the overall reduction of coal production of 800 million tons slated for 2020. Also pursued is production of "green petroleum coke" obtained from decaying coke in a coker unit with a high-sulfur variant, substituting for coal as a fuel and a low-sulfur, low-ash variant upgraded by calcining used in production of aluminum and steel. While cutting total carbon dioxide emissions by 60 to 65 percent by 2030 are essential if China is to comply with its commitments according to the Paris Climate Accords (2015), strong economic growth has, at least temporarily, increased coal consumption (5.3 percent in 2018), with the PRC, along with India and the United States, contributing substantially to the overall increase in global carbon levels in 2018. A waste product of coal burning and major contributor to air pollution, fly ash (*feihui*) is being increasingly used in construction materials for dams and other major projects replacing heavy carbon dioxide-producing concrete. The impact of coal burning on **health** is substantial with residents in the coal-heavy northern regions having a life expectancy rate three years below residents in the coal-light south. Similar threats to health are posed by continuous mining disasters, including underground fires that have flared for years in abandoned mines near the city of Wuda, Inner Mongolia.

COASTAL ECOSYSTEMS. With more than 14,500 kilometers (9,000 miles) of coastline, including 6,500 islands and bordering four major seas (**Bohai**, **East China**, **South China**, and **Yellow**), constituting more than 348,000 km^2 (215,000 mi^2), China's coastal ecosystem has undergone considerable degradation, especially since the era of rapid economic growth begun with the economic reforms in 1978–1979. Among the major areas experiencing serious contamination are salt marshes (*yanzhao*), **mangroves**, **forests**, sea grass beds (*haicao chuang*), tidal mud flats (*chao nitan*), and **wetlands**, which buffer vulnerable shorelines from storm damage and erosion, while also storing carbon (*tan*) and serving as crucial spawning grounds for commercial shellfish and finfish, along with biochemically processing terrestrial runoff. Additional damage has been caused to **coral reefs**, which have undergone serious degradation since the 1950s, in addition to dramatic increases in the frequency of algae plumes and red tides from fewer than 10 in the years before 1978, to between 70 and 120 since then. While the average body size of **coastal** marine **fisheries** has dropped, along with a thinning out of the 28,000 different species that inhabit China's nearby waters, surface seawater temperatures have increased, in part caused by ocean warming, coupled with significant increases in dissolved inorganic nitrogen in every sea except the northern-most Bohai.

Major factors bringing about this accelerated degradation largely derive from the rapid economic growth occurring in China's coastal areas, which has expanded eighteenfold since 1978–1979, and include overfishing, offshore waste dumping from **oil**-production facilities, coastal **sewage**, and industrial and agricultural wastewater carrying **fertilizer** and **pesticide** residue either discharged directly into coastal waters or via the seven major rivers that empty into the sea including the Yellow and **Yangzi**, which carry large amounts of discarded **plastics**, as well as heavy metals and polycyclic aromatic hydrocarbons from electronic and **chemical** industries. Also taking their toll have been coastal **reclamation** projects around **Shanghai** and Jiangsu province, with miles of coastline filled in with mud to build larger port facilities and conversion of coastal **habitat** to salt production (saltpans), which, along with other causes, have made degradation of the coastal ecosystem almost irreversible. Major efforts at amelioration include the establishment of marine protection areas, fishing moratoria with mandated catch limitations maintained at 1999 levels, and restoration of coastal mangroves, most of which is largely in the hands of local government officials, who are often reluctant to engage policies undermining economic growth.

COASTAL EROSION. Extending from the mouths of the Yalu River in the north bordering on the Democratic Republic of Korea (North Korea) to the Beilun River in the south bordering on Vietnam, one-third of China's coastline of 14,500 kilometers (9,000 miles) is subject to erosion more so in the low-lying areas north of the **Yangzi River**, especially in the muddy flats of the **Yellow River** estuary and its abandoned delta. Reflecting widespread regional variation, erosion is most serious along the coasts of the **Bohai Sea**, **Yellow Sea**, and **East China Sea** (46, 49, and 41 percent, respectively), and substantially less in the **South China Sea**, including Hainan Island (21 percent). Virtually all of the muddy areas of coastline, constituting 25 percent of the total, are eroded, while 70 percent of the sandy areas, constituting 75 percent of the total coastline, suffer similar eroded conditions, along with one to two meters lost per year of **coral reefs** and platform coasts, particularly in parts of Shandong province around Liaodong Bay bordering Liaoning province and Qinhuangdao in **Hebei province**.

According to large-scale marine surveys sponsored by the Chinese government, the primary causes of coastal erosion are attributed to both human and natural causes. The former come from excessive exploitation of coastal resources mainly by sand **mining** and land **reclamation**, plus significant reductions in sediment discharge from 200 to 100 million tons annually into the sea from **rivers** subject to significant dam and reservoir construction, while the latter consist of **sea-level rise** brought on by global **climate change**, which has produced increasingly frequent and intense typhoons, storm surges, and high tides.

Among the deleterious consequences of coastal erosion are land subsidence leading to saltwater intrusion into estuaries and rivers, as well as contamination of coastal aquifers (*hanshui ceng*) supplying urban water supplies. Deterioration of coastal **mangroves**, coral reefs, and **wetlands** in the country's tropical and subtropical zones has also occurred, along with shoreline retreat and downward cutting of shore face in the temperate zones. Major remediation efforts consist of such structural solutions as seawalls (*haidi*), jetties (*matou*), groins (*fugugou*), and offshore dikes (*diwei*), which, constructed primarily before the 1960s, protect two-thirds of the coastline. Nonstructural measures include beach nourishment, man-made dunes, planting of mangroves, and coastal shelter belts that dissipate tidal energy in front of the foreshore and have gained favor in more recent years. The Chinese government also implemented new integrated coastal management zone policy beginning in the 1990s, stressing the integration of terrestrial and marine components, with the State Oceanic Administration (SOA) taking the lead in functional zoning and more rational use of ocean space, including the creation of 20 mangrove and coral reef **nature reserves** along the coasts. Internationally, China's coastline is rated as a low-vulnerability area.

COMPANIES. China's increasingly aggressive pursuit of environmental amelioration policies beginning in the late 1990s has spurred the formation of domestic environmental firms, along with greater involvement by multinationals specializing in environmental protection and mitigation. Providing everything from environmental engineering equipment and services, to water conservation and treatment, to solid waste disposal, to public facilities management often through public–private partnerships (PPPs), these companies, of which there were about 70 Chinese firms in 2017, provide cost-effective solutions to ongoing environmental problems, for example, consultation on implementing tougher regulations regarding **health** and safety standards incorporated in the new Environmental Protection Law, enacted in January 2015. With investment in environmental protection slated to amount to RMB 15 trillion ($2 trillion) by 2020, and ten-year projects announced for remediating **air pollution** and **water pollution**, and improving **soil quality**, economic opportunities for environmental companies will continue to expand, especially with the strong emphasis on **renewable energy** and such projects as the **"sponge cities"** approach.

China has also produced its first "environmental billionaires," **Zhang Kaiyuan** of Beijing SPC Environmental Protection Technology Co. and **Ao Xiaoqiang** of Beijing SDL Technology Company, with several Chinese environmental companies committed to "going global" (*zou chuqu*). Generally small-scale with no apparent market leader, Chinese environmental companies rely heavily on imports to meet their equipment and high-technology needs from advanced countries like Germany, while joint programs have been established with such large multinationals as Microsoft, Veolia (France), and IBM in areas like pollution prediction apps for smartphone users.

COMPOSTING (*DUIFEI*). Applied to the treatment of approximately 20 percent of municipal solid waste (MSW) and **sewage** sludge produced in the People's Republic of China (PRC), composting is an alternative to landfill (*lese tian maichang*) and/or incineration (*fenhua*) that also provides nutrient material used for **agriculture**, **forests**, and **horticulture**. A method of waste **recycling**, composting also benefits the environment by pulling carbon dioxide (CO_2) from the atmosphere into ground storage. With China's major **cities** producing more than 150 million tons of solid waste annually, including enormous amounts of food waste, which in the richer cities constitutes 60 to 70 percent of the MSW, composting is considered a more environmentally viable approach to waste disposal as opposed to disposing of waste in landfills, which often leads to serious groundwater contamination and incineration, and has incurred widespread public opposition. Combined with wood chips, compost can also be used for the restoration of arid lands in rain-deficient areas in China's northwest, especially Ningxia province, and when produced at earthworm farms it can be applied to **soil** improvement schemes.

Major problems include the relatively high cost of composting derived from a complex sort-and-separation process, which, when introduced into many Chinese cities, for example, **Beijing**, in the early 2000s met with widespread local resistance, as poor people complained of the lack the time for such a tedious process, while the rich simply did not care. China also apparently lacks nationwide data on the use of composting and requires low-cost technology for advancing the practice. With the Chinese government mandating that offices, restaurants, and residences sort and separate household trash in 46 cities, use of composting is likely to increase.

CORAL REEFS (*SHANHU JIAO*). Formed by colonies of coral polyps bound together by calcium carbonate that grow best in warm, shallow, clear, sunny, and agitated waters, coral reefs cover 30,000 km^2 (77,000 mil^2), or 13 percent, of the global total, off the coasts of the People's Republic of China (PRC), primarily in the **South China Sea**, especially around the Spratly (*Nansha*), Paracel (*Xisha*), Hainan, and other islands, along with a scattering of mostly individual polyps off Fujian, Guangdong, and Guangxi provinces. Known as the "rainforests of the sea," coral reefs are a major breeding and feeding ground for thousands of fish and other marine species, including giant clams, turtles, mollusks, and crustaceans, with almost 600 different species, two-thirds of the global total, including staghorn, pillar, elkhorn, leptopsammia, and antipathies. Following years of widespread coral reef degradation, with a 70 percent loss in coral cover since the 1950s, significant restoration efforts have begun involving the Chinese government and academic agencies, joined by environmental nongovernmental organizations (ENGOs) and volunteer groups.

Natural factors causing extensive damage to coral reefs include, most importantly, **climate change** as increases in thermal ocean temperatures lead to "bleaching," the expulsion of algae living in coral tissue bringing about a whitening of coral color. Coral coverage is also undergoing serious decline especially in the Xisha Islands from the spread of "black band disease" involving complete tissue degradation. More important are the multiple human-induced factors inflicting damage, particularly to fringing reefs close to the shore. These include industrial pollution, overfishing, destructive fishing practices (poisons and blast explosives), land **reclamation**, and coastal development projects destroying **mangroves** and other protective vegetation. Most important is the highly controversial construction of military bases and ports in the internationally contested Spratly and Paracel islands and atolls beginning in 2014, with the PRC laying sovereign claim over disputed areas in the South China Sea, with 100 km^2 dredged to build artificial islands, oftentimes over coral reefs, serving as the foundation for construction of aircraft runways, buildings, and other facilities. **Steel** pipes driven into the seabed had additional deleterious effects on surrounding reefs, as the extent of shallow coral cover lost on occupied atolls was greater than unoccupied atolls, along with larger build-up of destructive sedimentation.

The newly created **Ministry of Natural Resources (MNR)** has committed the PRC to the long-term and difficult process of restoring lost coral reefs, with 1,300 hectares (3,211 acres) already added to reefs and atolls in the Spratly Islands chain. Despite an early survey of coral reefs conducted with the assistance of the **Soviet Union** in the 1950s, the continued absence of a comprehensive monitoring network in the PRC leaves the assessment of coral reefs sparse and uncoordinated, dimming the prospects for dramatic improvements, as transplanting juvenile coral from land-based nurseries to the sea is a tedious process with low survival rates.

CRIME (*FANZUI*). While the violation of environmental laws and regulations has generally been handled by administrative sanctions and fines, criminal prosecutions have become more frequent, especially following guidelines by the Supreme People's Court and Supreme People's Procurator for criminal prosecution in 2016, involving 18 conditions for "serious pollution of the environment." Incorporated into China's Criminal Code in 1997, actual criminal prosecutions of environmental offenders were for years a rarity, averaging only 20 cases annually between 1997 and 2003, as the legal threshold for initiating prosecution remained high. With growing demands from environmental **activists** and outspoken **journalists** for stricter government enforcement, including of local officials who often ignored or even conspired with violators, criminal prosecutions increased dramatically, reaching 1,188 in 2014, with a total of 4,636 cases between 2014 and 2016.

Key to invoking criminal prosecutorial action is evidence of actual physical harm done to people and/or natural resources by offenders involved in such actions as illegal dumping of toxins (*youdu*) or radioactive waste (*fangshexing feiwu*), or the releasing of pathogens (*bingyuan*) of infectious diseases that lead to substantial harm to public **health** and/or losses to public or private property. Also prosecuted are such acts as smuggling rare animals or **plant species** and falsifying environmental data that is crucial to the effective enforcement of laws and regulations. Special environmental courts have been established to handle the growing caseload of legal violations, with arrests sometimes numbering in the thousands, notably of corrupt or simply ineffective local officials for obstructing the administration of public order that the failure to enforce environmental laws and regulations can entail.

CULTURAL REVOLUTION (1966–1976). A largely urban phenomenon involving political infighting within the **Chinese Communist Party (CCP)** and mass violence carried out largely by young, radicalized Red Guards, the Great Proletarian Cultural Revolution (*Weida de Wuchanjieji Wenhua Geming*), while doing enormous damage to China's cultural and historical sites, and causing enormous human suffering, with the estimated death of 3 million people, along with 1 million incarcerated, had less impact on the Chinese environment. Continuing the **"war on nature"** inaugurated by CCP chairman **Mao Zedong** during the catastrophic **Great Leap Forward (1958–1960)**, China's environment suffered, especially in rural areas, from the formulaic campaign to "Learn from Dazhai" (*Xiang Dazhai Xuexi*), which by stressing the conquest of nature by hand and imposing a "one size fits all" set of agricultural practices intensified the deterioration of cropland with excessive terracing and other destructive practices. Equally, if not more, harmful to the environment was the virtual collapse of the administrative state, which halted the enforcement, already chronically weak and largely ineffective, of environmental protection laws and regulations.

"Sent down" (*xia xiang*) to the countryside by Mao's orders in 1968, the once-rampaging Red Guards, who had abused their victims among intellectuals and other targeted "counterrevolutionary" (*fan geming*) groups, now turned their ire on the local environment, for example, in the tropical **Xishuangbanna** region of **Yunnan province**, where primeval **forests** were wantonly cut down and even burned, making way for rubber tree plantations, which proved economically unviable. Yet, ironically, with the collapse of the **economy**, notably the highly polluting heavy industrial sector, overall **air quality** and **water quality** in China probably improved during the 10-year period, although little or no data was collected in the midst of the political chaos. While the massive **deforestation** during the Great Leap largely ceased, as the environmentally destructive **backyard furnaces** were abandoned, few efforts were pursued by the seriously weakened state apparatus to begin remediation of the damage, which would have to await the rebuilding of government authority and administrative competence beginning in the late 1970s.

D

DAI QING (1941–). Prominent female **journalist** and early environmental **activist** opposed to the construction of the **Three Gorges Dam** on the **Yangzi River**, Dai Qing has described the dam as the "most environmentally and socially destructive project in the world." Among Dai's many published works are *Yangtze! Yangtze!* and *The River Dragon Has Come!*, both collections of essays by scientists, one of whom is the outspoken **Huang Wanli**, and journalists opposed to the Three Gorges and other large-scale **water-control projects** in China, for example, the Sanmenxia dam on the **Yellow River**.

The adopted daughter of People's Liberation Army (PLA) marshal Ye Jianying, Dai attended the Harbin Military Engineering Academy, where she studied missile technology while subsequently working as a journalist for *Enlightenment Daily* (*Guangming Ribao*), where she was a columnist from 1982 to 1989. Attending a conference on the Three Gorges project in **Beijing**, where she witnessed open opposition to the venture, Dai was encouraged to publish the views of opponents to the project following a trip to Hong Kong, where opposition to the dam was freely expressed. Lobbying against international assistance to the construction of the Three Gorges project by such countries as Japan, she was arrested and imprisoned for 10 months following the military crackdown against the prodemocracy movement in Tiananmen Square in Beijing in June 1989. With the ascension to national power of prominent supporters of the project, notably Premier **Li Peng**, *Yangtze! Yangtze!* was banned, while upon her release Dai was prohibited from further publication in China, although other works, for instance, her memoirs of the prodemocracy movement (*Tiananmen Follies*), were published abroad.

Remaining committed to the environment with major investigative works, one of which is an account of the giant **forest** fire in the Greater Khingan Mountain region (*daxing'anling*) in Northeast China (May 1987), Dai has warned against the dire consequences in the form of intense sandstorms from the frequent dry-ups (*ganhe*) of the Yellow River, while becoming involved in a privately operated forest restoration project outside Beijing, where she still resides. Dai was recipient of the prestigious Goldman Environmental Award (1993) for her work on the Chinese environment.

DEFORESTATION (*SENLIN KANFA*). Dating back to ancient times beginning during the Zhou Dynasty (1045–256 BCE) and accelerating in the last 300 years, deforestation has been a major problem since the establishment of the People's Republic of China (PRC) in 1949. With **forest** land reduced to 5 percent of the total land mass (1949), plans for major reforestation of 100 million hectares (247 million acres) were announced in 1955, but quickly abandoned beginning with the **Great Leap Forward (1958–1960)**, the first of the "three great cuttings" (*san da fa*), when widespread and uncontrolled clear-cutting of forests was carried out, primarily to supply wood as a primary fuel for the 80,000 **steel**-making **backyard furnaces**.

During the **Cultural Revolution (1966–1976)**, the second "great cutting" occurred with virgin, old-growth forests, including rare Ginkgo trees, felled to make way for farmland and rubber tree plantations. The third "great cutting" came during the early years of economic reform (1978–1998) and has been described as the "most disastrous period of nationwide deforestation," largely to fuel a booming **timber industry**, with total undisturbed primary forest reduced by 50,000 km^2 (19,300 mi^2). Despite major reforestation campaigns beginning in the 1990s, forest loss continues, with 1,355 km^2 (523 mi^2) destroyed between 2000 and 2015.

Major causes of deforestation during the imperial era came from firewood collection (*chaihuo jihe*); charcoal-making (*mutan juece*) for heating; **reclamation** for conversion of forest lands to **agriculture**; brick kilns (*zhuanyao*); and timber for construction of housing, temples, grave sites for the nobility, imperial palaces, and such giant **water-control projects** as the Grand Canal (*Da Yunhe*). During the post–1949 period of central economic planning and extensive land reform (1953–1978), and following the introduction of economic reforms in 1978–1979, deforestation continued largely unabated, as major causes included the following: illegal **logging**; slash-and-burn (*daogeng huozhong*) agriculture; extensive **mining** operations; rapid expansion of the housing, furniture, and **paper** industries, along with disposable **chopsticks**; and the search for plants and animals used in **traditional Chinese medicine (TCM)**. Heavily forested regions in northeast China, especially Heilongjiang province, the largest source of commercial timber, along with Guizhou, **Sichuan**, and **Yunnan** provinces, were particularly hard hit, with large swaths of trees felled in the tropical rain forests of **Xishuangbanna**.

Major environmental effects of deforestation are as follows: **soil erosion** (5 billion tons annually) with attendant increases in **flooding** and siltation of reservoirs, **rivers**, and **lakes**; reduced precipitation and less **water** runoff as the absence of trees reduces the **recycling** of moisture from soils through vegetation and into the atmosphere (evapotranspiration), along with drops in water volume (desiccation) of such major rivers as the **Min** in Sichuan province; reduced carbon sequestration (*tan hui*); loss of **habitat** for **wildlife**, especially for such rare species as the **panda bear** and white-cheeked gibbon (*bai jia changbiyuan*); and increased **desertification**, with deserts (*shamo*) now encompassing 28 percent of the country's territory.

A ban on logging in natural, nonplantation forests was enacted in 2000. Yet, 8.85 million hectares (22 million acres) of forest cover was lost from 2001 to 2010, constituting a 5.4 percent decrease in forest cover, the greatest occurring in Guangdong province. With the country's forests in the southwest (Guizhou, Sichuan, and Yunnan) listed as one of the world's top 10 most threatened regions, China also continues to serve as a transshipment point for illicit tropical hardwood (*redai yingmu*) shipments from Southeast Asia to the United States.

DENG XIAOPING (1904–1997). Paramount leader of the People's Republic of China (PRC) from 1978 to the early 1990s, Deng Xiaoping inaugurated the dramatic shift in economic policy toward "economic reform and opening up" (*jingji gaige yu kaifang*) in 1978–1979, which greatly exacerbated the severe environmental degradation that had afflicted the country during the period of central economic planning (1953–1978). Like his predecessor, **Mao Zedong**, who served as chairman of the **Chinese Communist Party (CCP)** until his death in 1976, Deng showed little concern for environmental matters, as his policies, captured by the phrase "to get rich is glorious" (*zhifu guangrong*), led to a single-minded pursuit of economic growth, with deleterious effects on China's **water** and land resources, while intensifying the problems of **air pollution** throughout the country's many **cities**. Major exceptions included Deng's support for reforestation projects like the Three-North Shelter Forest Program (*Sanbei Fanghulin*), also known as the "Great Green Wall" (*Weida de Lü Qiang*), in 1978, which called for a 23 percent increase in forest cover by 2020, along with the country's first Environmental Protection Law (Trial) enacted in 1979.

While abjuring such environmentally destructive Maoist initiatives as the **Great Leap Forward (1958–1960)**, Deng instituted policies that unleashed economic forces through the evolving Chinese market, including the heavily polluting and largely unregulated township and village enterprises (TVEs/*xiangzhen qiye*). Decentralizing decision-making authority below the central level, implementation of new environmental laws and regulations was often stifled by local officials, for whom economic growth generally trumped environmental amelioration. With air pollution and **water pollution** intensifying throughout Deng's tenure, enormous environmental challenges would confront his political successors, from Jiang Zemin (1989–2002) to **Hu Jintao** (2003–2012) to **Xi Jinping** (2013–).

DESERTIFICATION (*SHAMOHUA*). Defined as the conversion of arid, semiarid, and subhumid regions by natural and human-made causes into desert-like landscapes, with wind-blown sand as the principal characteristic, desertification is one of the most severe and challenging environmental problems confronting the People's Republic of China (PRC). Dating back to the later years of the ancient Han Dynasty (206 BCE–221 CE), when agriculturally marginal lands abandoned by fleeing farmers withered and eroded into deserts and sandy lands, dry regions in contemporary China cover 2.6 million km^2 (1 million mi^2), constituting 27.4 percent of the nation's total territory. With an almost continuous belt of degraded land stretching for 5,500 kilometers (3,415 miles) from the northwestern provinces of Gansu, Ningxia, Qinghai, Shaanxi, **Xinjiang**, and Inner Mongolia into northeastern regions, including Jilin, Heilongjiang, and Liaoning provinces, 50,000 villages and major urban areas, notably **Beijing**, are afflicted by encroaching deserts and drylands.

Four levels of desertification exist in the PRC: "most severely" affected, with 50 percent or more of sand coverage, along with shifting sand dunes (*shaqiu*); "rapidly advancing," with 25 to 50 percent sand coverage; "ongoing," with 5 to 25 percent sand coverage; and "latent," with less than 5 percent sand coverage. In addition to such long-standing deserts as the massive Taklamakan in Xinjiang, lands have suffered from desertification for the last 100 years, along with less arid regions of patchy, desertified lands, with overall desertification increasing throughout most affected regions since the 1950s.

Halting the seemingly unbridled deterioration of **grasslands**, marshes, and vulnerable agricultural lands into deserts and sandy lands is a top priority of the Chinese government at every bureaucratic level. Increasing frequency of fierce sandstorms (*sha chenbao*) and drifting sand dunes moving toward Beijing and **Shenyang** (Liaoning province) threaten the livelihood of approximately 400 million people in 18 provinces and 471 counties susceptible to this almost-irreversible environmental degradation of land and agricultural **soil** resources. Official recognition of the problem came in 1959, with amelioration efforts begun in the late 1970s, including major tree planting along a 2,800-mile strip in the Three-North Shelter Forest Program (*Sanbei Fanghulin*), also known as the "Great Green Wall" (*Weida de Lü Qiang*), in 1978, and the adoption of the National Action Plan to Combat Desertification, in compliance with the United Nations Convention to Combat Desertification, held in 1994.

The rate of land loss to desertification grew from an average of 1,560 km^2 (602 mi^2) annually from 1950 to 1975, to 2,460 km^2 (950 mi^2) annually in the 1990s. While Chinese officials claim net reductions in desert lands between 1999 and 2014, evidence of continuing desertification includes massive sandstorms (*shachenbao*) hitting major urban areas (2010), as crucial reservoirs like Hongyashan in Gansu province, Asia's largest desert reservoir, has had one-third of its original volume filled with sand, leading to a major expansion.

Major types of desert in the PRC include the following: "aeolian" (*feng-cheng*), caused by wind erosion; water and soil loss resulting from the overuse of limited **water resources**; salinization (*yanhua*) from poor water management; and rock desertification, primarily affecting the Karst region in the southwest. Major deserts in the PRC, largely in the northwest, are as follows: Taklamakan in Xinjiang, which, at 337,000 km^2 (130,00 mi^2), is the second-largest shifting sand desert in the world, with 85 percent consisting of dunes; Gürbantünggüt in northern Xinjiang, covering 50,000 km^2 (19,000 mi^2); Badain Jaran spanning Gansu, Ningxia, and Inner Mongolia and known for its many colorful **lakes** and oases; Kumukuli, also in Xinjiang, and the highest desert in the world at 3,900 to 4,700 meters (12,800 to 15,500 feet); the gravelly Gobi in northern China and southern Mongolia, which is the fastest-growing desert on earth, expanding at a rate of 3.2 km (2 miles) a year; and the Tengger, Inner Mongolia and Ningxia, the fourth-largest desert in the country with massive sand dunes around Shapotou ("capital of sand") and Swan and Moon lakes. Other major deserts include the Kubuqi, Engebei, and Ulan Buh (Inner Mongolia), Muu-Us (Shaanxi, Gansu), Kumtaq and Lop (Xinjiang), and Qaidam Basin (Qinghai), along with large-scale sandy lands, such as Horqin, Hulun Buir, Hunshandake, and Orzindag, also in Inner Mongolia with excessively dry counties, such as Minqin (Gansu) losing upward of 70 percent of arable land to intruding deserts and sandy lands.

The primary causes of desertification include both natural and human-induced causes, with the former consisting primarily of **climate change**, which brings higher temperatures, stronger wind regimes, and reduced precipitation—with an annual drop of 25 percent in China's northwest region between the 1950s and 1980s. Human-induced desertification comes from overgrazing by livestock with attendant destruction of grasslands, overuse of groundwater resources, excessive **logging** with commensurate **deforestation**, and highly destructive open-pit **mining**. While the Great Green Wall was chiefly designed to stem the advancing Gobi Desert, the program's reliance on the planting of such water-consuming trees as pine and poplar in arid regions has actually accentuated desertification by decreasing valuable groundwater supplies, with only 15 percent of planted trees surviving beyond the first few years.

Other measures employed to stem desertification include limits on overgrazing by livestock and other destructive practices such as the draining of **wetlands** to create more pasture and farm land that end up desertified and the conversion of desert and sandy areas to vegetative lands. Examples of the last occurred in areas such as the 18,000 km^2 (6,940 mi^2) Kubuqi Desert in northern Inner Mongolia, once known as the "sea of death." The planting of desert-resistant vegetation included juncao wild grasses, desert willow (salix psammophila), shrubbery, and shelter belt forests consisting of willow trees, which consume less water, led to substantial recovery of lands from encroaching sands. Similar results have been achieved in the Ningxia-Hui Autonomous Region where encroaching sand dunes in the north from the Tengger Desert have been stemmed along with surface winds by the "Shapotou straw checkerboard sand barrier" consisting of grids of planted straw and shrubs developed by the renown Shapotou District Desert Research Center and recognized by the United Nations Convention on Desertification as a useful tool for halting desertification. Also developed by Chinese research scientists is a paste composed of modified sodium carboxymethyl cellulose solution, a non-toxic substance found in plant cell walls, that when added to water transforms sand into viable soils sufficient to grow crops but with application limited to relatively small areas. Conversion of desert lands to paddy rice production by such **companies** as Elion Resources Group has also been pursued leading to cultivation of so-called "desert rice" (*shami*). Overall, China's program to counter desertification has produced mixed results, with some desert areas shrinking while other areas continue to expand.

DEZHOU. A prefecture-level municipality of 5 million people, with 600,000 in the core city, located in the northwest of Shandong province, Dezhou is one of the designated "national **renewable energy** demonstration" **cities** in the People's Republic of China (PRC). **Solar** panels placed on rooftops, car ports, and even local statues provide 98 percent of **energy** in the central city, which is aided by the use of solar-powered electric buses, mobile generators, and cookers that altogether save an estimated 600,000 tons of **coal** annually. Dezhou is also the site of more than 100 **companies** involved in the production of renewable energy, most notably **solar power** and **wind power**, making it a low-carbon industrial base. City mandates call for new buildings to be equipped with solar panels and solar-powered water heaters, as 95 percent of buildings in the central city have solar energy applications, with 50 percent for the entire municipality, while city streets are illuminated by solar-powered lights.

Dezhou is home to the headquarters of the Himin Solar Energy Group, the largest manufacturer of solar-powered water heaters in the PRC. The company is headed by **Huang Ming**, known as China's "sun king" (*taiyang wang*), and located in the Sun and Moon Micro Emissions Mansion, the largest solar-powered building in the world, using only 10 percent of energy required in conventional buildings. Despite these advances, which have significantly reduced **air pollution** in the city and its environs, overall **air quality** and **water quality** in this formerly highly polluted area remains a challenge, as private **motor vehicle** ownership has grown, bringing with it increased levels of nitrogen oxide (NO_x), while the city is also the site of a **coal**-fired power plant and facilities for the manufacture of such goods as air-conditioners and automobile parts.

DIAN LAKE (*DIAN CHI*). Also known as "Kunming Lake," Dian Lake is located in Southwest China on the Kunming–Guizhou Plateau at an altitude of 1,886 meters (6,189 feet), close to Kunming city, the capital of **Yunnan province**. The eighth-largest **lake** in the People's Republic of China (PRC), at 298 km^2 (115 mi^2), and popularly known as the "great pearl embedded in a highland" (*qian zai gaoyuan de zhenzhu*), Dian Lake, while saved from being dammed in 1961, has become enormously polluted in recent years, as 90 percent of Kunming's wastewater and other industrial effluents from nearby factories and **mines** spewed untreated waste directly into its waters. Subject to major eutrophication (*shuiti fu yingyanghua*) and large-scale outbreaks of algae plumes (*zaolei yumao*), the lake received a **water quality** rating of grade V, "too polluted for any use," including **agriculture** and **industry**.

Also contributing to the lake's deterioration was the near-total destruction of marshes surrounding the body of water during the **Great Leap Forward (1958–1960)** in an effort to convert the area to rice production by trucking in dirt, leading to large-scale **deforestation**, **soil erosion**, and increased agricultural runoff as local farmers converted to excessive use of chemical **fertilizers**, and the planting of a deep-rooted type of water hyacinth (*shui hulu*), which clogged major portions of the relatively shallow lake at 4.4 meters (14 feet). By the late 1980s, pollution of the lake had gotten so bad that the entire surface was covered with a thin film of green algae, while the lake's waters suffered serious eutrophication, with a high concentration of nitrogen phosphates (*dan linsuanyan*), which reduced fish stocks by 55 percent. Equally destructive effects afflicted lake macrophytes, macroscopic plant life, and shoreline **wetlands**, which withered and died.

Remediation of lake conditions began in the 1980s and evolved throughout the years into a variety of methods to reduce pollution and restore the lake's waters to usable levels, especially for local agriculture and industry. In addition to the construction of water treatment plants for Kunming and local industrial/agricultural wastewater, with the first plant built in 1990, major diversions of fresh water (566 m^3 annually) from nearby **rivers**, one of which is the Niulan, 200 miles east of the Dian, are being carried out, with plans to add waters from the Jinsha (upper reaches of the **Yangzi River**) to literally flush the polluted waters of the lake into downstream rivers and ultimately the ocean. Moreover, restoration of nearby wetlands is being pursued, along with bans on new construction near the lake and the introduction of appropriately **invasive** water hyacinth, with prohibitions on fishing aimed at regenerating endemic and nonendemic fish species, which have been reduced from 10 to 1 and 25 to 2, respectively.

With water quality ratings in the lake improved to grade IV (suitable for industrial and entertainment use but not human contact), as ammonia (*an*) and nitrogen phosphate pollutants were lowered by 30 percent, the period of natural residence of effluents in the lake has also been reduced to three years from the more damaging four years, although the goal of making the lake once again "swimmable" by 2020 will require even more dramatic improvements.

DONGTING LAKE (*DONGTING HU*). Located on the middle reaches of the **Yangzi River** on the border of Hunan and Hubei provinces, *Dongting* ("Grotto Court") is the second-largest freshwater **lake** in China, at 2,820 km^2 (1,090 mi^2), expanding to more than 20,000 km^2 (7,720 mi^2) during the maximum **flood** season. Fed by four major rivers (Xiang, Zi, Yuan, and Li) and home to the **endangered** finless porpoise (*jiangtun*), Dongting has undergone severe environmental degradation, beginning in the 19th century, when **reclamation** of substantial segments of the shallow lake and **wetlands** was carried out to create new farmland, a process that continued following the establishment of the People's Republic of China (PRC) in 1949, especially during the **Great Leap Forward (1958–1960)**.

Major sources of increased degradation and pollution since 1949 include draining surrounding wetlands, some of the largest in China, for agricultural purposes and the dumping of an estimated 100 tons of wastewater containing a variety of heavy metals, many carcinogenic, including cadmium (*ge*), nitrogen (*dan*), lead (*qian*), zinc (*xin*), and chromium (*ge*), from shoreline **mining** and such industries as the **paper industry**. Other sources of contamination include seepage of human and animal wastes, the latter from nearby poultry and livestock farms, along with shoreline **garbage** piles and dumping of household wastes, especially by the thousands of fishermen living on boats. Equally destructive has been the dumping of inorganic **fertilizers** into the lake at a rate of 100 tons a day to enhance breeding of the lake's 114 different fish species and deposition by local **aquaculture** of harmful antibiotics (*kangshengsu*) including sulfonamides, trimethoprim, tetracycline, and quinolone.

Having shrunk in size by 50 percent in the last 70 years, from 4,350 km^2 (1,679 mi^2) to 2,820 km^2 (1,090 mi^2), Dongting Lake is now subject to serious remediation and pollution eradication efforts, begun in the early 2000s, with international assistance in addressing major **biodiversity** issues. In addition to shutting down factories lacking on-site wastewater treatment facilities, notably paper mills, and banning **chemical** fertilizer use in and nearby the lake, major goals call for improving **water quality** from the previous grade V (no use even for **agriculture** or industry) to grade III (allowing human contact), slated for the year 2020, through more than 600 local environmental protection projects in the four nearby **cities** of Yiyang, Yueyang, Changde, and Changsha.

With a focus on the restoration of the all-important wetlands that were added to the list of "Wetlands of International Importance" in 1992, the goal is to strengthen the lake as a major buffer against floods and restoring its longtime role as the "kidney of the Yangzi River" (*Changjiang de shenzang*), even as raw, untreated **sewage** continues to spew into the lake. Remediation efforts also include regulating large-scale sand excavation projects, while also removing illegal levees built by local fish farmers to carve out their own exclusive sections of the lake for fishing, which by impeding water circulation contribute to stagnation and incapacitate the lake's self-flushing capacity. Assisted by local government agencies and such environmental nongovernmental organizations (ENGOs) as the Ecological Environmental Monitoring Center and Green Hunan, these efforts are too often undermined by the collaboration of local government officials, including from **environmental protection bureaus (EPBs)**, with ongoing pollution from offending parties like paper mills and aquaculture, which are considered essential to the local **economy**.

DROUGHT (*GANHAN*). Along with **floods** and **earthquakes**, China has experienced devastating droughts throughout its history from ancient times to the present, primarily in the arid and semiarid (*ban ganhan*) regions of the north and northwest, and even in subhumid regions of the southwest and southeast. Defined meteorologically as an abnormal period of moisture deficiency relative to long-term averages for a given region, droughts have enormously devastating effects, including major **water** shortages for human and animal consumption, massive losses of crops that can bring about famine, deterioration of soil moisture and overall **soil quality**, and major drops in levels of groundwater. Serious droughts occurred in the 19th century, most severely from 1876 to 1878, and in the early 20th century, from 1928 to 1930, lasting 340 days, primarily in the north, and 1941 to 1942, in **Henan province**, spawning widespread famine.

Since the establishment of the People's Republic of China (PRC) in 1949, the country has endured 76 droughts from 1950 to 2006, covering at least 150,000 km^2 (58,350 mi^2) and lasting longer than three months, although 50 of these were shorter than six months. Included is the 1960 drought, which, coming in the midst of the highly destructive **Great Leap Forward (1958–1960)**, contributed to the Great Famine (1959–1961). With increasing dryness from 1990 onward, the country has been hit by a series of droughts, with concomitant reduction in precipitation, including reduced snow fall. Of greatest length and severity was the 1997–2003 drought, which extended for 76 months, affecting 3.9 million km^2 (1.5 million mi^2), 40 percent of the country's land mass, mainly in the north, central, and southwestern regions, peaking in October 1997, with severe water shortages, expanding **desertification**, and major dust and sand storms. Since then, major droughts have struck various regions: 2006–2007, along the **Yangzi River**, causing a significant drop in river water levels; 2008–2009, in North China; 2009–2010, in the southwest, primarily **Yunnan province**, with large-scale crop failures; 2010–2011, in eight provinces in the north and the Yangzi River Delta, the worst in 60 years; 2015, in Guangdong province, where 1,100 reservoirs went dry; and 2017, in North China, the worst in recorded history for that region. "Flash droughts" (*xunjian ganhan*) of extreme heat, low soil moisture, and elevated water evaporation rates have also become more frequent, resulting in significant shortfalls of potable drinking water. Changing weather patterns have led to a shorter rainy season, with 37 percent of the country becoming drier (and 26 percent wetter), causing significant crop damage, most notably in the southwest, to water-guzzling rubber (*xiangjiao shu*) and eucalyptus trees (*anshu*).

Among the major causes of the increasing frequency of drought, especially since the 1990s, is **climate change**, including rising temperatures and reduced precipitation, particularly in North China. With the recent weakening and southward shift of the East Asian monsoons, the primary source of precipitation for the country, along with the meteorological effects of El Niño and Arctic oscillations, environmental conditions in such **cities** as **Beijing** and **Shanghai** have become similar to desert regions in the Middle East. Exacerbating the situation is the prevalence of water-gulping industries like the **mining** of **coal**, **steel** production, and coal-fired power plants, along with wasteful practices by Chinese farmers and households, whose excessive water consumption is subsidized by the Chinese government, with water, like many natural resources in the PRC, priced well below market value.

Measures to alleviate drought conditions include provision of emergency water and grain supplies to areas suffering severe deprivation; government assistance in digging wells to relieve local water shortages; building pumping stations to provide **irrigation** to crops; and shooting silver iodine (*yindian*) into the atmosphere to generate rainfall. Longer-term plans include the completion of the **South-to-North Water Diversion Project (SNWDP)**, which will bring surplus waters to the north from the wetter and more humid south, where average rainfall is more than three times greater than the north, as prospects of a warmer and drier climate lead to enhanced evaporation, ensuring the persistence of periodic droughts as more and more soils lack the 120-day period of moisture necessary to grow crops.

E

EARTHQUAKES (*DIZHEN*). Located in one of the world's most seismically active zones, China has a long history of devastating earthquakes, most recently in 1976, in the eastern city of Tangshan (**Hebei province**), and in 2008, in eastern **Sichuan province**, causing 240,000 deaths in the former and 90,000 in the latter, including many children in poorly constructed school buildings. Environmental damage from both events was severe, especially in Sichuan, where the quake hit a heavily mountainous area, causing a flurry of **landslides** that blocked **rivers** and streams, with long-term effects to their **ecosystems** and harm to the nearby **Wolong National Nature Reserve** for the giant **panda bear**. A major release of air and water pollutants, although temporary, occurred, including 80 tons of liquid ammonia (*ye an*) from a damaged **chemical** plant in Sichuan.

Among the possible human-induced causes of earthquakes is the sheer weight of large-scale reservoirs produced by the construction of **hydropower** stations in Sichuan and Hubei province near the giant **Three Gorges Dam**, which some Chinese scientists believe may contribute to more active seismic activity beyond the purely natural causes from tectonic and other natural forces.

EAST CHINA SEA (ECS/*HUADONG HAIYU*). Covering an area of 1.24 million km^2 (478,000 mi^2), extending from the southern border of the Republic of Korea (ROK) to the northern tip of Taiwan, the East China Sea (ECS) has an average depth of 370 meters (1,210 feet), with a maximum depth of 2,715 meters (8,905 feet). With inflows of industrial and agricultural effluents from several major rivers, including the Qiangtang, Oujiang, **Min** (Fujian), and especially the **Yangzi**, the ECS suffers from serious **water pollution**, one of five major global seas, another of which is the **South China Sea**, infected by coastal eutrophication (*shuiti fu yingyanghua*). Attendant "dead zones" (*siqu*) in ECS waters contain significant levels of inorganic nitrogen phosphates (*dan linsuanyan*), **oil** hydrocarbons (*shiyou ting*), organic matter from **sewage**, heavy metals (*zhong jinshu*), and antibiotics (*kangshengsu*) from **pharmaceutical** plants. Major sources of these pollutants include agricultural run-off from excessive use of synthetic **fertilizers** with high nitrogen (*dan*) content and such highly polluting activities as dead seaweed from coastal porphyra (*zicai*) farms. The release of diesel fuel and petroleum into ECS waters periodically occur from shipping accidents, as occurred in January 2018, when a 128 mi^2 oil slick was created by a damaged Iranian oil tanker. Major **fisheries** in the ECS include cuttlefish (*wuzei*), mackerel (*qingyu*), shellfish (*beilei*), shrimp (*xia*), tuna (*jinqiang yu*), and the highly prized yellow croaker (*huangyu*), with many species suffering from overexploitation, although the sea remains a vital migratory route for humpback and gray whales (*zuotou, hui jingyu*). Crucial to the conversion in the **energy** sector away from **coal**-burning, ECS is site of a major underwater natural gas field located off Zhejiang province.

ECOCITIES AND ECOVILLAGES (*SHENGTAI CHENGSHI/NONGCUN*). With an urban population estimated at 813 million people in 2018, which, by 2030, will constitute 75 percent of the population of the People's Republic of China (PRC), construction of ecocities, also known as "green cities" (*lüse chengshi*), and ecovillages is central to Chinese government plans to create a less polluted, low-carbon intensive environment in both urban and rural areas. A major part of China's strategy for urban–rural development in the early 2000s, ecocities and ecovillages are designed to decouple economic growth from environmental degradation through a holistic, integrated approach to urban and rural development, including the following: the construction of energy-efficient buildings and retrofitting of existing structures, with greater reliance on **solar power**; the promotion of "green" transit—monorails, trams, and centrally located and highly accessible cycling and walking paths in major urban areas—reducing reliance on **motor vehicles**; carving out large areas of parks and vegetation; and **recycling** water and household waste by employing pneumatic municipal collection in place of vast numbers of **garbage** trucks with conversion into energy-producing methane gas through anaerobic biodigestion (*yanyang hengwu xiaohuoa*).

Among prefecture-level cities in China, 25 are undergoing construction as ecocities (Sino-Singaporean **Tianjin** as the most prominent), with 280 slated for future construction, many on existing toxic waste sites and under globally recognized LEED (Leadership in Energy and Environmental Design) certification. Considered an essential component of China's pursuit of global leadership in **sustainable development** and green **energy**, ecocities will substantially reduce China's overall **ecological footprint** via the facilitation of such programs as the Eco-City River Basin Management Program and with international involvement with facilities like the United Kingdom–China Research Center for Eco-Cities and Sustainable Development.

Early failures of planned ecocities include the Dongtan project on Chongming Island in **Shanghai** and Caofeidian in **Hebei province**, with critics, both domestic and international, noting excessively vague government standards allowing lax ecocity designation. Ecovillages are also being developed in rural areas involving the rebuilding of old houses to ensure retention of rainwater, access to solar power, and more favorable sunlight exposure of living areas, along with the construction of community parks on old garbage dump sites. In 2016, Hangzhou, Zhejiang province, was the site of an international forum on ecovillages involving the Sunshine Ecovillage Network and the Global Ecovillage Network, along with Chinese participants.

See also "FOREST CITIES" (*SENLIN CHENGSHI*); "SPONGE CITIES" (*HAIMIAN CHENGSHI*).

ECOFARMING (*SHENGTAI NONGYE*). Designed by agricultural scientists and opposed to modern farming practices involving heavy reliance on synthetic **fertilizers** and **pesticides**, and development of **genetically modified** foods, ecofarming is based on the concept of maintaining **soil** nutrients, employing traditional sources of fertilizer from animal manure, and using agricultural wastes like plant stalks and husks to improve overall **soil quality**. Increasing the variety of seed types and using biogas as a source of farm fuels are also major goals of the ecofarming movement, which is promoted in such major agricultural regions as the Shandong and Guangxi provinces.

See also ORGANIC FARMING (*YOUJI GENGZUO*).

ECOLOGICAL DISASTERS (*SHENGTAI ZAINAN*). One of the countries most affected by natural and man-made disasters, China has suffered from **earthquakes**, **floods**, **forest** fires, **landslides**, and typhoons, along with dam collapses and famines, producing death tolls in the hundreds of thousands into the millions. Since the establishment of the People's Republic of China (PRC) in 1949, areas suffering major ecological disasters—natural and man-made—have more than doubled, with the extent of destruction amplified by a lack of infrastructure and failure of the country's many poorly constructed reservoirs (*shuiku*) and buildings. Most devastating was the earthquake that struck the eastern city of Tangshan (**Hebei province**) in July 1976, killing 242,000 people, and the even more disastrous man-made Great Famine (1959–1961), which followed the ill-conceived **Great Leap Forward (1958–1960)** and caused an estimated 20 to 30 million deaths, largely in rural areas. Others have included the collapse of the Banqiao and Shimantan dams, the so-called "iron dams" (*tieba*) built with the assistance of the **Soviet Union** on the Ru and Hong rivers, major tributaries of the **Huai River**, in **Henan province**, during a massive typhoon in 1975, causing an estimated 170,000 deaths, and the massive forest fire in the Greater Khingan

Mountain region (*daxing'anling*) in the northeastern province of Heilongjiang from May to June 1987, burning 6,500 km^2 (2,470 mi^2) of forest and causing more than 200 deaths. More recently, an earthquake struck a mountainous area in eastern **Sichuan province** in 2008, killing 90,000 people, many of them children in poorly constructed school buildings, and causing substantial environmental damage, including blocked rivers and major landslides. **Chemical** spills and explosions have also occurred, the most serious in an illegal storage site in the city of **Tianjin** in August 2015, killing 173 people.

Dealing with such disasters and their aftermath is the responsibility of the Natural Disaster Reduction Center within the Ministry of Civil Affairs, with a Disaster Reduction Plan adopted in 2007, and a National Plan for Comprehensive Disaster Reduction incorporated into the 11th Five-Year Economic Plan (2006–2011), along with instructions to local governments to formulate commensurate disaster plans.

ECOLOGICAL ENGINEERING (*SHENGTAI GONGCHENG*). Defined as the design of **ecosystems** that integrate human society with the natural environment, ecological engineering was introduced into the People's Republic of China (PRC) in the late 1970s, where it has undergone both theoretical development and practical applications in a variety of fields, including **agriculture**, **aquaculture**, **wetlands**, **grasslands**, **coastal ecosystem** restoration, and pollution control. Synonymous with ecotechnology, ecological engineering combines base and applied sciences with engineering processes, relying on natural systems that are widely applied to agroecological engineering and environmental protection. Promoted by late professor **Ma Shijun**, known as the "father of ecological engineering" in the PRC, more than 2,000 sites in the country are engaged in practicing agroecological engineering, aquaculture, and wastewater control.

Major projects employing principles of ecological engineering include using water hyacinths (*shui hulu*) in waters near the city of Suzhou (Jiangsu province) for controlling **water pollution** and producing fodder; generating vegetation growth and carbon biomass stock to counter land degradation in Southwest China; manipulating **habitat** by growing nectar flowering plants (sesame) and trap plants (*xiangjing zhiwu*) to control rice pests with nonchemical methods; and generating **fisheries** synchronized with wetland production and harvesting on formerly barren coastline and developing marshlands for shoreline protection and food and fuel production. While a modern concept born out of biology, ecological engineering is consistent with classical Chinese principles of yin and yang, and the five elements—wood, fire, earth, water, and metal—to understand and describe phenomena in the cosmos.

ECOLOGICAL FOOTPRINT (*SHENGTAI ZUJI*). An index applied by Global Footprint Network (GFN) to more than 200 countries, with calculations based on data collected by the United Nations, the ecological footprint is defined as impact of human activities within a nation on the environment measured in terms of the area of biologically productive land and **water** required to produce the goods and services consumed and assimilate the waste generated. More simply, the ecological footprint is a measure of the amount of environment necessary to produce the goods and services required to support a certain lifestyle expressed through the global hectare (GHA) measurement unit.

A useful measure of biocapacity, GHA demonstrates how many people a certain territory can sustain, assuming current technologies and agricultural methods, with a global biocapacity per capita of 1.8. Components of the measure include carbon (*tan*), the direct and indirect consumption of fuel and electricity by households, the latter embedded in the production of consumer goods; the amount of cropland, grazing land, fishing grounds, and **forests**, the last for sequestering carbon dioxide; and built-up land for industrial and residential use. Also included in the ecological footprint are the Water Footprint and the Living Planet Index (LPI), the latter a measure of **biodiversity** measuring population trends of vertebrate species from terrestrial, freshwater, and marine **habitat**.

Three reports detailing the ecological footprint of the People's Republic of China (PRC) have been produced, in 2008, 2010, and 2012, with national rankings and basic GHA data published annually by the GFN. In 2008, China's per capita GHA of 2.1 was lower than the global average of 2.7, but higher than the sustainable global biocapacity of 1.8, indicating an overextraction of resources, with the highest and fastest-growing component consisting of carbon. With growing consumerism, especially substantial growth in the automobile market, by 2012 that figure had grown to 3.38, ranking China 71st in the world, and by 2018, to an even higher 3.7 (65th in the world), and with a total GHA by far the largest on the planet, at 5.2 billion, primarily reflecting the country's huge population and with a net deficit of 280 percent, exceeding capacity. Large territorial variations in per capita GHA exist in the PRC, as urban areas, with their higher levels of income and consumption of **energy**, outstrip the poorer rural areas, while highly developed Eastern China scores substantially higher than less developed western regions, but with the latter led by **Tibet**, enjoying much higher biocapacity per capita than the less well-endowed and more densely populated eastern region, particularly large urban areas like **Beijing**, **Shanghai**, and **Guangzhou**, where the carbon component constitutes 50 percent of their GHA score. That the number of provinces exceeding environmental capacity and operating in a deficit has grown from 19 to 26 indicates the deteriorating conditions growing consumerism is producing for China's ecological footprint.

ECOLOGICAL MIGRATION (*SHENGTAI YIMIN*). The movement of people either voluntarily or coerced from environmentally fragile and unsustainable areas, ecological migration and resettlement was begun in the People's Republic of China (PRC) in the 1980s and 1990s, and has involved some of the largest migratory programs in the world. Concentrated in the western regions of the PRC, the vast majority of migrants have come from the provinces of Gansu, Inner Mongolia, Ningxia, Qinghai, **Tibet**, and **Xinjiang**, particularly the highly distressed areas of Xihaigu (Ningxia), Dingxi (Gansu), the **Three Rivers Headwaters Area** (Qinghai), and the Karst region in Southwestern China (Guangxi, Guizhou, and **Yunnan**), where entire villages confronting environmental degradation have been emptied. By far the most significant of these ecological migrations occurred in the Ningxia-Hui Autonomous Region, where 1.14 million people have been moved, primarily out of the arid south, to the more prosperous and livable north, the largest environmental migration in world history. The most important driving force behind ecological migration is **desertification**, brought on by such human-induced actions as overgrazing and **climate change**, with commensurate drops in local **water resources**, as precipitation rates in Western China have dropped during the last several decades, with a concomitant rise in temperatures.

Major goals of ecological migration pursued by the Chinese government include mitigating environmental pressure on marginal lands from overgrazing and agricultural production, allowing for recovery and restoration of seriously degraded areas. Composed of populations with generally high **poverty** rates from some of the most inhospitable territories in China prone to natural disasters, ecological migrants are moved into areas offering better income-making potential, as well as better access to crucial services like **health** care, especially for the elderly. Included among the migrant population are people driven from their homes by large-scale **water-control projects** and **hydropower** ventures, for example, the gigantic **Three Gorges Dam** on the **Yangzi River**, which resulted in the displacement of 1.13 million people from villages and towns inundated by the reservoir (*shuiku*).

Results of government-led ecological migration in China have, by most measures, been mixed, as migrants have experienced improvements in income and access to crucial resources, primarily water, but with concerns about possible violations of human rights, along with unsustainable living arrangements. Oftentimes housed for free in hastily built villages in or near unfamiliar urban areas, with residences generally smaller than their native households and with limited access to farmland, migrants often confront limited employment opportunities, higher costs of living, and a less-relaxed lifestyle than was available in their ancestral homelands, with fears that government subsidies will at some point end. With many migrants being **minorities**, including Muslims, Tibetans, Mongolians, and Ewenkis (the last-known **hunting** tribe in the PRC), the government program of ecological migration is defended as an important measure in sustainable **ecosystem** management, while critics see it as simply a cover for enforced cultural assimilation and general human rights offensives in the interests of the country's Han majority.

ECOLOGY (*SHENGTAI*). A concept in biology originally developed by German scientist Ernst Haeckel in the late 19th century, involving relations of organisms to one another and their physical environment, ecology also describes political movements aimed at protecting environments, especially from pollution. An increasingly important academic discipline in the People's Republic of China (PRC) generally offered in life sciences and a component of **environmental sciences**, ecology is also the centerpiece of the commitment by the Chinese government to the creation of an "ecological civilization" (*shengtai wenming*) in the country's future.

Originally proposed in 2007, the concept of an ecological civilization was laid out in detail by Chinese President **Xi Jinping** at the 19th National Congress of the **Chinese Communist Party (CCP)** in October 2017. Stressing the "harmony between humans and nature," and calling on people to "respect and protect nature," Xi advocated that the PRC "encourage a simple, moderate, green, and low-carbon way of life" that would entail the restoration of the country's much-abused **forests** and **wetlands**, while pursuing a strategy of economic development with a greater reliance on **renewable energy**. Diametrically opposed to the **"war on nature"** promoted by CCP chairman **Mao Zedong** beginning in 1958, particularly during the environmentally destructive **Great Leap Forward (1958–1960)**, Xi's notion calls for the pursuit of further economic moderation within China's environmental capacity, as put forth in the concept of the **ecological footprint**.

Major components of "ecological civilization" include remediating existing environmental degradation and issuing guidelines for compensation from environmental damage; prioritizing pollution reduction and encouraging the more efficient use of natural resources; ensuring **food security**; instituting "green" financing; and promoting stricter enforcement. These became major topics for training CCP and government officials. Proposals were also aired to conduct audits of performance by officials even after they leave office and especially at the local level, where compliance with central laws and directives on the environment is weakest. Establishing clear-cut **property rights** to natural resources by state and private interests, including collectives and individuals, is also necessary for enabling a realistic valuation of environmental assets that have been historically undervalued and subject to overexploitation. Embedded in such ancillary government activities as the Yangzi River Economic Belt Program and the Greater Bay Area Plan for Guangdong Province, ecological civilization is also a topic for research by such popular organizations as the Ecological Society of China, founded in 1979. Greater attention to the lessons of indigenous ecological knowledge (IEK), traditional practices, often by **minority** groups that have maintained **biodiversity** and promoted **sustainable development**, especially in **agriculture**, have also occurred, with these practices often incorporated into government policies.

ECONOMY AND ECONOMIC DEVELOPMENT. Inheriting a national economy that, after years of internal disorder, civil war (1946–1949), and foreign domination, was in disarray, especially in terms of **agriculture**, and with a largely backward industrial base, top leaders in the People's Republic of China (PRC) pursued various policies of economic development and growth, with a major impact on the country's environment. Under **Chinese Communist Party (CCP)** chairman **Mao Zedong** (1949–1976), PRC economic policy began with central planning based on the model pursued in the **Soviet Union** and devoted to rapid development of heavy **industry**, especially **steel**, with little or no attention to environmental matters. Distrustful of markets and committed to large-scale reorganization of agriculture, Mao, despite relatively strong economic growth (6.2 percent annually, 11 percent in heavy industry), eventually abandoned the Soviet model in favor of radical plans for a **Great Leap Forward (1958–1960)**, which ultimately produced the Great Famine (1959–1961), followed by the economic chaos and disruption of the **Cultural Revolution (1966–1976)**, with seriously deleterious effects on the environment.

Shifting to major economic reforms promoted by paramount leader **Deng Xiaoping** beginning in 1978–1979, agricultural production was returned to family farms under the Agricultural Responsibility System, as a major role for markets was pursued in both industry and agriculture, along with dramatic expansion of foreign trade. With environmental protection still largely absent from the national policy agenda, rapid economic growth occurred in virtually every sector of the economy but with highly destructive effects on the country's **forests**, **rivers**, and **lakes**, as well as **air quality** and **water quality**, along with significant deterioration of **grasslands** and **coastal ecosystems**.

Economy and environmental protection were considered polar opposites, as the country experienced between 4 and 15 percent economic growth from 1980 to 2013, while even after the establishment of environmental institutions and the introduction of a series of environmental protection **laws**, the first in 1979, economics often trumped the environment, particularly among local CCP and state officials, as highly polluting township–village enterprises (TVEs) dominated industrial production.

With environmental protection taken more seriously under President **Hu Jintao** and Premier **Wen Jiabao** (2003–2013), economic growth remained strong, with limited initiatives taken on the environmental front as the breakneck speed of economic growth led to excess **energy** and resource demand.. Succeeded by the administration of President **Xi Jinping** and Premier **Li Keqiang** (2013–), a "war on pollution" was declared in March 2014, as many Chinese **cities**, especially in the north, experienced long periods, chiefly in winter time, of severe air stagnation and pollution, a so-called "Airpocalypse," raising the profile of environmental issues in the **media** and popular minds.

Rejecting the doctrine of "pollute first, clean up later" (*xianwuran houzhili*), economic growth was now linked to environmental protection, which had been incorporated into the 11th and 12th Five-Year Economic Plans (2006–2011, and 2011–2015, respectively), with an emphasis on **renewable energy** and **sustainable development**. With a rapidly growing and prosperous middle class of more than 200 million people increasingly concerned with quality of life issues, notably the environment, the PRC, in 2017, had a nominal gross domestic product (GDP) of RMB 13 trillion ($2 trillion), with plans to pursue more advanced industrial development based on low-carbon emissions and better allocation of natural resources to innovation and research and development, in line with the goals of Xi Jinping to achieve a "moderately well-off society" by 2021.

ECOSYSTEMS (*SHENGTAI XITONG*). Defined as a biological community of interacting organisms and their physical environment, ecosystems in China consist of **forests**, **wetlands**, **coastal** shorelines, and **grasslands**, the last covering approximately 400 million hectares (988 million acres), or 41 percent of the national land area, and constituting the largest ecosystem in China and second-largest grasslands in the world. Having undergone substantial degradation from biophysical changes, primarily **climate change**, and social-economic factors brought on by rapid industrialization and commercialization, in particular, beginning in 1953, major efforts at abatement and restoration were inaugurated in the early 2000s, with large-scale capital investments and government financing.

Assessments of the major ecosystems in China were conducted by national surveys carried out in 2000 and 2012, by the **Ministry of Environmental Protection (MEP)**, with the latter survey demonstrating real progress in the realms of carbon sequestration (*tanhui*) and retention, sandstorm prevention, **flood** mitigation, and **water** retention but with little change in **habitat** provision. Forty percent of land in China is incorporated into the Ecosystem

Function Conservation Area (EFCA), while China has hosted such international events as the Symposium on Ecosystem Monitoring and Evaluation. Needed improvements include sharing mechanisms for ecosystem monitoring and data collection, along with further development of legal and institutional innovations for assessment and abatement.

ECOTOURISM (*SHENGTAI LÜYOU*). A niche in the increasingly prosperous tourist market in the People's Republic of China (PRC), the third most visited country in the world, with 55 million foreign visitors in 2015, and an increasingly large number of domestic tourists, especially from the expanding middle class, ecotourism of the country's natural wonders was inaugurated in the 1990s and is still in the developmental stage. Defined internationally as "responsible travel to natural areas that conserves the environment and improves the well-being of the local population," ecotourism in China centers on the 2,750 **nature reserves** located largely in the western provinces of **Sichuan**, **Tibet**, Qinghai, and **Yunnan**, which constitute 14 percent of the country's land mass, with generally large **minority** populations and where ecotourist sites first took hold.

Along with ecoresorts, many at the luxury level, and carbon-neutral ecohotels, some in densely populated **cities** in the east, for example, **Shanghai**, ecotourist sites in China offer access to some of the most pristine areas in the PRC, with lush **forests** teeming with animal **wildlife**. These include **endangered species** like the Siberian/Amur tiger (*Xiboliya hu*); the Sika, or spotted, deer (*bandian lu*); the **golden snub-nosed monkey**; and the internationally revered giant **panda bear**. Major ecotourist sites include: Changbai Mountain, a dormant volcano with Tianchi crater **lake** near the border of North Korea (Jilin); Changqing National Nature Reserve with giant pandas and golden snub-nose monkeys (Shaanxi); Wulingyuan Nature Reserve with the famous sandstone "heavenly" pillars in the Zhangjiajie National Forest Park (Hunan); Hulun Buir Grasslands, the most well-preserved **grassland** in China (Inner Mongolia); Wanglang National Nature Reserve with pandas, black bear, and golden snub-nosed monkeys (Sichuan); and Jiuzhai Valley National Park with waterfalls (*pubu*) and colorful lakes attracting 1 million visitors a year (Sichuan). In addition to their striking natural vistas of dense forests, snow-capped mountains, and rare wildlife, availability of clean **air** and organic foods offered at these remote sites is also attractive to domestic tourists, especially those from densely populated, heavily polluted urban areas with ongoing issues of **food security** and contamination.

While international ecotourist standards call for avoiding any add-on to the local carbon footprint and preventing any additional harmful effects to the environment, relatively loose conditions for "ecotourist" designation in China has often involved lax regulations and weak enforcement, poor training of site personnel, and practices that have made some sites comparable to gaudy theme parks, with mass tourism replacing the more select and limited number of ecotourists. Unplanned and poorly constructed roads opening remote areas to **motor vehicles** are an additional threat to supposedly off-limit core and buffer zones in nature reserves. Also problematic is confused ownership, with managers often pursuing a quick profit, opening up reserves for a fee to local grazing and illegal **logging**, irresponsible tourist behavior reflecting the country's pervasive **"waste culture,"** and lack of public participation in management by locals, notably minorities.

Such poor practices afflict many ecotourist sites, particularly in nature reserves run by local and often corrupt governments that have turned "model" parks like Wanglang (Sichuan) into mass tourism sites, with the local Baima minority allowed to graze livestock and horses, and collect such plants and herbs as the prized Schisandra (*wuweizi*) for **traditional Chinese medicine (TCM)**, with detrimental effects on the park's **ecology**. Opening up remote areas can also be destructive to wildlife, as experienced by the Ordos Relict Gull Reserve, in Inner Mongolia, where pressure from too many misbehaving and ill-educated tourists led to the total destruction of gull (*ou*) nests. Ecotourism, especially in the wild rivers in Yunnan and Sichuan, is also threatened by the country's commitment to **hydropower** in areas like Tiger Leaping Gorge, the world's deepest on the Jinsha River, the upper tributary of the **Yangzi River** in Yunnan province, where revenue from sightseeing, rafting, and kayaking could, in the long run, outstrip electrical power production.

ELECTRIC VEHICLES (EVs). The largest market in the world for electric vehicle automobiles (*diandong qiche*), China in 2019 is home to 2 million units, up from a mere 10,000 in 2013, with total annual sales greater than Japan, Europe, and the United States combined, and constituting 2.2 percent of the entire 28 million-unit Chinese auto market. With automobile emissions becoming a major contributor to **air pollution** in Chinese **cities**, EVs are considered an important factor in providing for cleaner air, as well as enhancing **energy** security by reducing China's growing dependence on imported **oil**. Preceded by the production of electric bicycles, which effectively replaced motorized scooters, which many Chinese cities had banned, the shift to EVs began with the inauguration of the "863 EV Project" in 2001.

Channeling substantial investments into research and development of EVs, amounting to RMB 59 billion ($9.2 billion) from 2009 to 2015, the Chinese government also made available highly attractive financial incentives for consumers that dramatically reduced the price of the approximately 75 EV models now offered in the Chinese market. With the enactment of the Automotive Industry Restoration and Revitalization Plan (2009), priority has been given to producing electric and hybrid vehicles, as well as alternative-fuel vehicles, especially CNG/LNG (compressed natural gas and liquified natural gas), with a goal of EVs constituting 12 percent of the total automotive market by 2020, with a complete ban on internal combustion vehicles slated for before 2040. Major cities like **Beijing** and **Shanghai** already require Euro III **fuel emission standards**, with the former requiring GUO IV emission standards (Euro IV), the first city to implement such standards in the PRC. Along with the production in China of lithium-ion batteries, which power the EVs, China's State Grid Corporation is also building as many as 500,000 EV charging stations nationwide.

Major manufacturers of EVs in China include BAIC (Beijing Automotive Industrial Group), Brilliance, BYD, Chana, Changan PSA, Chery, Dongfeng, Geely, Guangqi Honda, NIO, SAIC (Shanghai Automotive Industry Corporation), and Senova, some in conjunction with foreign automakers from France and Germany, with plans by Tesla of the United States to open a production facility near Shanghai. The first mass-produced plug-in hybrid car produced in China was the BYD F3DM, along with an all-electric minivan, the Luxgen 7 MPV EV, and an all-electric long-range bus (500 kilometers), with academic institutions at major universities like Tsinghua University in Beijing also supporting research on EVs. Electric scooters and bicycles (e-bikes) have also replaced highly polluting gas-powered scooters in most major Chinese cities from the 1990s onward, with production of 20 million e-bikes units annually. With low emissions of CO_2, CO, and NO_x (eight percent of an average automobile), electric scooters and e-bikes provide enormous environmental benefits; the only major drawback is their utilization and ultimate disposal of highly toxic lead-based batteries.

Following the adoption of the National EV Promotion and Application Project (2003), multiple **cities** have quickly converted their public and private transportation systems to electric-powered vehicles. In Shenzhen, EVs constitute 95 percent of all taxis (21,000) and most buses, reducing carbon dioxide emissions by an estimated 850,000 metric tons annually. Environmental issues created by EVs include developing proper and environmentally safe disposal of used lithium batteries (17- to 20-year life cycle) to avoid landfills and other vulnerable sites.

ENDANGERED SPECIES (*BINWEI WUZHONG*). One of the most **biodiverse** countries in the world, the People's Republic of China (PRC) has a large number of animal and **plant species** designated as "endangered" or "protected," including, with percentages in parentheses, mammals (40), **birds** (7), reptiles (28), amphibians (40), and fish (7), some of which are endemic to China. Experiencing a relatively high rate of species loss, at 15 to 20 percent, versus the global average of 10 percent, China became a signatory to the Convention on International Trade in Endangered Species of Wild Fauna and Flora (CITES) in 1981, with the authority to designate a species as endangered and/or protected by the national, as well as local, governments.

Governed by the Law on the Protection of Wildlife (1988), municipalities like **Beijing** have the authority to designate threatened species with class I or class II protection, as registered in the *China Species Red List*.

The top 10 endangered animal species in 2010, with estimated numbers in the wild and/or captivity presented in parentheses, include, in reverse order, the following: Pere David's deer (*milu*/1,600); **Tibetan Antelope** (***chiru***/100,000); black-neck crane (*heijing he*/10,000); **Yangzi River** alligator (*eyu*/10,000); brown-eared pheasant (*zongse yeji*/2,000); crested ibis (*zhu lu*/1,400); South China tiger (*Huanan hu*/98, all in captivity); **golden snub-nosed monkey** (16,000); Yangzi River freshwater dolphin (*baiji tun*/now declared functionally extinct); and giant **panda bear** (*da xiongmao*/1,864). While the poaching of wildlife for purposes of **traditional Chinese medicine (TCM)** and exotic culinary fare continues, several species have been pulled back from near-extinction, for example, Pere David's deer and the South China tiger, the latter no longer in the wild. Various species of snakes (*she*), frogs (*qingwa*), turtles (*gui*), pheasants (*yeji*), partridges (*zhugu*), pika (*shutu*), toads (*chanchu*), paddlefish (*shiwenxun*), and salamanders (*rong*) are also listed as endangered and/or protected, along with the Chinese pangolin (*chuanshanjia*). The last female species of the Yangzi River giant soft-shell turtle (*juxing jiayu*) died in April 2019. Major endangered plant species on the verge of extinction include a rare cypress tree, with only one left in the wild in northern **Sichuan province**; the *baishanzu* fir, with only three left; and many angiosperm (flowered plants with seeds).

ENERGY (*NENGYUAN*). The largest contributor to greenhouse gasses (GHG) in the world, with serious **air pollution**, especially in urban areas, the People's Republic of China (PRC) relies heavily on the burning of **coal**, which constituted more than 60 percent of fuel for power generation in 2017. Clean sources, mainly natural gas (*tianranqi*) and various forms of **renewable energy**, contributed 20.8 percent of total energy production as part of the Chinese plan to decarbonize the national **economy**. The world's second-largest consumer of **oil** behind the United States, China is the largest oil importer, at 8.4 million barrels a day, largely to provide fuel for its burgeoning **motor vehicle** sector, including an explosion in private automobiles, whose emissions as a source of urban air pollution, notably nitrogen oxide (NO_x), have grown substantially.

In 2004, China's end expenditure of energy was 13 percent of the gross domestic product (GDP), double the figure in the United States, while energy consumption per RMB 10,000 of the GDP is 10 times higher than Japan and 3.4 times the world average, indicating enormous waste and inefficiencies, which China is addressing through the gradual elimination of energy price subsidies. The unit energy consumption for 33 major products, for instance,

steel, is 46 percent higher in China than the international average, with demand for total energy continuing to grow and expected to reach 2.3 billion tons coal equivalent (BTCE) in 2020.

In response, China is pursuing various scientific and technological innovations in the production of conventional sources of energy, including coal, oil, natural gas, and **nuclear power**, along with renewables, especially **wind power**, **solar power**, **geothermal power**, and **hydropower**. Losing substantial electrical power over porous long-distance transmission lines, China is also planning a $600 billion upgrade to a proposed nationwide electric power grid. New technologies in coal include carbon capture as eight facilities are planned for the country including the Yanchang Integrated Carbon Capture and Storage Project located in coal-rich Shaanxi for converting 360,000 tons per annum of coal and natural gas and residual heavy **oil** into chemicals products including fuels and **fertilizers**. Other evolving technologies include coal-to-gas and clean coal, the latter in conjunction with several foreign enterprises, particularly from the United States, that specialize in green thermal power generation. Energy consumption to GDP has dropped 19 percent (2005–2010) and 18 percent (2010–2015), with a target of 15 percent (2015–2020), although in individual years, notably in 2007, total emissions have increased, primarily because of increased coal burning. China's development of a Tokamak nuclear-fusion reactor also promises unlimited clean energy but with commercial application still decades in the future.

Dramatic increases in air pollution from coal burning shrouding many of the country's biggest **cities**, including **Beijing**, **Chongqing**, Harbin, and **Lanzhou**, led China to ramp up the import of cleaner-burning natural gas, while older and generally smaller coal-fired plants with outdated and heavily polluting technologies are gradually being closed down. The country has signed long-term contracts with natural gas suppliers in Qatar, Australia, **Russia**, and Indonesia, and is building a pipeline through Myanmar that is partly intended to tap offshore gas supplies. Domestic sources are also being pursued, especially through introduction of hydraulic fracking technology for production of shale gas. Total reserves in the PRC are estimated at 36 trillion M^3 but proven reserves are only 800 billion M^3, concentrated in remote and often **mountainous** regions of the **Sichuan**, Tarim, and **Yangzi River** basins. Embedded in deep, heavily faulted and deformed rock formations that are hard to access and costly to produce, the shale gas industry in China has expanded with projects the likes of the first shale gas liquefaction plant in Sichuan (2014). However, the industry suffers from multiple problems, including water scarcity, population density, a dearth of advanced technology and **transportation** infrastructure (especially adequate pipelines), and low natural gas prices. In a burst of international natural gas deals, China Petroleum and Chemical Corporation (Sinopec) has piloted numerous mergers and acquisitions, including Hong Kong–based China Gas. Natural gas is slated to provide 10 percent of the country's total energy needs by 2020.

In the oil sector, offshore drilling has been promoted, with 3,000-meter deep-water semisubmersible drilling platforms, along with a variety of new technologies through petroleum research and exploration. New technologies are insufficient, however, to prevent continued growth in oil imports, which began in 1993, and are expected to grow to 40 percent of consumption in the 2020s. China is also engaged in research and development of so-called "green petroleum," the conversion of wet biomass into crude-like oil through a process of hydro-thermal liquefaction (HTL) under moderate temperature and high pressure through bio-reactors with carbon dioxide (CO_2) treated as a raw material. Natural gas production constitutes about 3 percent of the national energy consumption and was used primarily as a feedstock for **fertilizer** plants until 1997, when the completion of the Shaanxi–Beijing natural gas pipeline shifted use to electricity production, along with household cooking and heating. Natural gas reserves located largely in Western and North-Central China are estimated at 53.3 trillion FT3, while natural gas production in 2002 was 1.15 trillion FT3. China plans to expand the consumption of natural gas as a substitute for coal by embarking on a major expansion of its gas infrastructure, including the "West-to-East" pipeline from **Xinjiang** to **Shanghai**. Other pipelines connect the vast Ordos gas field in Inner Mongolia to Beijing, while a link into the Russian natural gas grid in Siberia is undergoing construction, with gas expected to start flowing to China in 2019, under a 30-year contract.

In terms of nuclear power (*hedian*), which by not contributing to air pollution is considered a clean, renewable energy in China, 17 nuclear power reactors are in operation at six separate sites, for example, the Daya Bay, Ling'ao, and Qinshan nuclear power stations, with 32 undergoing construction. Safety concerns and potential environmental damage from the March 2011 nuclear disaster at Daiichi in Japan led to a full review of the Chinese nuclear program, which has never experienced a major accident, with China focusing on developing new nuclear technologies, including pressurized water reactors, very-high-temperature reactors like the pebble bed reactor, and the even more advanced fast neutron reactor. In terms of safety, China has announced that beginning in 2017, new construction will be limited to Generation III plants, with plans to raise the percentage of China's electricity produced by nuclear power from 2 percent to 6 percent by 2020.

Wind power, solar power, and hydropower, along with biogas power, tidal power, and geothermal power, have been targeted for development since the adoption in 1994, of the Development Plan of New and Renewable Energy, 1996–2010. Yet, the contribution to total energy production from these sources in 2000 was less than 1 percent, with nonfossil fuels contributing 12 percent of energy consumption in 2015. Institutional fragmentation and lack of coordination between research institutes and **companies** continue to hinder the development of new technologies in these fields. Some in China have also advocated the creation of a Green Dragon Fund, consisting of $200 billion to be devoted to the testing and production of clean-energy technologies in the PRC and throughout the world. The idea is to channel long-term investments to advance potential leapfrog technologies and buy equity in promising start-ups and large companies on the cutting edge of these technologies and investment funds, for instance, Khosla Ventures. Ironically, as one of the world's largest polluters and carbon emitters, China is uniquely positioned to make the massive investment required to jump-start the clean energy revolution by relying on a relatively efficient, timely, political, and economic **policy-making** system, as well as the country's large foreign exchange reserves. Advocates of the Green Dragon Fund are calling for China to divest 20 percent of its more than $1 trillion worth of U.S. Treasury bill holdings and reallocate the money into an energy-focused sovereign wealth fund.

In the midst of **climate change** and dwindling supplies of fossil fuels worldwide, massive clean-energy investments will bring immediate aid to the country, allowing it to continue growing its energy-hungry economy, even as the global economy suffers the consequences of surging energy demands and skyrocketing raw material prices. With these megatrends converging, China will become a world leader in the nascent clean energy revolution, turning the country into a leading international force for renewable energy.

China has enormous hydropower potential, with thousands of fast-flowing **rivers** over steep runoffs from the Qinghai-Tibetan Plateau and heavy rainfall in the south and central regions that have been harnessed by 80,000, mostly small-scale **water-control projects** constructed since 1949. In 2004, these projects provided 21 percent of the nation's total power capacity, a figure that is slated to increase to 24 percent by 2020, as such major projects as the **Three Gorges Dam** on the **Yangzi River** and the Ertan Power Station on the Yalong River come fully online. As China has embraced the model of **sustainable development** with its emphasis on renewable energy resources, plans are also afoot to draw on wind, solar, and geothermal energy. But currently less than 1 percent of energy comes from these diverse sources, as the regions of high solar and wind generation are in the far west, at a considerable distance from the country's heavily populated eastern areas. Pilot solar and wind projects are ongoing in Gansu province, **Tibet**, and Inner Mongolia, where a wind farm of 96 turbines is slated to become the largest in Asia, in conjunction with joint ventures like the German–Chinese solar collector plant in Jinan, the largest such project in the world.

In addition to drafting a Renewable Energy Promotion Law (2004), China has joined the European Wind Energy Association and **Greenpeace** in their "Wind Force 12" project, with plans to grow by 50 to 75 percent a year, so that by 2020, wind will generate 200,000 megawatts of electricity, with 10 percent of all energy in China coming from renewable sources, for example, small hydroelectric dams. Characteristic of many relatively underdeveloped countries with a large percentage of its population in the countryside, China relies on biomass consisting of crop stalks, straw animal manure, and solid waste that is converted into energy by the simple, low-cost process of pyrolysis (use of heat in absence of oxygen) into bio-gas, bio-oil, bio-char, and bio-coal (especially in the rural sector, where overall consumption is dramatically lower than in urban areas).

According to the 10th, 11th, and 12th Five-Year Economic Plans (2001–2005, 2006–2011, and 2011–2015, respectively), China is shifting its focus from the expansion of energy production, the centerpiece of China's five-year development plans since 1980, to an emphasis on the production of clean energy by a massive technological upgrading of the energy **industry**, along with energy efficiency incentives to industry and other users. This will include the construction of a more efficient national electricity grid with more market reforms in the power industry, where prices of energy resources, except coal, are fixed or partially fixed by the state. China's heavy reliance on cheap labor and high consumption of energy and minerals is slated to last no more than 15 to 20 years, necessitating dramatic improvements in energy efficiency by lowering the dependency on coal and imports of mineral resources, improving industrial efficiency, and completing price liberalization. Enormous regional imbalances will continue to exist, however, between the energy-rich eastern, **coastal** regions and the energy-deficient western, thinly populated regions, with increasing public pressure for remediation, especially by more frequent **social protests**.

ENVIRONMENTAL AWARENESS (*HUANJING DE YISHI*). Virtually nonexistent during the period of central economic planning (1953–1978) and the early years of economic reforms beginning in 1978–1979, environmental awareness by the Chinese public has increased dramatically, beginning in the 1990s and extending through the 2010s, with enhanced knowledge of the country's environmental problems and a perceived need for major mitigating actions by government and society. With surveys showing that almost 50 percent of the Chinese people believe **air pollution** is a "very big problem" (2013), while petitions and complaints regarding environmental problems by the public to the National People's Congress (NPC) have exploded in number, 2013 was the evident tipping point in the public mood, when **Beijing** and other major cities were hit by severe "Airpocalypse" events, demonstrating threats to public **health** and well-being, and provoking public debate on environmental issues.

Increased reporting by **journalists** in major national and provincial newspapers and via the **internet**, along with greater **media** attention on television and in the movies, and publicity events on World Environment Day (5 June), have also strengthened public awareness, especially from 2013 onward. Widespread popularity and use of "greenspeak," namely such hot-button words as **biodiversity** (*shengwu duoyangxing*), green consumption (*lüse xioafei*), smog (*yanwu*), and animal rights (*dongwu quanyi*), along with globalized terms like **sustainable development**, have also elevated public attention. Increasingly affected by air pollution and **water pollution**, which impact daily life, people living in both urban and rural areas, many involved in volunteer activities like tree planting and **garbage** collection, are becoming increasingly aware of the multiplicity of environmental problems confronting the country, including **desertification**, greenhouse gas emissions (GHG), and **deforestation**, aided by government transparency involving the publication of air pollution and water pollution levels in urban areas. Two nationwide surveys of environmental awareness were conducted in the PRC in 1998 and 2007, the latter showing increased awareness and knowledge but with continuing skepticism about mitigation efforts by the Chinese government.

ENVIRONMENTAL DIPLOMACY (*HUANJING WAIJIAO*). A term coined in 1989, by the then-chairman of the State Commission on Environmental Protection (SCEP), "environmental diplomacy" pursued by the People's Republic of China (PRC) has gone through a dramatic transformation from general opposition, to global agreements on environmental protection, to assuming an international leadership role, especially on the important issue of **climate change**.

Aligning with a broad coalition of developing countries in the group of 77 (G-77), including, most prominently, Brazil, South Africa, and India (known as BASIC), each of them high emitters of greenhouse gases (GHG) and other pollutants, China took a leading role in maintaining the decision embedded in the Kyoto Protocol (1997) to give the most responsibility for global environmental remediation, notably financing, to developed countries. Arguing for the priority of economic development and national sovereignty, China, along with India and Brazil, failed to agree on targeted goals for global pollution amelioration aimed at halting impending climate change at the United Nation Climate Change Conference in Copenhagen, Denmark, in 2009. Confronting a slowdown in its domestic **economy** and the impending end to the Kyoto

Protocol, China quickly switched positions, assuming an increasingly high-profile role in pursuing **international agreements**, most notably the Paris Accord on Climate Change (2015).

Primary motives for China's embrace of a proactive approach to environmental diplomacy include attracting foreign assistance for domestic environmental amelioration, primarily from the **World Bank** and such supportive developed countries as Canada; building institutional and human capital modeled on leading developed countries like the United States; protecting domestic stability by taking the initiative on environmental protection in the face of increasing and sometimes violent **social protests**; curbing unsustainable economic growth, which was generating significant environmental damage; and burnishing China's global reputation, seeing international environmental leadership as a source of "soft power." Following the abrupt withdrawal of the United States from the Paris Accords in 2017, China made it clear the country was, according to former environmental executive **Xie Zhenhua**, "capable of assuming leadership on global climate change."

ENVIRONMENTAL EDUCATION (*HUANJING JIAOYU*). Considered a vital component in the national environmental strategy of the People's Republic of China (PRC), environmental education is designed to increase public knowledge of and participation in the many issues of environmental protection confronting the country. Begun in the early 1970s following China's participation in the United Nations Conference on the Human Environment (1972), programs in environmental education at the primary, secondary, and university levels, along with adult education programs, are aimed at fostering public stewardship of the environment and promoting ecological sustainability among a population, both young and old, caught up in a **"waste culture"** of extravagant exploitation of scarce environmental resources, especially **water** and food, with 40 percent of the latter wasted.

Environmental nongovernmental organizations (ENGOs) like **Friends of Nature (FON)** have joined with Chinese government authorities in the environmental educational enterprise by sending environmental specialists into rural schools to teach children the importance of environmental protection while establishing easily accessible websites to further public **environmental awareness**. Major government actions include the Platform for Action on Nationwide Environmental Publicity and Education (1996), which is revised every five years, along with the Ministry of Education (2003) mandating the infusion of environmental education content in every subject and at every level of the Chinese public-school system.

Academic disciplines like environmental engineering and **environmental sciences** are offered at most major universities, along with BELL program for integrating environmental education into MBA programs. Large-scale training of primary and secondary schoolteachers in environmental education, along with the creation of "nature school" learning centers and "green schools" promoting campus-wide conservation and **recycling** are also part of the coordinated educational effort. Increased attention to environmental issues has occurred in government-controlled **media**, including in newspapers and on television and radio shows, while such programs as the Environment-Friendly Youth Ambassadors Action Programs are active in enlisting youth participation in environmental protection.

Government and private institutions involved in China's promotion of environmental education include the Center for Environmental Education and Communications in the **Ministry of Environmental Protection (MEP)**, now the Ministry of Ecology and Environment; the **Beijing**-based Environment Literature Research Society, which oversees publications with strong environmental protection themes; the Chinese Society for Environmental Education (CSEE); and such international environmental groups as **Greenpeace**, the World Wildlife Fund (WWF), Ecologia, Friends of the Earth, and Pacific Environment. Many multinational corporations operating in China, such as Volkswagen and British Petroleum, have also contributed to environmental education programs. The Environmental Education Initiative (EEI) is a 10-year joint environmental education program to advance the transition to a low-carbon **economy**.

Gauging the actual impact of environmental education and enhanced publicity efforts is difficult to determine in improving environmental awareness of the Chinese public and especially in altering the "waste culture" that pervades so much of China's increasingly consumer-oriented society. Too often the Chinese public exhibits a casual and indifferent attitude toward disposing of **garbage** and the ever-present Styrofoam lunch boxes, tossing them everywhere and anywhere in pristine **nature reserves and national parks** or into the nation's vulnerable **rivers** and **lakes** with little or no regard for the environmental consequences.

ENVIRONMENTAL ETHICS. *See* RELIGION (*ZONGJIAO*).

ENVIRONMENTAL HISTORY. The history of the environment in China prior to the establishment of the People's Republic of China (PRC) in 1949 was one of great destruction and damage to the natural world, along with significant but limited conservation practices by rulers and the general populace alike. Begun in the ancient period of the Bronze Age (2,000–1,000 BCE), environmental degradation extended through the early (1,000 BCE–300 CE), middle (300–1,300 CE), and late imperial eras (1,300–1800 CE) into the first half of the 20th century.

The primary factor behind this long history of environmental deterioration was the massive extension of sedentary, small-scale **agriculture** into much of the empire, especially marginal and environmentally fragile lands in the south and southwestern regions, under the aegis of a strong centralized state. Nascent development of **industry** also occurred, notably iron smelting, beginning as early as 500 BCE, for production of weapons for frequent wars, especially during the Warring States (475–221 BCE), along with **mining**, primarily of silver (*yin*) and copper (*tong*), which were refined and used for cash in an increasingly freewheeling commercial **economy**, at considerable cost to the environment. Large-scale canals and **water-control projects** were also constructed, largely for **irrigation**, most notably Dujiangyan on the **Min River**, which remains in use today. Following the establishment of a centralized and increasingly powerful state under the Qin Dynasty (221–206 BCE), the ability to reengineer the environment, especially by "controlling the waters" (*zhishui*), became the raison d'etre of state power and the emperor's "mandate of heaven" (*tianming*), supported by agricultural taxes that grew with ever-expanding farm land.

Major victims of this seemingly inexorable process of agricultural expansion and economic development with concomitant increases in the empire's population were the extensive **forest** lands, which were under pressure from the earliest dynasties, notably the Shang (1600–1046 BCE). For the North China Plain, the geographical cradle of Chinese civilization, disruption of forest and vegetative cover accelerated during the expansionist Zhou Dynasty (1045–256 BCE) and had virtually disappeared by the middle of the Han Dynasty (206 BCE–221 CE), turning the region into one gigantic farm, with most **wildlife** also extinguished. Large-scale human migration brought similar results to forest lands in the south and southwestern regions, especially modern-day **Sichuan province**, where small-scale agriculture was introduced into remote hillsides and mountainous regions cleared of substantial tree life by the 11th century.

Dramatic increases in **timber** production were also carried out for shipbuilding and major construction projects like the Great Wall and the Grand Canal, along with more than 4,400 walled **cities** built by the 17th century, many serving as imperial capitals, for instance, Luoyang (690–705, 923–936), Xi'an (221–206 BCE, 618–907 CE), and **Beijing** (1271–1912), featuring large palaces made of wood. **Reclamation** of new farmlands from **lakes**, including **Tai** and **Dongting**, along with consolidation and diking of the **Yellow River** and the construction of seawalls (*haidi*), especially around the city of Hangzhou (imperial capital, 1127–1276), was also carried out, along with the proliferation of earthen works ("polders") in the lowlands of the **Yangzi River**, which produced millions of new cultivatable lands by the 12th century.

Undergoing a "medieval industrial revolution" from the 9th to the 13th centuries, mining and burning of **coal** for industrial purposes, along with household heating and cooking, led to early cases of **air pollution**, where in **cities** like Kaifeng (imperial capital, 960–1127), with a population of 1 million people, Chinese urban society became perpetually separated from the natural world. The expansion of markets for agricultural and timber products increased **deforestation** in increasingly remote regions, often populated by **minorities** whose indigenous lifestyles, largely beneficial to the local environment, became targets of campaigns of "ecowars" destroying such **habitat** as forest land, swamps, lakes, and **grasslands**, which were permanently replaced by farmland and the destructive effects of the plough for what became a largely agrarian empire. While rainforest and subtropical regions, as in **Yunnan province**, remained largely untouched, resources and especially wildlife were threatened by widespread **hunting and poaching**, including by the imperial household. Despite maintaining wildlife reserves and protected forest land, emperors often projected their authority and dominion over nature by the wholesale slaughter of especially ferocious animals like tigers. By the early 1700s, serious shortages and extinction of wildlife, for example, the saltwater alligator, were evident, along with the virtual disappearance of **wilderness areas** throughout much of the empire, notably in **Tibet**, where forest lands were felled in favor of grasslands and pasture for livestock grazing.

With the introduction of such New World crops as maize (*yumi*) and sweet potatoes (*ganshu*), the most marginal and fragile lands, oftentimes on steep hillsides, were subject to cultivation, which, by the 19th century, created a broad environmental crisis, most importantly the widespread loss of **soil** nutrients, detrimental to agriculture. With the extension of increasingly unproductive agriculture into the northwest and other arid and semiarid regions, much of the remaining forest lands came under increased pressure, while major **river** basins—for example, the Han, **Huai**, and Wei—suffered serious degradation.

Along with the major loss of riverine **fisheries** by mid-century, continued sediment buildup led to even more course corrections by the Yellow River (1887, 1897) among many in Chinese history (26 since 595 BCE), producing serious **flooding**. With the imperial state increasingly preoccupied with foreign intervention, especially after the two Opium Wars (1839–1842, 1856–1860), traditional statecraft involving flood control, dike construction, and maintenance of the Grand Canal faltered, with serious environmental and economic effects involving **resource depletion**, particularly in the north. With further plundering of the country's natural resources by foreign invaders, primarily the Japanese in Manchuria (1931–1945), China confronted widespread ecological impoverishment, posing enormous challenges to the new **Chinese Communist Party (CCP)** rulers of the PRC from 1949 onward.

Traditional concepts and principles used to sanctify the plundering of the environment during the imperial era included such Chinese language idioms (*chengyu*) as "draining the lake for fish" (*ba hushui pai gan*) and "killing the hen for eggs" (*shasi muji de dan*). *See* MAPS.

ENVIRONMENTAL IMPACT ASSESSMENT (EIA/*HUANJING YINGXIANG PINGGU*). Adopted by the National People's Congress (NPC) of the People's Republic of China (PRC) in October 2002, and implemented in September 2003, the Environmental Impact Assessment (EIA) Law requires Chinese enterprises—state-owned and private—to investigate the effects of projects on communities and the environment before inaugurating construction. Formulated in the context of the Trial Environmental Protection Law enacted in 1979, EIA establishes a four-stage assessment process involving the design of the investigation, evaluation of background environmental quality, prediction of environmental impact, and analysis and assessment of environmental impact. Seeking to reach a balance between the frequently contradictory goals of economic development and ecological conservation, EIA incorporates provisions and guidelines for conducting environmental impact assessment from the evolving process of ecological impact assessment (ECOIA), which began in the 1970s with a formative stage of development involving the application of methods from abroad and experimentation in different ecological environments. Generally restricted to major national infrastructure construction projects, assessments focused on potential geological and hydrological effects, as well as the impact on local **wildlife** and **plant species**.

Strengthened by the adoption of a series of technical guidelines and shifting from static to dynamic analysis, the Chinese EIA process is the most advanced among developing countries, with substantial involvement by environmental nongovernmental organizations (ENGOs), although with little direct public participation. Theoretically required prior to inaugurating project construction, the law has been frequently circumvented with the collaboration of local government officials, as EIAs were often submitted belatedly in the form of "makeup" assessments, allowing enterprises to bypass the formal process with only minor fines imposed on offenders. Revised in 2016, with substantially stiffer fines of as much as several million RMB, public input was virtually nonexistent, while the amended law allows for other project approvals to precede the EIA, making any project veto on EIA grounds very costly. Responding to criticism of an excessively opaque process, steps were announced in 2016, for increasing public input, while Chinese universities are heavily involved in theoretical and practical research on the assessment process.

ENVIRONMENTAL INSPECTIONS (*HUANJING JIANCHA*). An integral part of environmental enforcement in the People's Republic of China (PRC), environmental inspections are carried out by 18,000 inspectors organized into 200 teams, with investigations focused on enterprises and local government officials who fail to enforce the country's environmental statutes. Following the declaration of a "war on pollution" by Premier **Li Keqiang** (March 2014), inspector authority was enhanced by the 2015 Environmental Protection Law, including imposing substantial fees and closing seriously offending facilities, along with taking punitive action against ineffective and weak local officials. Major targets of inspection include the **energy** and **transportation** sectors, along with land-use policies, as inspections in 30 provinces and regions have led to the shutdown of 30,000 enterprises and penalties against 5,700 officials. As part of the three-year antipollution plan announced in 2018, new rounds of inspections were announced for the **Yangzi River** Delta and Shanxi and Shaanxi provinces, with an emphasis on reducing the release of volatile organic chemicals (VOCs), estimated at 25 million tons in 2017.

ENVIRONMENTAL JUSTICE (*HUANJING ZHENGYI*). Defined as the fair treatment and meaningful involvement of affected populations regardless of status in the development, implementation, and enforcement of environmental **laws and regulations**, environmental justice is increasingly a major topic of concern and debate in the People's Republic of China (PRC), especially among environmental **activists**. With the emergence of environmental nongovernmental organizations (ENGOs), including public interest groups like **Legal Assistance to Pollution Victims** willing to engage in legal action and **lawsuits**, an increasingly autonomous and investigative **media**, and an overall more assertive **civil society**, especially among the emerging middle class in major urban centers, the public space devoted to issues of environmental protection has expanded, particularly on matters of land usage, urban planning, and **health**. Legitimation of public input into environmental **policy-making** by decisions of the ruling **Chinese Communist Party (CCP)** at the 17th National Congress of the CCP in 2007, plus the enactment of laws requiring **environmental impact assessments (EIAs)** and greater transparency in measures involving environmental protection have opened up a previously opaque process with greater recognition of public interests and environmental rights, and more attention to legal and administrative procedures.

As public complaints (*gongzhong tousu*) regarding ill-conceived plans for waste treatment facilities and lax enforcement of environmental regulations of offending industrial facilities, particularly in the **chemical** sector, have ballooned, with the number of complaints in 2006 numbering 600,000, and those in 2009 equaling the entire period from 1997 to 2007, government environmental agencies, chiefly at the local level and in heavily middle-class, urban areas, can no longer operate in a purely top-down, command-and-control fashion with impunity; however, most popular input comes after laws and regulations are enacted as opposed to the process of policy-making, which remains highly insular.

Numerous cases of successful public opposition to the construction of waste incinerators have occurred, for example, in **Beijing** municipality (2012) and Guangdong province (2015), although in most cases the potentially affected communities are prosperous middle-class areas with a generally well-educated population employing scientific evidence, supported by the media, and with informal contacts in relevant government agencies. For poorer areas, particularly in rural communities in Western China, with large numbers of marginal groups, including **minorities**, achieving environmental justice is much less likely, as popular understanding of environmental rights is considerably weaker, with few, if any, organized groups representing local interests and less attention in the media, as was the case with many of the country's so-called **cancer villages**. Formal channels for protest, for instance, local arms of the People's Congress and **environmental protection bureaus (EPBs)**, are often impervious to public complaints from "backward" communities, which fail to transcend local specificities and are often treated as enemies of economic "progress," as even the courts can gum up lawsuits interminably with the willing connivance of local officials. **Social protests**, which can often turn violent, then become the last avenue for direct action, with the distinct possibility of arrest and prosecution for activists and average participants alike.

ENVIRONMENTAL PROTECTION BUREAUS (EPBs/*HUANJING BAOHU JU*). Established as enforcement agencies of the national environmental bureaucracy consisting of the State Environmental Protection Agency (1987–1998), State Environmental Protection Administration (1998–2008), and **Ministry of Environmental Protection** (2008–2018), environmental protection bureaus (EPBs), numbering 3,000, exist at the subnational provincial, municipal, district, county, and township levels. Assisting in drafting local regulations to supplement central policies and directives, EPBs work directly with local factories and especially polluters, along with industrial bureaucracies and government actors, but with dual responsibilities to both higher authorities and local leaders, creating inevitable jurisdictional conflicts.

Established in some locales from the late 1970s onward, EPB heads were generally sympathetic to the macrogoals of rapid economic growth, while funding for the widely understaffed bureaus, primarily from local sources, is often insufficient, especially in the poorer areas of the country. EPBs have also suffered from a lack of routine monitoring of polluters, with, for example, a mere 30 percent of enterprises in **Shanghai** providing data. Operating under a variety of central ministries without independent coercive enforcement authority, EPB personnel often rely on intervention by local **Chinese Communist Party (CCP)** leaders to resolve difficult environmental issues, as the bureaus were demoted in the bureaucratic hierarchy in the 1980s and 1990s. Strengthened by the enhanced commitment to environmental protection by the central government in the early 2000s, with significant expansion of staff to 170,000 nationwide in 2007, EPBs now operate with considerable institutional independence, carrying out diversified functions, with sufficient outside funding, adequate staff, and a routinization of **policy implementation**. Considered one of the powerful "hegemonic powers" (*bawang*) at the local government level, EPBs confronted with resistance from offending parties turn to the courts, including the newly established environmental courts, as the last resort for enforcing key decisions and rulings.

ENVIRONMENTAL PROTECTION LAW. *See* LAWS AND REGULATIONS.

ENVIRONMENTAL SCIENCES (*HUANJING KEXUE*). An interdisciplinary field integrating physical, biological, and information sciences, environmental sciences at the university level and in research institutes have flourished in the People's Republic of China (PRC), especially since the vigorous pursuit of environmental protection and conservation beginning in the early 2000s. Introduced as an area of academic study at Nankai University (**Tianjin**) in 1983, environmental sciences are an absolute necessity in the training of the technically competent personnel charged with handling such complex problems as alternative **energy** systems, pollution control and mitigation, natural resources management, effective waste disposal systems, and effects of global **climate change**, along with providing systematic solutions to environmental problems. As of 2017, 18 higher education institutions in the PRC offer programs in the field, many of them some of the country's most prestigious universities, including Peking, Tsinghua, Fudan, and Zhejiang universities, and Huazhong University of Science and Technology and Harbin Institute of Technology, along with foreign universities operating branches in the PRC, for example, the University of Copenhagen.

Major research institutions include the Chinese Research Academy of Environmental Sciences (publisher of the *Journal of Environmental Sciences*), affiliated with the **Ministry of Environmental Protection (MEP)**, and the Research Center for Eco-Environmental Science, Chinese Academy of Sciences (CAS), where many institutes dealing with environmental matters are housed. Affiliated areas of study include social sciences–oriented environmental studies and the more technologically focused environmental engineering. Subjects touched upon by environmental sciences include atmospheric sciences, biology, **botany**, chemistry, **ecology**, geology, physics, **plant** and **soil** sciences, and zoology.

ENVIRONMENTAL STATISTICS AND ASSESSMENTS. Essential to an effective process of **policy-making**, the collection and assessment of statistics in the People's Republic of China (PRC) has undergone dramatic transformation, replacing the old system of bottom-up reporting through an elaborate bureaucratic hierarchy subject to manipulation and distortion to a modern technological system based on remote-sensing (*yaogan*) by satellites and ground-based radars, providing for more accurate data. During the period of central economic planning (1953–1978), China's statistical system was underdeveloped and suffered from serious flaws, most prominently during the **Great Leap Forward (1958–1960)**, when data collected at the local level by officials competing to impress their superiors suggested massive increases in grain production that proved enormously overblown, contributing to the Great Famine (1959–1961). Similar, although less catastrophic, flaws have plagued the administrative system governing environmental protection and restoration, especially in the case of new **forest** acreage and official measurements of **water quality**, which have proven wildly overstated. Inadequate and unsystematic data collection, along with a lack of transparency by highly secretive government authorities, hampered environmental protection work for decades.

Systems and networks to monitor emissions and other data began in 2006, with a network of high-definition (*gaofen*) earth observation satellites launched beginning in 2013, and numbering 10 in 2019. With high spectral resolution, this system is capable of assessing smog, particulate matter concentrations, and **air pollution** from space, along with enhanced agricultural and forestry surveys, and is supplemented by ground-based radars used to detect the distribution and spread of airborne pollutants. Covering seven provinces and major **cities** in the north, the satellite-radar network is slated to expand nationwide, providing data less subject to manipulation and distortion by local officials accused of such tactics as diverting traffic away from monitoring stations and spraying water on the streets to improve **air quality**.

Additional problems with China's environmental statistics and assessment include the lack of a comprehensive nationwide and standardized system of data collection; excessively low targets of remediation set according to previous standards for measuring progress in cleaning up contaminated areas and achieving improvements in air and water quality; exceptionally upbeat reports by agencies with flexible and easily modified procedures in which data is subject to bargaining and negotiation between local and higher-level officials, as political goals frequently trump statistical accuracy; and limitation of assessment to a few major industries, excluding **agriculture**, and with targeted entities like polluting enterprises determining sample selection, leading to underreporting of contamination levels and generally low-quality data, especially from rural areas. That the national survey of provinces, autonomous regions, and municipalities in 2014 indicated that virtually every unit had reached its targeted goal for environmental amelioration indicates continuing problems of political bias in environmental statistics and assessment. China has also relied on aerial drones specially equipped with cameras and sensors for monitoring air and water quality, begun in 2011, and serving as a supplement to the existing monitoring system.

ENVIRONMENTAL TECHNOLOGY (*HUANJING JISHU*). Environmental protection and remediation require new, advanced technologies, as well as policies, **laws, and regulations**. Monitoring equipment to measure **air quality**, **water quality**, and **soil quality**, along with real-time data collection and transmission devices like drones and optical sensors, and sampling apparatuses, are essential to the "green tech" that make dealing with the enormous problems of industrial effluents and wastewaters and efficient management of **fertilizer** and **pesticide** application such a challenging task. Relying on substantial imports from more advanced economies in Europe, Japan, and North America, along with new domestic startup **companies** specializing in such areas as computer apps for tracking air quality and factory emissions, China is moving on several fronts to build the technological infrastructure for effective pollution management and environmental restoration of blighted water and lands.

Major arenas of environmental amelioration, with examples of necessary technologies listed in parentheses, include the following: **air pollution** control (continuous emissions monitoring systems; ambient air quality monitoring equipment; fuel vapor control systems; source emissions measurement technology); industrial air pollution reduction (wet/dry scrubbers; carbon injection systems; NO_x, mercury, CO_2, and PM 2.5 monitoring systems; catalytic and noncatalytic reduction controls; pumping and fuel-handling equipment; leak detection devices); power plant emissions reduction (dry sorbent injection technology; flue gas desulfurization equipment); solid waste and **recycling** (sorting and composting equipment); recycling discarded electronics (crushing and grinding machines; materials handling equipment); hazardous waste management (wastewater treatment technology; **brownfield** site removal equipment); municipal water and wastewater treatment (advanced filtration and anerobic digestion equipment; biologic denitration); sludge treatment (sludge disinfection equipment; biogas recovery); groundwater monitoring pollution prevention and renewal (sparging; bioremediation; in situ air stripping equipment); water efficiency and reuse (nonrevenue water control software; membranes); processing and producing water (advanced filtration); soil remediation (engineering and design); and environmental engineering and consultation (**environmental impact assessment [EIA]** procedures).

New technologies are also being employed in developing precision **agriculture** in which weather forecasting is combined with data from sensors, aerial photos, and soil properties for monitoring and improving efficiency of fertilizer and pesticide application utilizing crop-dusting drones, and use, along with the **internet** of things (IoT) and SAP S/4 hana enterprise resource-planning software. Calls by national leaders, one of whom is President **Xi Jinping**, to implement a "robotics revolution" are made to improve production and reduce **energy** usage, while the restaurant **industry** is pursuing methods for tracking and cutting back on food waste. With a capacity to handle massive amounts of big data, quantum computers are developing highly complex environmental modeling and carbon-capture technology, along with new, more efficient batteries for **electric vehicles (EVs)** and energy grid management.

E-WASTE (*DIANZI LESE*). The largest dumping ground for electronic waste material from both international and domestic sources in 2013, the People's Republic of China (PRC), with its relatively cheap labor and lax environmental regulations, became a major importer of e-waste beginning in the 1970s, growing to 70 percent of global supply. Consisting of a variety of discarded products, including scrapped air conditioners, refrigerators, television sets, and particularly cellphones and laptop computers, foreign e-waste, despite official bans on imports, continues to pile up in China, primarily through illegal channels via Hong Kong, ending up in such small Chinese towns as **Guiyu** and Qingyuan, Guangdong province, and Taizhou, Zhejiang province. Dirty and dangerous work that is damaging to the local environment, the dismantling and separation of e-waste, especially in small, family-run businesses with unskilled and ill-equipped manual labor, involves the application of hydrochloric acid (*yansuan*) to extract valuable metals like copper (*tong*) and **steel**, along with the burning and incineration of electronic components like circuit boards and the chipping and melting of **plastics** into pellets, which are sold back to textile and other manufacturers. Among the toxic materials and heavy metals released into the local environment are lead (*qian*), beryllium (*pi*), cadmium (*ge*), mercury (*gong*), and hydrocarbon ash (*tan qing huahewu hui*), which make their way into the local **air**, **water**, and **soil**, and, ultimately, crops, including rice and vegetables, creating problems of **food security**. With the growth of a middle class in China wedded to such prized consumer products as cell phones, whose constant technological upgrade leads to a high level of turnover, the PRC is now the second-largest producer of e-waste after the United States.

In addition to the official ban on e-waste imports enacted in 2002, in conjunction with the United Nations Basel Convention on the Transboundary Movement and Dispersal of Hazardous Wastes from developed to developing countries, China has instituted several measures with the support of major multinationals like Nintendo to stem the tide of accumulated domestic e-waste, which amounted to 3 million tons in 2011. Included are stricter regulations imposed by the **Ministry of Environmental Protection (MEP)** in 2008 and 2011, along with the replacement of the myriad small-scale recyclers by the establishment of industrial parks, which, in places like Guiyu, have improved **air quality** and reduced human exposure, especially among children, to toxic substances in dust, water, and soil. More aggressive efforts include pilot e-waste **recycling** programs in major urban areas, including **Beijing**, Hangzhou, and Qingdao, along with an Extended Producer Responsibility (EPR) system adopted from the European Union, requiring manufacturers of electronic goods to carry out extended management of their products even after being discarded, with full implementation slated for 2025. With total e-waste growing by 20 percent annually, reaching 6.7 million tons in 2015, and popular resistance to recycling programs in favor of selling used electronic products to illicit buyers, China's e-waste problem is likely to persist, although poorer regions of the world, especially in Africa, have become the new preferred dumping sites.

EXPENDITURES. Public and private spending and investment on environmental protection and programs, especially for the development of **renewable energy**, have grown substantially in the People's Republic of China (PRC) following the devastating **floods** along the **Yangzi** and other major **rivers** in 1998. By the year 2000, environmental spending, primarily by the government, amounted to RMB 79 billion ($12 billion), or 1.2 percent of the country's gross domestic product (GDP). By 2013, this figure had grown to RMB 420 billion ($64 billion), with RMB 30 billion ($4.6 billion) in direct spending and RMB 390 billion ($59.6 billion) in long-term investments, although this figure represented a 10 percent decline from 2012. In 2015, the total figure grew to RMB 480 billion ($68 billion), with total spending of RMB 5.1 trillion ($770 billion) from 2011 to 2015. At 1.7 percent of the GDP and $6,800 per capita, China, in 2013, fell below more developed nations, for instance, neighboring Japan (2 percent), as most experts agree the PRC should be spending between 2 and 4 percent of its GDP annually on the environment, amounting to RMB 8 to 10 trillion ($1.2 to $1.5 trillion) from 2015 to 2020, as opposed to the slated RMB 4.5 trillion ($680 billion), if full restitution of existing environmental damage and degradation is to be achieved. Plans for spending on renewable energy come to RMB 2.3 trillion ($360 billion) for the same period from 2015 to 2020.

F

FERTILIZERS (*FEILIAO*). Confronting agricultural lands denuded by centuries of overcultivation, the government of the People's Republic of China (PRC) maintained direct control over the allocation of fertilizer, overwhelmingly organic, during the period of central economic planning (1953–1978), largely to high-priority crops, especially basic grains. With agricultural productivity lagging, following the introduction of economic and agricultural reforms in the late 1970s and early 1980s, the production of inorganic **chemical** fertilizers increased dramatically, spurred by the purchase from abroad of 13 large nitrogenous-fertilizer plants. With substantial subsidies to the chemical fertilizer **industry** and agricultural prices increasingly set by the market, overuse of consistently underpriced fertilizers by Chinese farmers quickly occurred, with consumption in the PRC amounting to 30 percent of the global supply on just 9 percent of the planet's arable land.

Along with the equally excessive application of agricultural **pesticides**, chemical fertilizer use in China has grown at a rate substantially greater than the expansion of agricultural production, with one-half of the country's agricultural regions exceeding the international standard by 225 kilograms (495 pounds) per hectare (2.47 acres), with 8.8 million tons deployed in 1978, expanding to 59 million tons in 2013, a 5 percent increase per annum, with a concomitant reduction in the use of animal waste. A mere 30 percent of fertilizer applied to cultivated lands makes it into crops, with 70 percent ending up as agricultural run-off consisting primarily of nitrogen-based compounds, such as ammonia and nitrogen oxides (NO_x). This run-off eventually flows into the country's **rivers**, **lakes**, and groundwaters, creating significant problems of **water pollution** and eutrophication (*shuiti fu yingyanghua*).

Among the major causes of widespread and long-term inefficiency in fertilizer use is the pursuit of high yields by Chinese farming households, encouraged by the country's policy of agricultural self-sufficiency, with limited reliance on imports, particularly in basic grains (*liangshi*). Numbering 200 million in 2017, with 98 percent cultivating small plots averaging less than 2 hectares (4.94 acres), Chinese farmers are confronted with soils made infertile by centuries of overcultivation, along with other problems, including alkalization (*jianhua*), salinization (*yanhua*), and waterlogging (*lao*). The overuse of fertilizers also results from the high and generally unsustainable costs of sophisticated management techniques, which are beyond the reach of the average capital-poor Chinese farmer, along with an overall shift to fertilizer-guzzling vegetable and fruit production.

Efforts to reduce fertilizer overuse include mandating China's agricultural extension service to include environmental goals in their services to Chinese farmers, as pilot programs demonstrate that with greater crop variety and better planting times, along with more rational application of seeds, fertilizer use drops an average of 15 percent, even as crop output increases. China is also shifting to controlled-release and slow-release fertilizers (CRFS and SRFS, respectively), as tougher environmental regulations imposed on the highly subsidized nitrogen fertilizer industry have limited supply, with prices rising largely as a result of market liberalization, which evidently contributed to a decline in fertilizer use in 2017.

FISHERIES (*YUYE*). The largest producer and exporter of fish and fish products in the world, the People's Republic of China (PRC) consumes approximately one-third of the global fish catch, with domestic sources from inland waters and **coastal** seas. With 176,000 km^2 (67,000 mi^2) of **rivers** and **lakes**, plus 98,000 reservoirs and four major seas (**Bohai**, **East China**, **Yellow**, and **South China**), China deploys a fishing fleet of 187,000 vessels, some of which are more than 2,500 distant-water ships, yielding an annual catch of 32 million tons. Prior to 1963, during the period of central economic planning (1953–1978), when fishing was not a priority, much of the freshwater catch came from wild inland fisheries, which, since the introduction of economic reforms in 1978–1979, have declined significantly, a result of overfishing, including the use of explosives and dam construction and land **reclamation** of major lakes. Following the enactment of the Fisheries Law (1986), which called for the rapid development of fishing in Chinese waters and allowed for the privatization of fishing vessels and growth of a domestic fishing market, production increased 200 percent between 1986 and 1995, with commensurate damage to freshwater and marine fisheries.

Among the 1,323 freshwater species (877 endemic to China), 199 are threatened or **endangered**, including the paddlefish (swordfish), giant catfish, barbless carp, and Sichuan taimen (salmon family), with the largest at-risk number found in the species-rich **Pearl River**. Marine fish species found in Chinese waters number 3,000, with 150 fished commercially, some of which are chub mackerel, black scrapper, anchovy, crab, krill, and shrimp. A total of 270 marine fauna species are listed as endangered, with 19 labeled as "critical," notably the popular yellow croaker (*huangyu*), horseshoe crab, Chinese shrimp, and spiny lobster, with four declared extinct. The loss of as much as 50 percent of coastal **wetlands**, 73 percent of coastal **mangroves**, and 80 percent of **coral reefs** has put many marine spawning grounds at risk, even as demand for luxury seafood like crab (*pangxie*), lobster (*longxie*), salmon (*sanwenyu*), scallops (*shanbei*), oysters (*muli*), and abalone (*baoyu*) by the country's expanding middle class continues to increase. Consumption of the traditional luxury item shark fin (*yuchi*) has been heavily criticized and subject to an international boycott in a nationwide **media** campaign.

Major damage to inland fisheries has occurred as a result of large-scale **chemical** spills and widespread **water pollution** of rivers and lakes, while coastal waters have suffered from overfishing, for example, in 2016, when the total Chinese catch came to 13 million tons, surpassing the sustainable coastal catch level of 8 to 10 million tons per year. Confronting the reality of diminishing supply from both inland and coastal fisheries, with catches containing increasingly large amounts of small "trash fish" (*lese yu*) unsuitable for human consumption, the Chinese government has issued a series of policies beginning in 2006, aimed at stabilizing and restoring domestic fish stocks and limiting the annual catch. Included is the Program of Action on the Conservation of Living Aquatic Resources (2006), which addressed the problem of deteriorating aquatic environment, including the decline of fisheries and the increasing number of endangered species. This was followed by more aggressive policies. These included adopting the "China Plan" in 2017, extending seasonal closures of fishing grounds, begun in 1999, to four and one-half months in the summer; banning such damaging fishing practices as trawling with giant nets and the use of explosives and fish poisons, along with new rules on maximum net size; establishing fish quotas for 11 coastal provinces, notably the major fishing centers of Jiangsu and Anhui; cutting fuel subsidies and tax breaks, on which many private fishing firms rely to stay in business; and reducing the number of fishing vessels by 20,000 and commensurately reducing the catch by 15 percent by 2020. Yet, as China's domestic waters will hopefully benefit from these restrictions, increased Chinese presence with the largest distant water fleet in the world in fishing grounds off West Africa is having a deleterious impact on local fisheries crucial to the economic health of poor countries like Senegal and Sierra Leone, where illegal "pair-trawling" by Chinese fishing boats is devastating local fish stocks.

Other controversial fishing practices include the conversion of captured "trash fish" to fishmeal used for feeding larger fish, poultry, and livestock. Relying on highly destructive bottom-trawling, large quantities of small fish are caught, estimated in a **Greenpeace** report at 7 million tons annually, threatening the food chain with disruptive overfishing that has produced a vicious cycle of increasingly scarce quantities of fish pursued by more and bigger fishing vessels, which the Chinese government, despite the efforts at regulation, seems unable to control. Enacting legal remedies like the Fisheries Law (1986/amended 2000), China also signed on to such **international agreements** as the International Plan of Action to Prevent, Deter, and Eliminate Illegal, Unreported, and Unregulated Fishing (2001). Adopting science-based catch management strategies, including total allowable catch (TAC) and fish improvement projects (FIP) for threatened species the likes of the red-swimming crab, China has also participated in major **international conferences**, for instance, the Our Ocean Conference, held in Chile in 2015.

See also AQUACULTURE (*SHUICHAN YANGZHI*).

FLOODS (*HONGSHUI*). With 50,000 **rivers** and 14,500 kilometers (9,000 miles) of coastline, China lives under the constant threat of major floods, especially on the **Yangzi River** and **Yellow River**, with headwaters situated in the high, snow- and ice-covered altitudes of the **Qinghai–Tibetan Plateau** and, in the case of the Yellow River, carrying enormous amounts of sediment (*chendian*). Perhaps the most devastating flood in world history occurred in China in 1931, when massive floods, combing melting snow and ice, joined with heavy spring rains after a long **drought**, overflowed the banks of the Yangzi River and **Huai River**, plus many tributaries, to inundate 180,000 km^2 (69,000 mi^2), affecting eight Chinese provinces and killing an estimated 3.7 to 4 million people.

Since the establishment of the People's Republic of China (PRC) in 1949, major floods have occurred in 1954, 1991, 1998, and 2010, the last in south China, with dramatically smaller death tolls of 30,000, 3,000, 3,856, and 3,185, respectively, causing a loss of 1 percent of the gross domestic product (GDP). Largely a result of vigorous flood control and flood management policies, the Chinese government relies heavily on a network of River Basin Commissions (seven in all) operating under the authority of the Ministry of Water Resources (MWR). Along with such natural phenomena as typhoons and heavy snow/ice melt, largely in the spring flooding season, major causes of floods in China derive from man-made factors, especially **deforestation** and attendant **soil erosion**—particularly in **mountainous** areas, **reclamation** of land from such major flood detention areas as **lakes** and **wetlands**, and the reduction or even the elimination of major flood plains by the expansion of farmland and urbanization.

Concentrating on structural measures for flood control from 1949 to 1998, China extended the river dike system to 280,000 kilometers (173,000 miles). The country built 85,000 dikes, largely during the **Great Leap Forward (1958–1960)** and **Cultural Revolution (1966–1976)**, and 97 large flood detention areas. With many of the dikes (*diwei*) and reservoirs (*shuiku*) suffering from poor construction and subject to failure, as occurred during the devastating 1998 flood, which affected 223 million people and left 15 million homeless, China has since shifted policy to a greater reliance on nonstructural measures. These include making changes in land use from farming to reforestation and the restoration of lakes; reducing the number of people living in flood prone, low-lying areas or removing them altogether, with stricter building codes and better drainage systems; and enhancing efforts at environmental protection, especially to address the potential consequences of **climate change**, for instance, intense precipitation events sparking flash floods. Measures to address urban flooding in cities such as **Shanghai** include constructing roads and paved areas with pervious lightweight concrete that absorbs and diverts rainwater into specialized pipes and storage tanks, avoiding accumulation on surface areas.

In gaining assistance from international agencies like the Asian Development Bank (ADB) and employing such new technologies as digital terrain models, remote sensing, and flood hazard mapping systems, China has adopted an integrated flood management policy that includes a National Flood Management Strategy (2005), drawn up by MWR, although floods remain a constant threat, as occurred on the Yangzi River in July 2018, causing major **landslides**, destroying crops, forcing the evacuation of 1 million people, and killing dozens of people, especially in high-**poverty** areas.

FOOD SECURITY (*SHIPIN ANQUAN*). Concerned with issues of food availability and safety, food security has been a major issue in the People's Republic of China (PRC), especially since the Great Famine (1959–1961) following the disastrous **Great Leap Forward (1958–1960)**, along with more recent and highly publicized cases of tainted food, for example, the contamination of milk and infant formula by melamine (*sanjuqing'an*), copper sulfate on preserved duck eggs, **pesticide** residue on ginger, and cadmium-laced rice. Since the introduction of agricultural reforms beginning in the early 1980s, China has witnessed an enormous increase in food production and supply, achieving 95 percent self-sufficiency in major grains (*liangshi*), including wheat (*xiaomai*), corn/maize (*yumi*), and rice (*mifan*), with concomitant increases in vegetables, meat, and other high-protein products.

While there are still ongoing concerns with food supply, including conversion of agricultural lands to nonfarm uses, growing water shortages, destructive effects of **climate change**, and deteriorating **soil quality**, a general shift has occurred in terms of issues of food safety, particularly as the rate of malnutrition in the country has dropped from 23 percent (1990–1992) to 9.3 percent (2014–2016). Major causes of food contamination include soil and **water pollution** from the overuse of **fertilizers** and **pesticides**, with elevated lead (*qian*), cadmium (*ge*), formaldehyde (*jiaquan*), arsenic (*shen*), and borax (*pengsha*) levels found in rice, noodles, and other foods, excessive antibiotics (*kangshengsu*) and other toxic substances in animals and **plants**, and illegal **chemical** additives and growth hormones by the generally poorly regulated food processing **industry**, along with weak enforcement of food safety standards.

With 80 percent of the Chinese public indicating grave concern about food safety and enormous distrust of the domestic food industry, the Chinese government has issued a slew of **laws and regulations** beginning with a Food Hygiene Law (1995, with a Trial Food Hygiene Law in 1982), strengthened in 2007, and a Food Safety Law (2015), both committed to ensuring hygienically safe food processing. New government bodies have also been established at the national and subnational levels for addressing food security issues, including the State Food and Drug Administration (2013), the National Institute of Nutrition and Safety, the Food Security Committee under the State Council, and the National Center for Food Safety Risk Assessment. Ranked 40th in the world out of 113 nations in terms of food quality, with fewer than 20,000 cases of food poisoning in a 10-year period, the Chinese government confronts a huge problem in regulating a highly fragmented food industry composed of 450,000 producers, while dealing with soil contamination afflicting 15 percent of agricultural lands due mainly to toxic runoff.

"FOOLISH OLD MAN WHO REMOVED THE MOUNTAINS" (*YUGONG YISHAN*). Ancient Chinese mythology on the benefits of perseverance and willpower invoked by **Chinese Communist Party (CCP)** chairman **Mao Zedong** in a 1945 speech, the story regarding the indefatigable efforts of a 90-year-old man to remove two **mountains** penning up his house stone by stone was used to support radical plans for transforming the natural world, especially during the **Great Leap Forward (1958–1960)** and **Cultural Revolution (1966–1976)**. Largely discarded from China's national narrative since the adoption of policies emphasizing environmental protection and conservation, the tale is still sometimes used to glorify the dogged efforts of elderly Chinese farmers to convert barren hillsides into productive farmland and even by environmental **activists** lauding local villagers in their unending quest to pressure large **chemical** producers to remove toxic wastes sites, despoiling their land and **water** supplies. Associated with large-scale engineering projects that often bring environmental harm, the myth with the message of moving mountains remains alive and well in the form of huge land-creation projects in such **cities** as **Lanzhou** (Gansu province) and **Chongqing**, where entire mountains are being flattened and valleys covered with landfill to make way for urban expansion, with potential deleterious environmental impacts like **landslides** and increased **air pollution**.

FOREIGN WASTE (*YANG LAJI*). Once the world's largest importer of waste, constituting 56 percent of the total global supply in 1988, including 70 percent of **e-waste** most from developed nations like the United States, Japan, and the European Union, the People's Republic of China (PRC) instituted a ban on 56 different types of foreign waste beginning in January 2018. Lacking many raw materials and commodities crucial to industrial production, especially of consumer goods, which took off in China in the early 1980s, China imported vast amounts of recyclable materials, for example, **paper**, **plastic**, metals, and other solid wastes, which were then converted into inputs to its evolving industries, including textiles, automobile parts, and electronics, particularly smartphones.

While an entire **industry** of waste processing and **recycling** emerged, poor management and lack of adequate oversight led to increasing problems of domestic **air pollution**, **water pollution**, and **soil** pollution, leading to tougher restrictions on imports under the so-called Green Fence and National Sword initiatives (2013 and 2017, respectively), which imposed a stricter inspection regime on imports of foreign waste, including scrap metal, low-grade plastics, and paper, followed by a total ban in 2018. Declaring much of the imports "dirty" (*zang*) and "hazardous" (*weixian*), China currently accepts as imports only high-quality recyclable wastes with no more than 0.5 levels of contamination, while relying increasingly on the country's own growing waste product, especially of plastics (20 billion pounds in 2010, including 73 billion plastic water bottles), paper, and metals, to generate inputs into manufacturing.

While smuggling of prohibited items still occurs, imposition of the ban is seen as sparking further development and improvement of China's own domestic recycling system, which handled 30 percent of potentially recyclable wastes in 2017 (versus 35 percent in the United States), with prices of discarded items rising. While foreign governments have vociferously protested the ban as a violation of World Trade Organization (WTO) rules, notably the United States, which, with wastes as its sixth-largest export to the PRC, at $5.6 billion, called for its "immediate" termination, China has steadfastly resisted such appeals and even expanded the list of banned items to include automobile parts, certain wood products, and old ships, while urging foreign exporters to solve their own waste problem.

"FOREST CITIES" (*SENLIN CHENGSHI*). An architectural concept designed to convert **cities** into pollution-eating areas, forest cities are a product of a decision by the State Council of the People's Republic of China (PRC) in 2016, calling for new urban construction in the country to be "economic, green, and healthful." Based on an Italian design with construction planned for the city of Liuzhou in north central Guangxi province (once known as the "capital of **acid rain**" because of highly polluting heavy **industry**), the forest city will be built on a 350-acre self-contained site for a planned population of 30,000 people, consisting of 70 buildings, which will be covered with 40,000 trees and 1 million plants that will absorb 10,000 tons of carbon dioxide (CO_2/*eryang huatan*) and other pollutants annually, while producing 900 tons of oxygen and reducing the surrounding air temperature. **Energy** for the buildings will be supplied by **solar** panels installed on rooftops and from **geothermal** sources, with intracity **transportation** provided by a rail line for **electric vehicles (EVs)**. Surrounding parks and open areas will provide **habitat** for **birds** and other **wildlife**, with construction beginning in 2018, and slated for completion in 2020.

See also "SPONGE CITIES" (*HAIMIAN CHENGSHI*).

FORESTS (*SENLIN*). The most **biodiverse** forests in the world, with 2,800 different species, and, at 310 million hectares (765 million acres), covering 21.6 percent of national territory (2018), the People's Republic of China (PRC) is engaged in a major campaign of reforestation. With forest cover at 8.6 percent of the national territory in 1949, with some estimates as low as 5 percent, major forestry areas exist in the northeast in Heilongjiang, Inner Mongolia, and Jilin provinces, in the south including Fujian and Guangdong, and in the southwest in Guizhou, Guangxi, **Sichuan**, and **Yunnan** provinces. Massive **deforestation** occurred during the **Great Leap Forward (1958–1960)** and **Cultural Revolution (1966–1976)**, and in the early years of economic reform (1978–1979), with scientific forest management and data collection widely neglected. Beginning with the Three-North Shelter Forest Program (*Sanbei Fanghulin*), also known as the "Great Green Wall" (*Weida de Lü Qiang*), in 1978, targeting an area from desert regions in **Xinjiang** in the far west to the Greater Khingan Mountains (*daxing'an ling*) in Heilongjiang in the northeast, major efforts at reforestation were pursued, with goals of soil conservation, desert stabilization, **grassland** expansion, and crop protection. Similar campaigns began in the 1980s, one of which was the Great Plains Project, aimed at increasing forest land on the North China Plain and in various regions of the **Yangzi River** and **Pearl River** basins; the Greening Taihang Mountains Program on the border of **Hebei** and Shanxi provinces; the Coastal Protective Forest Project; and the Obligatory Tree Planting Project, mandating that Chinese citizens older than 11 plant three to five trees a year or engage in other forestation projects.

Reforestation programs were expanded from the 1990s onward, especially following devastating floods on the Yangzi River (1998), which stemmed in good part from loss of forest cover, particularly in headwater regions. With total forested areas in the country reaching an all-time low of 17 percent, six key forestry programs were inaugurated, most importantly the National Forest Conservation Program (1998), which banned **logging** in natural forests, and the **Sloping-Lands Conversion Program**, also known as the Green-for-Grain Program (1999). Offering substantial financial incentives to farmers to reforest barren and surplus cropland, these programs also addressed widespread problems of **soil erosion**, that required nationwide programs of reforestation and effective forest management, along with major devotion of resources amounting to RMB 680 billion ($90 billion). Included is Saihanba forest, Hebei province, begun in the 1960s, on sandy lands, and which, at 750,000 km^2 (285,000 mi^2), is the largest man-made forest in the world.

Importing seeds from abroad and planting more than 1.5 billion trees from the early 1980s onward, China increased tree cover throughout the country from Liaoning province in the north down to the Yangzi River in Central China. Consisting almost exclusively of single, uniform poplar trees (*baiy-angshu*), fast-growing trees for the booming **timber industry** and quick to bulk up as a protective barrier for crop fields, the new forests absorb more carbon than slower-growth varieties, assisting in combatting **climate change**, while also reducing China's exploitation of forests in poorer countries in Southeast Asia and Africa.

Since the late 1990s, total forest cover in the country doubled, adding 69 million hectares (170 million acres) by 2013, with an annual average increase of 2 million hectares (5 million acres). The PRC ranks sixth in forest cover in the world behind, in rank order, Brazil, **Russia**, the United States, the Democratic Republic of Congo, and Canada. Still far below the global average of 31 percent total cover, China ranks even lower on a per capita basis, at one-fourth the world average. Chinese forests, particularly in the new plantations, are also of generally low quality, with highly uneven regional distribution and young growth outstripping mature growth by two-to-one. Plans call for national forest cover to reach 23 percent by 2020, with 6.6 million new hectares (16 million acres) slated for 2018.

Forest types in the PRC include cold-temperate coniferous (*zhenye*); temperate coniferous; temperate coniferous and broad-leaved (*kuoye*) mixed; warm coniferous; deciduous (*luoye*) broad-leaved; subtropical evergreen (*changlu*) and deciduous broad-leaved mixed; sclerophyllous (hard-leaf) evergreen broad-leaved; and tropical monsoon rainforests (*redai jifeng yulin*). More than 1,000 tree species in the country have economic value, providing, in addition to timber, such products as foods, edible oils, fruit and nuts, and woody vegetables, along with crucial food sources for **wildlife**. Known as the "kingdom of bamboo" (*zhuzi wanguo*), with an estimated 4 million hectares (9.8 million acres), China has bamboo forest consisting of 500 different species, one-half the global total, including arrow and square bamboo, the staple diet of the giant **panda bear**. With greater emphasis on planting and cultivating mixed forests, as opposed to the previous reliance on single species such as poplar and Korean pine, China is also engaged in developing new, salt-tolerant tree species for the many generally barren coastal areas, such as in Shandong province, afflicted by highly saline soils.

The largest forests in China are Saihanba (Hebei), Greater Khingan (Heilongjiang), Changbai Mountains (Jilin), Linzhi and Bomi (**Tibet**), Tianshan and Altai (**Xinjiang**), Wuzhi (Hainan), Wuyi Mountains (Fujian), Shennongjia (Hubei), and Qinling Mountains (Shaanxi). Top scenic forests include

Tianshan Xueling Spruce Forest (Xinjiang), Shu'nan Bamboo Forest (Sichuan), Changbai Mountain Mixed Forest (Jilin), Jianfengling Tropical Rainforest (Hainan), Baima Snow Mountain Forest (Yunnan), Linzhi Spruce (Tibet), **Xishuangbanna** Tropical Rainforest (Yunnan), Arxan (Inner Mongolia), and Huayang Poplar Forest (Xinjiang). Trees native to China include Camphor (*Zhangshu*), China Fir (*Zhongguo shanmu*), China Tulip (*Yujinxiang*), Dawn Redwood (*Liming hongmu*), Dove (*Gezi shu*) or Davidia, Fujian Cypress (*Fujian bai*), Golden Larch (*Jinse luoyesong*), Gutta Perch (*Yajiao*), Happy (*Xishu*), Hall's Crab (*Chuisi hatang*), Horse Chestnut (*Qiyeshu*), Nanmu Evergreen, Osmanthus (*Guihua*), Parasol (*Wutang*), Weeping Banyan (*Rongshu*), Wingceltis (*Tan*), and Yunnan Cypress (*Yunnan bai*).

Rare and threatened trees, many unique to China, include the following: Catkin Yew (Cat *hongdoushan*), with as many as 400 years of growth and used in **traditional Chinese medicine (TCM)**; Dove, considered a "living fossil," with as many as 800 years of growth; Katsura (*gui*), with a girth measuring as much as 55 meters; Baishan fir (*Baishan lengshan*), the rarest conifer in the world, found in southern Zhejiang province; Dragon Spruce (*Long yunshan*), with as much as 700 years of growth; and the Ginkgo (*Yinxing*), or Maidenhair, with as much as 1,400 years of growth. The ginkgo, known as a "living fossil" because it has remained unchanged for 200 million years, has leaves that turn a golden color in the autumn.

Institutions involved in the training of personnel for forestry management include the Beijing Forestry University and Sichuan Agricultural University. International groups participate in reforestation projects, such as the planting of two million trees in the Horqin Sandy Lands in Inner Mongolia by Timberland Earthkeepers and Green Network. A system of forest police stations under the People's Armed Police is charged with enforcing laws and regulations against pilfering forest products—especially fuelwood for heating and cooking. Research by Chinese scientists indicates that forests of mixed tree species absorb twice as much carbon as single species forest land.

"FOUR PESTS" (*SIHAI*). Targeted for elimination during the campaign launched in the mid-1950s as part of the National Program for the Development of Agriculture, rats (*dashu*), flies (*cangying*), mosquitoes (*wenzi*), and sparrows (*maque*) were considered threats to public **health** and human well-being. With sparrows, technically the Eurasian tree sparrow, singled out by **Chinese Communist Party (CCP)** chairman **Mao Zedong** based on the belief the **birds** consumed large amounts of grain seeds—enough to reputedly feed 60,000 people—millions of sparrows were killed, along with the destruction of nests and the extermination of young chicks. With warnings by scientists and experts of the detrimental consequence of eliminating a major insect predator totally ignored, sparrows were hounded by slingshot-totting youngsters and gong-banging crowds, driving the birds to exhaustion, as so-called "free-fire zones" were set up for mass shootings.

Near-total destruction of the birds left the countryside open to a massive infestation of grain-consuming locusts (*huangchong*), contributing to the Great Famine (1959–1961), which followed the **Great Leap Forward (1958–1960)** and led to the death of between 20 and 30 million people. Bordering on extinction, sparrows were imported from the **Soviet Union** to rebuild the domestic population in China. An example of excessive tampering with the **ecosystem** promoted by Mao, who knew little to nothing about animals, while ignoring expert advice, the campaign to "Eliminate the Four Pests" was an integral part of the Chinese government's **"war on nature."** Wreaking havoc on the Chinese environment, the campaign was replaced in the late 1990s, with an emphasis on environmental protection and restoration, invoking the traditional Chinese concept of "heaven and humanity in harmony" (*tian ren heyi*).

FRIENDS OF NATURE (FON/*ZIRAN ZHIYOU*). The first environmental nongovernmental organization (ENGO) founded in the People's Republic of China (PRC), in 1994, by a group of intellectuals headed by **Liang Congjie**, Friends of Nature (FON) is a cautious, nonconfrontational environmental action group. Focused primarily on raising the **environmental awareness** of the Chinese public, FON engages in educational and publicity programs through workshops, field trips, and the training of teachers, especially in rural areas. Registered with the Ministry of Civil Affairs and prohibited by Chinese law from organizing nationally, the organization is based in **Beijing**, working in conjunction with other Chinese ENGOs, including Green Earth Volunteers, the Nature Conservancy, and the **Institute of Public and Environmental Affairs (IPEA)**, headed by **Ma Jun**, as well as such international groups as **Greenpeace**, the World Wildlife Fund (WWF), and the International Fund for Animal Welfare. Concerned with the fate of China's **wildlife** amid widespread **deforestation** and animal poachers seeking a quick profit, FON has taken on such tasks as protecting the **Tibetan antelope (*chiru*)** and the **golden snub-nosed monkey**, while also joining in public-interest **lawsuits** against offending enterprises, now permitted under Chinese law.

Establishing the Green River Network, FON joined in the largely successful effort to block the construction of **hydropower** projects on the **Nu River** in **Yunnan province**. Major ongoing projects include addressing such issues as urban waste, promoting low-carbon and **energy**-efficient living in Chinese households, and taking on the pervasive **"waste culture"** rife in China's increasingly consumer-oriented society. The recipient of numerous national and international awards, including the Asia Environment Award (1995) and the Ramon Magsaysay Award for Public Service (2000), given to Liang

Congjje, FON has 10,000 members, including professional teams that assist localities in dealing with environmental problems. The organization also publishes a quarterly newsletter.

FUEL AND EMISSION STANDARDS (*RANLIAO PAIFANG BIAOZHUN*). With **air pollution** in major urban areas becoming one of the most important national problems of environmental protection, the People's Republic of China (PRC) has gradually imposed increasingly stringent fuel standards for on- and off-road vehicles, including automobiles, buses, and trucks, while banning the production or import of **motor vehicles** in violation of these standards. Avoiding such rules in the early years of domestic automobile production to assist Chinese auto producers in catching up with their foreign counterparts, for example, General Motors and BMW, which manufacture in the PRC, the first set of fuel standards were implemented in January 2000 (China I), requiring lead-free gasoline. The second set (China II) was implemented in July 2005, allowing sulfur levels of 500 parts per million (PPM), while the third (China III) was implemented in December 2009, lowering allowable sulfur standards to 150 PPM, with the fourth (China IV) and the fifth (China V) sets coming in December 2013 and December 2017, respectively, further reducing levels to 50 and 10 PPM, respectively. The sixth set (China VI) will include additional limits on benzene (*ben*), aromatics (*fangting*), olefins (*xiting*), and temperature levels, which will reduce emissions of hydrocarbons, carbon dioxide, **ozone**, and particulate matter, slated for introduction in **Beijing**, **Tianjin**, and 26 other large **cities** in the northeast provinces of **Hebei**, Henan, Shanxi, and Shandong by December 2019.

Generally consistent with European precedent and in accord with China's Air Pollution Control Action Plan (2005), fuel standards in the PRC are set by the Standardization Administration of China (SAC), with provinces and cities allowed to develop their own, more strict standards with the approval of the National Development and Reform Commission (NDRC). Trucks and buses unable to meet comparable standards for diesel fuel, numbering 400,000, have been retired and replaced by newer models, while China has also set policies governing new **energy** vehicles, especially **electric vehicles (EVs)**, including automobiles.

G

GARBAGE (*LESE*) AND GARBAGE DISPOSAL. Producing 520,000 tons of garbage and rubbish per day, 203 million tons annually, 1 kilogram (2.2 pounds) per capita, the People's Republic of China (PRC) is confronting major problems involving waste disposal and management, especially with the emergence of an increasingly consumer-oriented and waste-producing **economy** in both urban and rural areas. By province, the largest producer of waste, both household and industrial, is Guangdong, while large- and medium-sized **cities** account for 185 million tons annually, with the four largest cities in rank order being **Beijing**, **Shanghai**, **Chongqing**, and Shenzhen. Beginning in 2009, designated as "Year One of the Garbage Era" (*laiji yuannian*), the Chinese government, along with many environmental nongovernmental organizations (ENGOs), moved dramatically to improve waste management, incorporating it as a major task in the 12th Five-Year Economic Plan (2011–2015), and by strengthening the Solid Waste Pollution and Control Law (2015).

Garbage collection and disposal vary substantially throughout the country, with the most effective waste management implemented in relatively high-income urban areas like Beijing, where 10,000 garbage trucks operate daily but with poorer municipalities (and even parts of Beijing) relying on an army of pickers and scavengers, mainly migrant workers, some of which are children. With rubbish, particularly from urban areas, composed primarily of household food wastes and inorganic materials, for instance, **coal** ash (*meihui*), improved solid waste management (SWM) is crucial to environmental amelioration.

Nationwide, increasing amounts of waste are burned in new, environmentally friendly, and high-technology incinerators, with the heat used to generate electricity, although public opposition to these facilities and landfills runs high. With 41 percent of solid waste nationwide (235,000 tons a day) incinerated and a goal of increasing this figure to 50 percent by 2020 (591,000 tons), landfills (*lese tian maichang*) remain the largest disposal sites, especially in poorer communities. Problems of these facilities include groundwater contamination, especially by hazardous waste, such as used lithium-ion batteries from **electric vehicles (EVs)**. Local governments that burn their trash often rely on cheaper and dirtier incinerators.

Among major problems involving garbage disposal in China is the persistent problem of illegal dumping, with an estimated 500 mi^2 in the country covered by wastes and piles of refuse in the countryside and even major cities. Other major and also illegal dumping sites include **rivers** and **lakes**, with estimates that the **Yangzi River** receives 2,000 tons of illegal waste per day. With most waste management in the hands of private companies with a financial incentive to offer the lowest per-ton prices of garbage disposal, sometimes as low as $4 per ton, especially to many cash-strapped local governments, waste is collected, sold to middlemen for a hefty profit, and ultimately dumped illegally, with nondegradable material, especially **plastics**, ending up in the oceans.

Plans also exist to charge households for garbage collection, while mandating **recycling** systems for 46 major cities, by 2020, as the Chinese government is also putting greater emphasis on the development of "green" industries, particularly **renewable energy**. Despite these efforts, **World Bank** estimates are for the generation of 1.4 million tons of garbage per day by 2025, which includes hundreds of thousands of discarded bicycles from newly established bike-sharing companies. Programs are also being pursued by the Chinese government and such environmental nongovernmental organizations (ENGO) as **Friends of Nature (FON)** to bring about significant waste reduction on a national and per capita basis.

GENETICALLY MODIFIED ORGANISM (GMO/*ZHUANJIYIN SHIPIN*). With 20 percent of the world's population, at 1.4 billion people, and only 7 percent of the planet's arable land, **food security** is a perennial problem in the People's Republic of China (PRC). Yet, the country has not embraced widespread production and human consumption of genetically modified organisms, especially foods (GM food). While Chinese scientists have developed a GMO variety of rice that was approved at a technical level for commercial planting, environmental anxieties, along with social concerns regarding food safety, blocked its deployment. Despite the promise of enhanced yields, no GM foods, except for papaya (*fan mugua*), are grown in China and made available for human consumption, with strict bans against GM rice and other domestically grown grains (*liangshi*), for example, corn (*yumi*) and wheat (*xiaomai*). While China does allow the import of GM soybeans (*huangdou*), corn, and rapeseed (*caizi*), especially from the United States, stringent biosafety certifications are required, with the imports going toward animal feed and processed products like cooking oil.

Extensive research is conducted in China on GMO crops at institutions like Shandong Agricultural University, with Shandong province growing the most GM cotton in the country. While Chinese government officials testify to the harmlessness of GM foods, public opinion in the PRC is generally opposed to their introduction into the Chinese market, with stiff punishments meted out for their illegal production and/or sale. China has also blocked imports of foreign grains suspected of containing unacceptable GM varieties, as in 2014, when China rejected U.S. corn imports containing insect-resistant MIR 162. Even with qualities undoubtedly appealing to China, for instance, higher yields, reduced water and **pesticide** requirements, and tougher resistance to hostile growing environments, the country, and especially a general public highly sensitive to food security issues, is in no hurry to adopt GM foods for daily consumption, even as President **Xi Jinping** has declared that the PRC should "dominate the high points of GM technologies."

GEOENGINEERING (*DIQIU GONGCHENG*). Defined as large-scale intervention in the earth's natural systems to counteract the effects of **climate change**, geoengineering was introduced to scientists and policy-makers in the People's Republic of China (PRC) in 2011, drawing on years of experience by many local communities engaged in weather modification efforts. Major research projects began in 2014–2015, involving the study of possible methods of altering weather and **atmospheric** conditions aimed at mitigating China's enormous vulnerability to a changing climate. Potential geoengineering projects include the spraying of particles into the stratosphere to scatter sunlight (in effect replicating the cooling effects of large volcanic eruptions) and making **coastal** clouds more reflective, both designed to offset, at least temporarily, increases in global temperatures, slowing the impact of climate change and providing countries like China more time to decarbonize their **economy**. Critics of these potential projects note the possible environmental side effects of such major **ecosystem** changes, including **ozone** depletion and reduced rainfall, by possibly altering Asia's monsoon precipitation.

Other major Chinese geoengineering projects focus on weather modification, primarily efforts in the dry, arid regions of North China, like Gansu and Shaanxi provinces, along with northeast Jilin province, seeding clouds with silver iodide (*dianhua yin*) or liquid nitrogen (*yedan*) to thicken water droplets, where they become heavy enough to provoke rainfall or snowfall. Endorsed by **Chinese Communist Party (CCP)** chairman **Mao Zedong** during the **Great Leap Forward (1958–1960)** and influenced by similar practices adopted in the **Soviet Union** but largely rejected as ineffective in Western countries, cloud seeding (*rengong jiangyu*) was carried out largely by the Chinese People's Liberation Army (PLA) either by firing **chemical**-filled shells from antiaircraft guns on the ground or dropping them via aircraft into clouds.

Similar projects include a plan to build thousands of chemical furnaces in **Tibet**, burning rocket fuel that would bring about the release of the cloud-seeding agent sulfate iodide (*liusuan dian*), with its crystalline structure, into the air, generating increased rainfall and substantially expanding freshwater supplies on the arid **Qinghai–Tibetan Plateau**. Known as "Sky River" (*Tianhe*) and developed by the China Aerospace Science and Technology Corporation originally for military purposes, the project would increase China's **water** security and even slow the melting of Himalayan **glaciers**, although silver iodide is a potential contaminant. Engaged with the international community from the United States and Europe, also involved in geoengineering, the field involves the use of sophisticated climate models like the Geoengineering Model Inter-Comparison Project, which deals with the possible climatic effects of various projects, for example, solar reduction management schemes. Major research institutes include the Geo-Engineering Investigation Institute in Shanxi province.

GEOTHERMAL POWER (*DIRE*). The extraction of naturally hot water or steam from below the earth's surface and pumped through pipelines largely for heating of homes and businesses, geothermal power is a type of **renewable energy** slated for substantial development in the People's Republic of China (PRC), chiefly as a substitute for highly polluting **coal** and other fossil fuels. With exploration begun in the 1970s and large untapped reserves located specifically in the Himalayan region in **Tibet**, 25 megawatts (MW) of geothermal energy are harnessed in China, with plans calling for production to increase to 100 MW by 2020 (70 million tons coal equivalent), the largest in the world. In addition to the Yangbajain field in Tibet, which is the only geothermal site in China used to generate electricity totaling 100 gigawatt hours supplying 30 percent of demand in the capital city of Lhasa, major geothermal power is used in the city of Xianyang (Shaanxi province), with plans outlined in the 13th Geothermal Development Plan (2016–2020) to develop shallow and medium-depth geothermal sites, particularly in the northeast region, including **Beijing** and **Tianjin** municipalities, as well as **Hebei province**. Working in tandem with companies from Iceland, where 30 percent of energy comes from geothermal sources, China received a $250 million loan from the **World Bank** to tap underground hot springs, which are slated to supply clean **energy** to 336 Chinese **cities**.

GLACIERS (*BINGCHUAN*). Home to 12 glacierized regions with a total in 2017 of 48,571 glaciers covering 51,840 km^2 (20,010 mi^2), 90 percent in **Tibet** and **Xinjiang**, the People's Republic of China (PRC) contains the largest ice mass in the world outside the two poles and Greenland, with 82 percent of glaciers undergoing substantial melting and retreat. Covering 0.6 percent of the national territory, glaciers provide 2 percent of the annual discharge into the country's **rivers**, including major waterways with headwaters in glacial areas, including the **Yellow**, **Lancang**, and **Yangzi** rivers, the last fed by 753 glaciers, providing 5 to 7 percent of the total water volume. Situated at altitudes ranging from 2,800 to 6,000 meters (9,184 to 19,680 feet), glacial regions are found in major **mountain** ranges including Altai, Tian, Qilian, Pamir, Karakoram, Kunlun, Tanggula, Nyainqentanglha, Gangtise, Himalayas, and Hengduan, with Kunlun having the largest total glacier cover, at 12,000 km^2 (4,632 mi^2), and the Karakoram site serving as the longest glacier in the world (42 km/26 miles) outside the north and south poles. Located in the five provinces of Gansu, **Sichuan**, Tibet, Xinjiang, and **Yunnan**, mountain glaciers are of three types—maritime, continental, and subcontinental—determined by differences in regional climates, with all three represented in the country.

Surveys of glaciers in the PRC were carried out by expeditions beginning in 1959, which led to such ill-conceived policies as blackening the ice flows by dumping **coal** dust and soot to accelerate melting to feed into major rivers. More scientifically sound procedures have been followed in recent years, including two major glacier inventories (1979–2002 and 2002–2014) conducted by the Cold and Arid Regions Environmental and Engineering Research Institute, Chinese Academy of Sciences (CAS), with data collected from an extensive system of monitoring stations, some sited on glacial surfaces.

The most scenic and frequently touristed glaciers include Rongbuk (Tibet), at the foot of Mt. Everest (*Qomolangma*) in the Himalayas; Tuoermu (Xinjiang), located at an altitude of 7,435 meters (24,900 feet); Hailuogou (Sichuan), situated at the base of Gongga mountain, highest in the province, and near virgin forests; Telamukanli (Xinjiang), 28 kilometers (17 miles) in length, constituting a "solid water tower" (*guti shuita*) with protruding ice peaks and 2.2 billion cubic meters converted water yield; Midui (Tibet) at 2,400 meters (7,872 feet) above sea level the world's lowest; "July 1st" (*Qiyi*) glacier (Gansu) so named for the date of discovery by a Chinese–Soviet team in 1958 at 5,150 meters (16,897 feet) and located near the Gobi Desert; and Transparent Mengke (Gansu), located in the Qilian Mountains, is one of the country's longest glaciers at 10 km (6.3 miles). Baishui, in Yunnan province, is one of the world's fastest-melting glaciers, having already lost 65 percent of its mass, while glaciers on the **Qinghai–Tibetan Plateau** have sustained a 15 percent loss. Similar rapid glacier shrinkage has occurred in the Tianshan ("Heavenly") Mountain range in Xinjiang, doubling from an annual loss of 5,000 m^2 (1962–1986) to 10,600 m^2 (1986–2017). Declared as a "wake-up call" on global **climate change** by **Greenpeace**, glacial melt in Western China has also produced sudden glacial debris flows contributing to major **flooding** in downstream rivers.

GLOBALIZATION (*QUANQIUHUA*). Defined as the interactions and integration of nations, multinationals, and peoples stemming from improvements in **transportation** and global communication centered on international trade and cultural exchange, globalization has had an enormous impact on the People's Republic of China (PRC), notably the environment. Beginning with the adoption of the economic reform policies (1978–1979), China's environmental conditions were profoundly affected by the rapid growth of international trade with the global **economy** and the impact on Chinese popular attitudes and behavior by international contacts, both economic and cultural. The growth in exports entailed greater user of hazardous materials and **chemicals**, with facilities devoted primarily to selling goods abroad accounting for more than 30 percent of carbon dioxide (CO_2) emissions, amounting to 1,700 million tons (2005). The import and production of **motor vehicles**, especially automobiles, along with the proliferation of increasingly large household units into **biodiversity** hot spots, have also degraded the environment, with autos and trucks becoming a major contributor to **acid rain** and **air pollution**.

The lure of a Western lifestyle, with a strong emphasis on personal consumption, has had a dramatic impact on both urban and rural populations, the latter often living in environmentally sensitive areas where the accumulation of **garbage** and other refuse has become a major blight on the land and in **rivers** and **lakes**, as environmentally sound collection and disposal are seriously underdeveloped. China has also had deleterious effects on neighboring countries, as air pollution, sandstorms, and pollution of major rivers emptying into the seas and generating eutrophication (*shuiti fu yingyanghua*) have impacted other northeastern nations, particularly the Koreas and Japan, and the entire East Asian marine system, notably international **fisheries**. Even such well-meaning environmental protection polices as the ban on most domestic **logging** (1998) have led the PRC to become a major importer of **timber** (third largest in the world), which effectively involves exporting problems of **deforestation** to poorer neighbors in Southeast Asia, for example, Papua New Guinea and Africa.

Positive effects of globalization on the environment are evident, especially in the diffusion of environmental norms into the country by international trade, along with the role of multinationals operating in China, introducing advanced technologies of environmental control and amelioration. **Environmental awareness** among the Chinese public has also been strengthened with the growing availability of ecofriendly products sold by foreign firms and the inclusion of environmental commitments in trade agreements and international commercial contracts, which often require certification of environmentally acceptable production processes by Chinese suppliers to such foreign firms as Walmart and Apple. Compared to Chinese state-owned enterprises (SOEs), with their antiquated and highly polluting production processes, international firms often employ cleaner technologies and more **energy**-efficient operations, while Chinese firms "going global" (*zou chuqu*) confront similar realities when seeking approval to operate abroad, particularly in the developed world.

Contact with international environmental groups like **Greenpeace** has also strengthened the legitimacy and role of comparable Chinese institutions, for instance, **Friends of Nature (FON)** in the PRC. That the State Council mandated Chinese industrial enterprises to meet national environmental standards by the end of 1996, and then shut down tens of thousands of highly polluting factories in October 2017, occurred in a context of an increasingly internationally exposed economy and society brought about by globalization, with China asserting a global leadership role on **climate change** and **renewable energy**.

GOLDEN SNUB-NOSED MONKEY (*JINSE HOU*). Endemic to China and also known as the **Sichuan** golden-hair monkey (*jinmao hou*) and the snub-nosed leaf monkey (*yehou*), the golden snub-nosed monkey is found in the **mountainous** regions of Guizhou, Hubei, Sichuan, and **Yunnan** provinces. Numbering 16,000 in the People's Republic of China (PRC) and considered **endangered**, largely because of loss of **habitat** to **deforestation** and **hunting and poaching**, now outlawed, the species is divided into three types—gold (Sichuan), gray (Guizhou), and black (Yunnan)—with most dwelling in **nature reserves** like Baima Snow Mountain (Yunnan) in dark pine coniferous **forests** under protected status at altitudes of 1,500 to 3,400 meters (4,900 to 11,100 feet), where, in winter, the species withstands snow cover and colder average temperatures than any other nonhuman primate. Classified as "old world monkeys" medium to large in size with nongrasping tails, the monkeys are overwhelmingly arboreal in the wild, feeding primarily on lichen and gathering in groups of 20 to 30 for mutual protection from predators like wolves (*lang*) and eagles (*ying*).

GOVERNMENT STRUCTURE. The institutional structure of the environmental protection bureaucracy in the People's Republic of China (PRC) has developed through several stages toward increasing size and authority beginning with a Leading Group on Environmental Protection under the State Council, established in 1974 and culminating in the formation of the Ministry of Ecology and Environment (MEE/*Shengtai Huanjing Bu*) in 2018. Institutional arrangements during the intervening years, when environmental protection became an increasingly important priority of the Chinese government, included the establishment of the State Environmental Protection Agency (1987–1998), followed by the State Environmental Protection Administration (1998–2008) and the **Ministry of Environmental Protection (MEP)** (2008–2018), succeeded by the MEE, along with the Environment and Resources Protection Committee in the National People's Congress (NPC).

Major administrative players in the environmental protection sector of the Chinese government include **Qu Geping** (1987–1993), **Xie Zhenhua** (1998–2005), and Li Ganjie (2017–), the last minister of the newly created MEE. Functional tasks carried out by departments in the MEP include policies, **laws, and regulations**; science and technology and standards; pollution control; natural **ecosystem** protection; **environmental impact assessments (EIAs)**; international cooperation; nuclear safety; and **environmental inspections**. In addition, there are five regional offices (East, South, Northwest, Southwest, and Northeast), with the implementation of laws and policies carried out by **environmental protection bureaus (EPB)** at the provincial level and below. Monitoring stations numbering 300 were set up to collect **environmental statistics**, while the MEP also oversaw several "key laboratories" and technical research centers dealing with such topics as urban air particles pollution and industrial wastewater pollution control.

The consolidation of **policy-making** and enforcement of environmental protection into the Ministry of Ecology and Environment (MEE) reflected the enhanced commitment by President **Xi Jinping** to creating an "ecological civilization" (*shengtai wenming*) in the PRC and general concern with excessive fragmentation and overlapping jurisdictions between the former MEP, the National Development and Reform Commission (NDRC), and other central government agencies. Environmental policy areas shifted to the MEE from other agencies listed in parentheses include the following: previous responsibilities of the MEP; **climate change** (NDRC); groundwater pollution (Ministry of National Land and Resources); watershed protection (Ministry of Water Resources); agricultural pollution control (Ministry of Agriculture); and marine conservation (State Oceanic Administration). With ministry staff increased to 500 personnel, other changes involve separating management and supervisory functions and employing a more holistic approach to environmental issues through policy coordination with the newly established **Ministry of Natural Resources (MNR)**.

GRASSLANDS (*CAOYUAN*). The largest national resource in the People's Republic of China (PRC), grasslands cover approximately 41 percent of the country's entire land mass. Totaling 400 million hectares (988 million acres), grasslands in China extend from Inner Mongolia to the **Loess Plateau** in Shaanxi, Gansu, and **Xinjiang** provinces in the northwest to the **Qinghai–Tibetan Plateau** in the southwest and west. Third largest in the world, China's grasslands consist of four main steppe types—meadow (*caodi*), typical, desert (*shamo*), and alpine (*gaoshan*)—with the first consisting of moist and fertile **soils** capable of sustaining lush vegetation and forage, and supporting the world's largest flock of sheep and goats, along with livestock, for grazing. Suffering from years of overgrazing, **overpopulation**, unregulated **mining**, and **reclamation** into farmland, along with the deleterious effects of **climate change**, 36 percent of the nation's grassland, estimated at 113 million hectares (273 million acres), is seriously degraded, leading to **desertification**, with hard, impermeable ground surfaces retaining little **water** and subject to serious **soil erosion**, outbreaks of rodent and insect infestation, and deterioration into drifting sand dunes (*shaqiu*).

Legally defined as state property in the 1982 State Constitution, grasslands, under the Grassland Law (1985), have been leased in limited contracts to collectives and individuals involved in **nomadic pastoralism**, while livestock grazing on vulnerable areas was banned in 2006. Offered state subsidies to restrict and limit their livestock, herders have engaged in widespread violations of the ban, while strongly opposing government resettlement programs. Monitoring stations have been set up throughout the grassland regions, which began in 1976, along with various programs, for example, the adoption of a new Grasslands Law (2002) to encourage the more rational use and scientific planning of grassland resources and the Herds for Grass Program (2005), adopted to restore degraded areas, along with ongoing research by the China Ecological Research Network (CERN).

Dramatic improvements have been hampered by the lack of an integrated approach to range management, which would also draw on local knowledge, improvement in pasturing methods, and effective pest control, as the low productivity of most grasslands is below the level required for supporting current livestock production, creating a downward spiral of more animals feeding on less land. The largest and most well-preserved grassland in the country is Hulun Buir (Inner Mongolia), which covers 250,000 km^2 (96,000 mi^2), is known as the "grass kingdom" for its rich variety of grasses, including alkali, esparto, alfalfa, and wheatgrass in a largely pollution-free environment with 3,000 **rivers** and 500 **lakes**. Others include: Sangke (Gansu), Bashang (**Hebei**). Ordos, Qahar, Ulan Qab, and Xilin Gol (Inner Mongolia), Qilian Mountain (Qinghai), Ruoergai (Sichuan), Gannan, Naqu, and Qiang Tang (**Tibet**), Shangri-la (Yunnan), and Bayanbulak, Narat, and Yili (Xinjiang).

GREAT LEAP FORWARD (*DAYUE JIN*/1958–1960). A radical attempt in the People's Republic of China (PRC) to overcome economic backwardness and achieve the stage of "communism" (*gongchanzhuyi*) in one fell swoop through mass mobilization, the Great Leap Forward was a bold plan adopted largely at the behest of **Chinese Communist Party (CCP)** chairman **Mao Zedong** that ultimately produced a major economic and environmental disaster. Relying on the country's almost unlimited manpower to substitute for the severe lack of capital goods, dramatic increases were foreseen in the production of both **agriculture** and **industry** but at enormous costs to the country's agricultural land and especially **forests**, where 400- to 500-year-old Korean pine (*Hong song*), camphor pine (*zhangnao song*), pear tree (*lishu*), and Northeast China ash (*baila shu*) were felled by uncontrolled and widespread **logging**.

Reorganizing Chinese agricultural units into massive, multivillage people's communes (numbering 26,000), highly questionable agricultural practices were adopted, many based on the questionable agronomic theories of Trofim Lysenko from the **Soviet Union**, including deep plowing and close planting of seedlings with the effect of pulling nutrient-poor subsoil to the surface and depriving crops of necessary sunlight and adequate **water**. Poorly conceived **irrigation** projects also led to waterlogging (*lao*), with concomitant increases in soil salinization (*yanhua*) and alkalization (*jianhua*), reducing crop yields. With plans calling for dramatic increases in agricultural production, especially grains (*liangshi*), from 190 to 350 million tons, with little or no additional capital inputs, for instance, **fertilizer**, the quality of soils deteriorated rapidly, with concomitant declines in overall output, even as the state was dramatically increasing grain procurements based on false claims of a surge in national output. Massive land **reclamation** projects were also executed, converting **grasslands** and **lake** beds into largely useless farmland, which throughout time led to **desertification**.

Equally, if not more, destructive was the environmental catastrophe wrought by the resort to the rural **backyard furnaces**, designed to dramatically increase **steel** production from scrap metal, which in the absence of **coal** for fuel led to massive **deforestation**, with entire hillsides and **mountains** denuded of tree cover. While the Great Leap would result in the Great Famine (1959–1961), with an estimated death toll of 20 to 30 million people, mostly in the countryside, China's environment, especially forest land, would take years to recover, as the country remains saddled with hastily and poorly constructed **water-control projects**, including dams and reservoirs prone to failure, which contribute to the constant struggle against persistent and widespread **flooding**.

GREEN BUILDINGS AND HOUSING. One of the largest markets in the world for cost-effective and environmentally friendly structures, the People's Republic of China (PRC) has made the construction of **energy**-efficient "green" (*lüse*) buildings a major priority in the country's rapidly expanding building and housing construction **industry**. Plans call for 520 million m^2 (5.6 billion ft^2) of green buildings through new construction and retrofitting of existing buildings in 336 Chinese **cities**, that, by 2030, will constitute 28 percent of all structures, with some cities, for example, **Beijing**, **Shanghai**, Wuxi, and Changde, requiring new construction to meet green certification. The second-largest market in the world for LEED (Leadership in Energy and Design) certification, introduced into the PRC in 2008, energy efficiency and conservation in the construction industry began in the 1980s, after years of neglecting environmental building standards, especially during the 1950s, when massive apartment blocks were built modeled on housing in the **Soviet Union** with little or no regard for insulation or other ecological and **biodiversity** considerations.

Tighter building standards and codes were first issued in the 1980s, with increased attention to "green" quality construction, especially before such major international events as the 2008 Beijing Summer Olympics. Measures for improving energy efficiency and conservation were also gradually introduced, including the installation of gas, water, and electrical power meters in new buildings, along with new ventilation and air conditioning technologies. Drawing on contacts with such international bodies as the World Green Building Council and with the involvement of domestic environmental nongovernmental organizations (ENGOs) like **Friends of Nature (FON)**, along with government subsidies provided to "green" builders, China is second only to Brazil in terms of annual growth in green building construction for both residential housing and business. Relying on its own three-star rating system and LEED certification, entirely new cities are being built according to green certification, with one example being Suzhou Taihu New City, Jiangsu province. Major universities like Tsinghua and research institutes have also constructed model ultra-low energy consumption ecobuildings in **ecocities and ecovillages**, with involvement by specialized foreign enterprises. China is also looking into development of new forms of construction cements which with addition of chemicals including titanium dioxide turn harmful pollutants such as carbon dioxide and nitrogen oxide into harmless materials.

GREEN CONSUMER GOODS. With an increasingly consumer-oriented **economy** led by an estimated 420 million members of the rapidly expanding middle class, 31 percent of the population in 2012, up from 4 percent in 2002, demand for green consumer goods, especially "organic" (*youji*) foods and **health** products, many imported, has expanded exponentially. Motivated primarily by health and safety considerations, Chinese consumers have undergone a rapid transformation, demanding more "green" products, even with their generally higher price. Included are foods certified by the Green Food Development Center of the Ministry of Agriculture with the "green food" (*lüse shipin*) logo; household products and **energy**-saving appliances; personal care products, especially women's cosmetics; clothing; furniture and building materials; and products for babies and children.

According to national surveys, 71 percent of middle-class households in China express a willingness to purchase green products, a figure highly correlated with income, education, and age, with millennials (26 to 35 years old) and women more than men showing the greatest interest. Promoted by the Chinese government in the 13th Five-Year Economic Plan (2016–2020) and supported by state subsidies to producers and consumers in many urban areas, green consumption is also advocated by many environmental nongovernmental organizations (ENGOs), notably **Friends of Nature (FON)**, with its education programs highlighting environmentally friendly and energy-saving products with less packaging, while disparaging such environmentally destructive items as disposable diapers. Government regulations have also been imposed on **chemical** usage in the cosmetic industry, prohibiting 145 chemicals labeled as harmful, while also calling for less packaging and urging the use of reusable shopping bags instead of throwaway **plastic** bags, with financial incentives for turning in old appliances for new energy-saving models. Avoiding products with **pesticide** residue, formaldehyde (*jiaguan*), phosphorous (*lin*), and other harmful substances known to damage the environment, "green" consumers in China rely on accurate labeling of such items as **genetically modified** foods, which are generally unpopular with the Chinese consumer.

GREEN GROSS DOMESTIC PRODUCT (GDDP/*LÜSE GUONEI SHENGCHAN ZONGZHI*). Developed during the 1970s by Western economists and designed to factor into the conventional gross domestic product (GDP) the environmental consequences of economic growth, including **resource depletion**, loss of **biodiversity**, environmental degradation, and restorative costs, the green gross domestic product (GDDP) was first adopted in the People's Republic of China (PRC) in 2004, dropped in 2009, and re-adopted in 2015. Strongly supported in 2004, by then-head of the State Environmental Protection Administration (SEPA), **Pan Yue**, and central leaders, one of whom was **Chinese Communist Party (CCP)** general secretary and president **Hu Jintao**, calculation of the GDDP was carried out by SEPA, in conjunction with the National Bureau of Statistics, yielding an estimated cost to the Chinese **economy** of RMB 511 billion ($77 billion), 3.05 percent of the GDP. Strongly opposed by provincial and local CCP and government officials, whose status and future promotions often depend solely on economic performance, as measured by the standard GDP, following the one report issued in 2006, the GDDP was abandoned three years later, with criticisms that it was based on speculative estimates of **sustainable development**.

Following years of increasingly severe **air pollution** and **water pollution**, with costs, for example, **health** care, rising fourfold from 2004 to 2013, and with central CCP leaders emphasizing the importance of "greening" the Chinese **economy**, the GDDP was reintroduced in 2015, to allow central and local leaders to assess the progress of conservation efforts, which are now considered a major criterion in judging the performance of government officials and organs. Part of the effort to reach sustainable development by 2020, and supported by important leaders, most notably President **Xi Jinping**, research, along with pilot programs into the GDDP, began in 2016.

Major problems in calculating the GDDP include quantifying in physical units of pollution levels, for instance, tons of carbon dioxide (CO_2) emissions, total wastes, **timber** consumption, and other physical quantities extracted or returned to the environment, followed by environmental valuation in which a monetary value is assigned to the quantitative figures. With the PRC as the only major nation to employ GDDP, as such countries as Germany consider the calculation severely flawed, other similar indexes include the Sustainable Development Goals (SDG) Index and the Genuine Progress Indicator (GPI), along with Environmental Performance Indexes produced by Yale and Columbia universities, on which China has scored poorly.

GREENPEACE. One of the largest international environmental nongovernmental organizations (ENGOs) in the People's Republic of China (PRC), Greenpeace East Asia has offices in **Beijing** and **Guangzhou** (2002), as well as Hong Kong (1997), and is headed by Ms. Li Yan, originally from Gansu province, with a degree in **environmental sciences** from Peking University. Oftentimes working in tandem with the Chinese government on such tasks as providing consultations to the National People's Congress (NPC) in preparation of a draft law governing the development of **renewable energy** (2006), Greenpeace also issues reports on the state of the Chinese environment, which frequently take issue with official policies on how best to reduce **water pollution**, while also criticizing environmental offensives by state-owned enterprises (SOEs) like the giant Shenhua Coal Company, largest **coal** producer in the PRC. Subjects addressed in periodic reports on China's environment include **e-waste** in Guangdong province; overdependence on coal and the need to speed up development of renewable energy; the importance of importing environmentally sound **timber**; **glacier** meltdowns in the Himalayan **Mountains**; pollution of **rivers** from the textile industry, especially in Zhejiang province; and rising **ozone** levels in major Chinese **cities**.

Greenpeace also lauds government policies where proven effective, for instance, the significant reductions in **air pollution** in major Chinese cities like **Beijing** and **Tianjin** (2017–2018). Other campaigns by Greenpeace, often in conjunction with such local Chinese environmental nongovernmental organizations (ENGOs) as **Friends of Nature (FON)**, include the following: promoting reusable **chopsticks**; revealing toxic substances in children's toys and clothing, including international brands; opposing the role of large **chemical** multinationals like Monsanto in Chinese **agriculture** and the food **industry**; examining the destructive impact of giant fish farms and **aquaculture** on local **fisheries**; and opposing China's role in advancing illegal **logging** in such poorer areas as the Congo region in Central Africa.

GUANGZHOU. The third-largest city in the People's Republic of China (PRC), with a population of 14 million people, Guangzhou ("Canton" in English) leads the other major Chinese **cities**, including **Beijing**, **Tianjin**, **Shanghai**, and **Chongqing**, in achieving significant environmental improvements, especially in ambient **air quality**. With such major industries as textile and automobile production, along with large numbers of **motor vehicles** operating in the city's perennial humid air, Guangzhou, like most major cities in China, suffered from persistent smog and haze that periodically led the city to declare emergency measures, including mandating emission reductions by local factories and restricting driving days for private automobiles, which ultimately resulted in a 42 percent reduction in fine particulate matter (PM 2.5) levels from 2012 to 2016. Plans to replace the entire fleet of fossil-fueled buses and taxi cabs with 30,000 **electric vehicles (EVs)**, featuring 10,000 public buses, by 2020 promise to continue air quality improvement, with more than 300 days of clear skies in 2016, labeled the "Canton blue" (*Guangzhou lan*) phenomenon. Emblematic of the city's commitment to green technology is the 71-story Pearl River Tower, an **energy**-efficient building powered solely by **wind power** and **solar power**.

Contamination of local **water** supplies, particularly in the **Pearl River** estuary, by industries, agricultural runoff, and easily disposable **plastic** bags has also beset the city, with as many as 3 million people consuming dirty water, while rainfall levels have dropped precipitously. Also affected are local **coastal** waters, which, in 2009, suffered 11 cases of red tide algae plumes. Measures at amelioration of water pollution have included widespread shutdowns of the most seriously polluting factories in Guangzhou and surrounding Guangdong province, most notably in October 2017. Guangzhou also confronts, as does all of Guangdong province, **sea-level rise**, which amounted to 3.4 millimeters per year from 1980 to 2014, along with the impact of **climate change**, which is greater in Guangzhou than for any city on the planet, with growing threats of **coastal erosion**, frequent and more intense typhoons, and salt tides bringing the intrusion of saltwater into freshwater **rivers**.

With a passion for the exotic, especially in cuisines and **traditional Chinese medicine (TCM)**, Guangzhou has been a hub for illegal trade in **endangered species**, with local markets teeming with such rare and threatened creatures as salamanders (*rong*), leather-backed turtles (*pibeigui*), monitor lizards (*jianzhi xixi*), and pangolin (*chuanshanjia*), a large anteater with highly prized scales used in TCM and as meat for expensive southern dishes.

GUIYU. Located in Guangdong province south of the capital of **Guangzhou** on the coast of the **South China Sea**, Guiyu is a township of approximately 150,000 people that became known as the "electronic graveyard of the world." With a workforce consisting largely of **poverty**-stricken migrants from China's interior, large numbers of computers, cell phones, and other electronic products were dismantled, often by hand, producing enormous amounts of **e-waste**, severely damaging the local environment.

The largest e-waste site in the world in 2005, covering 52 km^2 (20 mi^2), as many as 60,000 workers processed discarded electronics from as many as 100 trucks per day, totaling 1.5 million pounds a year, often using toxic acids to strip away parts like computer chips, while burning circuit boards and other components to free up precious metals, including gold (*jin*), silver (*yin*), and copper (*tong*), for resale. With electronic refuse dumped into pools of toxins or thrown into **garbage** piles left uncollected throughout the township, toxic substances leached into local groundwater, making crops, primarily rice, and **water resources** unfit for human consumption. Major toxins to which workers were exposed included cadmium (*ge*), polyvinyl chlorides (*julu yixi*), brominated flame retardants (*xiuhua zuranji*), chromium (*ge*), and mercury (*gong*), considered possible causes of brain damage, cancer, and kidney failure.

Among the most deleterious effects of e-waste is the high rate of lead poisoning (*qian zhongdu*), found primarily in children, as much as 80 percent in one study, with the township judged in 2009, as having the highest level of toxins in the world. Drawing the attention of both domestic and international **media** organizations, along with the Basel Action Network, governing the international export of toxic waste, the provincial Guangdong government relocated the e-waste worksite into a nearby industrial park beginning in 2013, while merging the highly fragmented workshops into larger enterprises but with the township remaining contaminated by toxic remnants. China's ban on the import of **foreign waste**, enacted in January 2018, will likely reduce the e-waste supply, although this may be counteracted by the growth of domestic consumption and subsequent casual discarding of electronic products.

H

HABITAT (*QIXIDI*). Crucial to the survival of **wildlife** and the preservation of **biodiversity**, habitat in the People's Republic of China (PRC) has suffered substantial destruction and deterioration, especially from the 1950s to the 1990s, with major efforts at preservation and restoration beginning in the late 1990s, primarily through the creation and expansion of 2,750 **nature reserves** (2017). The major cause of habitat destruction and fragmentation in the PRC has been extensive **deforestation** brought on by **overpopulation**, urbanization, excessive and illegal **logging**, road development, and encroachment of **agriculture**, which have made many previously primeval **forests** and **wetlands** unable to support substantial numbers of wildlife species like tigers (*hu*), elephants (*daxiang*), and the iconic giant **panda bear**. While several nature reserves have been created to protect the panda, existing habitat consists of no more than mere slivers of **mountainous** terrain set aside in northern **Sichuan province** and extending into neighboring Gansu and Shaanxi provinces, with intrusion into these areas by livestock and wild horses, along with continued and often illegal logging by impoverished local residents.

China's population of wild elephants, estimated at 300, are confronted with the same threat, as the giant animal's rainforest habitat in **Xishuangbanna, Yunnan province**, has been supplanted in large areas by expanding rubber tree plantations with threats from local farmers, who often see the giant animals as predators and "pests" (*haichong*). In such areas as the **Pearl River** Delta, rapid urbanization, growing at more than 8 percent annually, has destroyed 35 percent of the local natural habitat and more than 40 percent of wetlands. Periodic national surveys begun in 1995, employing modern technologies, for example, remote sensing (*yaogan*), track habitat conditions, with the Chinese government pledging in 2010 to halt virtually all **biodiversity** loss by 2020, even as **climate change** remains a potential threat to habitat, notably the bamboo forests, on which giant pandas depend.

HE BOCHUAN. One of the first authors in the People's Republic of China (PRC) to sound a public alarm on the emerging environmental crisis in the country, He Bochuan published *China on the Edge: The Crisis of Ecology and Development* in 1988, during the height of political reform and liberalization in the PRC, with translation into English in 1992. A former lecturer in the philosophy of science at Zhongshan (Sun Yat-sen) University in **Guangzhou**, Guangdong province, He blamed much of the country's deteriorating environment, as well as other social and economic problems, on the path of economic development and reform enhancing the role of markets pursued since 1978–1979. Exploding population, inefficient **water** allocation policies, and a confused land ownership system that denied farmers the right to own the land, thereby disincentivizing long-term investment in **soil quality** and other environmentally friendly policies, were, for He, major factors in leading to poor environmental management and overall environmental deterioration. Issued by the relatively obscure Guizhou People's Publishing House, an outlet for other dissenting works, including those by environmental **journalist** Ms. **Dai Qing**, He's book sold more than 400,000 copies in the PRC before being banned by the Chinese government in the period following the military crackdown against prodemocracy demonstrators in Tiananmen Square in **Beijing** and other **cities** in June 1989.

HEALTH. Rapid economic growth in the People's Republic of China (PRC) has led to major environmental problems, with significant deleterious effects on public health among both the urban and rural population but with substantial progress in alleviating serious, especially waterborne, diseases. Among the major environmental causes of health risks in the PRC are **air pollution** and **water pollution**, as well as **soil** contamination, especially from heavy metals and other toxins, with air as the biggest problem from the lethal mixture of **coal**-burning and **motor vehicle** emissions in the **cities** and poor indoor air quality in rural areas from coal/biomass-fueled stoves and heating systems. With contaminated air responsible for strokes and heart disease among the public, the World Health Organization (WHO) estimates that 300,000 Chinese annually die prematurely, a figure affirmed by Chinese medical researchers, but some estimates were as high as 1.6 million people in 2013.

While China has made remarkable progress in bringing piped-in water to 94 percent of urban areas in 2007, up from 30 percent in 1985, which has significantly reduced such waterborne diseases as cholera (*huoluan*), dysentery (*li*), typhoid (*shanhanzheng*), and schistosomiasis (*xuexichongbing*), two-thirds of the country's rural areas still lack such facilities, particularly among the poorer populations, which remain exposed to groundwater often containing **chemicals**, heavy metals (e.g., lead [*qian*]), and other toxins. Among the major health problems are the so-called **cancer villages**, with high rates of the deadly disease, along with clusters of children suffering from lead poisoning in the countryside, where notoriously polluting sources include unregulated **chemical** factories, township and village enterprises (TVEs/*xiangzhen qiye*), and such **e-waste** sites as **Guiyu** township in Guangdong province.

Chinese government action to improve environmental protection and reduce health risks has grown since the late 1990s, most notably with the release of the National Environment and Health Action Plan (2007–2015), which called for a nationwide surveillance network to determine the nature and extent of environmental pollution and public health effects. Along with a program to ascertain the health impact of urban air pollution and smog, begun in 2013 (unreleased as of 2018), the document "Healthy China 2030," released in 2016, also called for heightened monitoring of environmental health risks, which was strengthened by an enhanced technical system proposed in 2018. Yet, even as life expectancy at birth in China improved between 1990 and 2001, the country still lacks a comprehensive analysis of environmental indicators and health risks as new problems arise, for instance, the "multicontaminant" nature of air pollution toxins combing the effects of industrialization and urbanization, primarily from motor vehicles, especially in medium-to-small cities in the PRC, with still-unknown consequences for public health.

HEBEI PROVINCE. Located in North China, with a population of 74 million, and surrounding the two centrally administered municipalities of **Beijing** and **Tianjin**, *Hebei* ("north of the [**Yellow**] river") covers 188,000 km^2 (72,900 mi^2), the 12th-largest administrative territory in the People's Republic of China (PRC). Situated in the heart of the North China Plain, one of the most agriculturally productive regions in the country, Hebei is bordered by the Bohai, Taihang, and Yan **mountains** to the east, west, and north, respectively. Drained by several **rivers**, including the Chao, Dai, Hai, Liao, Luan, Ming, Yongding, and Ziya, the province has two major **lakes**, Hengshui and Baiyandian, the latter the largest freshwater lake in North China, known as the "kidney of north China." Home to the man-made Saihanba **forest**, with 618 vascular plants and a variety of trees, for example, larch, spruce, pine, and birch, **plant species** in the province include the following: the Asiatic dayflower; the Chinese pink; the Chinese violet cress; the common hibiscus; the northern maidenhair fern; the Sichuan pepper; the winter jasmine; the yellow sweet clover; and a rare, recently discovered cliff flower, with more than 200 varieties, including medicinal and industrial plants declared **endangered** or vulnerable. Threatened **wildlife** include large mammals like the Amur leopard (*Amu'er bao*) and the brown-eared pheasant (*zongse de yeji*), while tigers (*hu*) have been declared extinct in the wild, with small mammals in the province largely preserved, along with such **birdlife** as swans (*tian'e*), with many endangered animal species ensconced in Qinghuangdao Wildlife Park, the largest wildlife preserve in the country.

An industrial **economy** based on **mining** and heavy-sector manufacturing, Hebei ranks as the most polluted of the 28 provinces in the PRC, with six of the top 10 most polluted urban areas in the country: Baoding, Handan, Hengshui, Tangshan, Xingtai, and **Shijiazhuang**, the last rated the most polluted city in the entire nation in 2017. Major polluting industries include cement (*shuini*), **chemicals**, **coal** (*mei*), coke (*jiaotan*), glass (*boli*), petroleum (*shiyou*), textiles (*fangzhipin*), and **steel** (*gang*), with mills in several **cities**, including Langfang, Zhanjiakou, Shijiazhuang, and Tangshan, the largest steel-making city in the world. Afflicting the province are high levels of **air pollution** and **water pollution**, along with **acid rain**, contaminated agricultural runoff, and industrial **sewage**, the latter dumped into local rivers and ultimately ending up in neighboring **Bohai Sea**. With pollution more serious in the southern than the northern region and greater during the dry than the wet season, pollution from Hebei spills over into Beijing and Tianjin, with the former claiming that one-third of urban smog originates in Hebei, especially from coal-fired power and steel plants.

Measures introduced by Hebei provincial authorities since the inauguration of the country's 11th Five-Year Economic Plan (2006–2011), and especially following the declaration of a nationwide "war on pollution" by Premier **Li Keqiang** (March 2014), include the following: production caps imposed on major polluting industries, for example, cement, glass, coal, coke, and steel, the last reduced from 320 million tons in 2011, to a target of 200 million tons by 2020, with plans to consolidate production in 70 mills down from a high of 87, along with shifting facilities from major cities to newly established industrial parks. Campaigns have also been launched to clean up polluted groundwater (*dixiashui*), with the goal of reducing grade V (unfit for any use) to 25 percent by 2020, along with expanding "ecological red zones" (*shengtai hongse quyu*), land declared off limits to development. Targets for 2020 also include raising nonfossil fuel consumption in the province to 10 percent, with heavy environmental taxes authorized by the National People's Congress (NPC) on pollution-offending industries, including coal-fired power plants and antiquated industrial boilers. Included in the plan for a regional approach to economic and environmental issues, known as the Beijing–Tianjin–Hebei Integration Project, Hebei authorities have resisted attempts by its two large and powerful neighboring municipalities to shift outdated, polluting industries into the province, while the city of Baoding, still seriously polluted, has emerged as a major center for **renewable energy**, with local **companies** specializing in **solar power** and **wind power**.

Hebei is also encouraging the development of low-carbon service and tourism industries, while engaging in major reforestation projects for enhancing carbon sequestration. Improvement in ambient **air quality** has occurred in most major cities throughout the province, a benefit of the national campaign to reduce air pollution in the north, but with recorded fine particulate matter (PM 2.5) levels still exceeding by substantial margins the annual average standards of 10 micrograms established by the World Health Organization (WHO). State prosecutors are also bringing legal action against local authorities in the province, along with officials in Beijing and Tianjin, for failing to fully implement existing environmental statutes.

HENAN PROVINCE. The third most populated province, at 100 million people (2017), in the People's Republic of China (PRC), *Henan* ("south of the [**Yellow**] river") covers 167,000 km^2 (64,500 mi^2), consisting of fertile lands that make up the North China Plain and in close proximity to the notoriously unpredictable Yellow River, subject to massive **flooding** and, more recently, periodic dry-up. Throughout the course of modern Chinese history beginning in the 20th century, the province was the site of several major **ecological disasters**, many human-created, including the massive floods and loss of life brought about by the intentional destruction of Yellow River dikes (*diwei*) by the Nationalist government of Chiang Kai-shek to halt the advance of the invading Japanese army (1938). Labeled as the "largest act of environmental warfare in history," resulting fatalities totaled between 400,000 and 500,000, with millions made homeless, along with the long-term deleterious impact on surrounding agricultural lands. With the hydraulic and agricultural systems of the province severely disrupted, Henan was struck, three years later, by a major **drought**, which led to widespread famine (1942–1943) concentrated in its western regions, with little or no effort at alleviation by the Nationalist government.

Following the establishment of the PRC in 1949, Henan suffered again, most notably during the **Great Leap Forward (1958–1960)**, when misguided policies were vigorously pursued by provincial **Chinese Communist Party (CCP)** leaders and the province was hit by serious drought and a resulting famine. Relying on highly inflated grain production statistics, with local leaders operating in total denial, claiming "severe drought, excellent crops" (*yanzhong ganhan, youliang zuowu*), the province was once again hit by widespread famine, leading to approximately 3 million deaths, of which there were 1 million fatalities in the single prefecture of Xinyang. Other equally destructive events include the widespread flooding of the **Huai River** (1950), followed by the cascading collapse of dams, many constructed during the **hydropower**-building frenzy of 1958, on the Ru and Hong rivers, Huai River tributaries (August 1975), which resulted in an estimated 170,000 fatalities. In more recent years, Henan has suffered from widespread **air pollu-**

tion, especially in major **cities** like Zhengzhou, the provincial capital, with some of the highest levels in the province, and the ancient capital of Luoyang, site of the Longmen Buddhist grottos, suffering damage from exposure to air contaminants, notably **acid rain**.

Relying on such heavy-polluting **industry** as aluminum, **chemical**, **paper**, **tanneries**, and **coal** production following the inauguration of economic reforms (1978–1979), Henan joined **Xinjiang** as one of the two most polluted provinces in the PRC in 2016, with **air quality** undergoing significant decline in 2018. Major pollutant discharges, many illegal, of industrial wastewater and chemicals, for example, highly toxic dyes (*ranliao*), into local waterways have also been recorded, turning crystal-clear rivers like the Ying and Yun, tributaries of the Huai, into virtual cesspools. Agricultural runoff from the excessive use of **fertilizers** and **pesticides** by the province's prolific grain growers has also had deleterious effects, most prominently the emergence of **cancer villages**, for instance, Duying, Huangmengying, Mengzhi, Sunying, and Xiditou.

Amelioration measures include shutting down the most highly polluting **coal**-burning factories and power plants; shifting **energy** production from coal to natural gas; promoting greater reliance on the province's rich **geothermal** resources; and alternating use in major cities of **motor vehicles**, especially private automobiles, now numbering 3.7 million in Zhengzhou and contributing 25 percent to local air contamination. Henan authorities have also opposed policies pursued by more developed regions like **Beijing** and **Shanghai** of shifting highly polluting factories into the province, while local officials accused of slack enforcement of environmental protection law have, since 2016, lost their positions to increasingly vigorous **environmental inspections** by central government officials.

HORTICULTURE (*YUANYI*). Defined as the cultivation of ornamental **plants**, flowers, fruits, trees, and vegetables, along with landscaping, gardening, and **soil** management in greenhouses and other such facilities, horticulture in the People's Republic of China (PRC) is a major industry in **agriculture**, with 11 million hectares (27 million acres) under cultivation. In a highly diverse climate ranging from tropical and subtropical conditions in the south to semiarid and arid regions in the north and west, China is the world's largest producer of vegetables and fruits, constituting 50 percent of global total, including everything from kiwi (*mihoutao*), citrus (*ganju*), and apples (*pingguo*) to a rich variety of flowers (*huahui*), some with air-purifying functions, demanded by China's growing middle class. Dominated for years by small-scale, family operations with low productivity and lack of climate control, horticulture in China has undergone major structural modernization during the last several years, developing large-scale and more diverse production units with modern technology and facilities, for example, green horticultural production houses.

Areas with horticultural production include Shandong, **Sichuan**, and **Yunnan** provinces, plus the northeastern city of Dalian, along with the desert city of Jinchang (Gansu province), once a heavily polluting **mining** center known as the "nickel capital of China" (*Zhongguo de niedu*), has been transformed into a major flower-growing region. Major foreign involvement and investment from advanced horticultural nations like The Netherlands and New Zealand is also occurring, although the sector is still afflicted by ongoing problems of **water** scarcity, excessive use of **fertilizers** and **pesticides**, particularly by small producers, and **soil erosion** and degradation. Major universities offering academic programs and research in horticulture include China Agricultural University, Sichuan University, and Zhejiang University, with promotional efforts like China Green Week (Horti China) held in **Shanghai** in November, offering a trade platform for new seeds and plants. The manufacture and trade of seeds is governed by the China Seed Law (2000), which, amended in 2016, allows for easier and less cumbersome approval of new seed varieties for the market, as China moves to produce more diverse horticultural products.

The PRC has also hosted the International Horticultural Exhibition on three occasions, 1999 (Kunming, Yunnan Province), 2006, (**Shenyang**, Liaoning Province), and 2019 (**Beijing**) where President **Xi Jinping** opened the gathering featuring 1,200 plan varieties with a commitment to "green development." Manufacture and trade in seeds is governed by China Seed Law, which, amended in 2016, allows for easier and less cumbersome approval of new seed varieties developed by **biotechnology** onto the market as China moves to produce more diverse horticultural products.

HOUSEHOLD WASTE (*JIATING LESE*). The largest generator of household waste in the world, amounting to 220 million tons annually (2018), with projections of 533 million tons by 2030, the People's Republic of China (PRC) has a goal of **recycling** 35 percent of all such waste by 2020. With landfills (*lese tian maichang*) as the most common form of disposal (79 percent), contaminating groundwater and beginning to reach capacity, while incineration (18 percent) poses a threat to public **health**, a functioning recycling system is an absolute necessity in the PRC, where growing urbanization and increased urban consumption has led to waste in some areas littering the streets or ending up in giant uncollected **garbage** and solid-waste dumps.

Consisting of **paper** (30 percent), organic materials (27 percent), **plastics** (18 percent), glass (15 percent), and metal (5 percent), household waste in China's urban areas, which is three times larger in volume than in the countryside, is much easier and less costly to dispose of when first sorted out in bins or other containers by residents prior to collection, a practice many municipal governments have encouraged since 2000, but with little effect on Chinese consumers, as proper sorting is often carried out by unregulated scavengers, many of them children. With pilot programs established in multiple **cities** in a span of almost two decades, with minimal impact on the population's deeply engrained **"waste culture"** of casual discarding and dumping of waste, including animal and human feces, cities are taking a tougher stand, imposing fines on violators discovered in some urban areas through the use of bar-coded trash bags.

In Shenzhen, in Southern China, mandatory recycling of furniture, home appliances, and other large unwanted items has been introduced, the first city in China to institute such a program, while in **Beijing** prizes and other awards are being handed out to households for proper sorting. Nationwide mandatory recycling is slated for implementation by 2020, with a commitment by the central government of RMB 190 billion ($29 billion) to establish the programs, which should also be advanced by the PRC decision to ban most **foreign waste** (January 2018).

HU JINCHU. A specialist on **forest** growth and **habitat** for **wildlife**, especially the giant **panda bear**, Hu Jinchu is a professor in the School of Life Sciences, West China Normal University, Nanchong, **Sichuan province**. Head of the Chinese team cooperating with representatives from the World Wildlife Fund (WWF) in tracing and monitoring giant pandas with fitted radio transmitters in the **Wolong National Nature Reserve**, China's largest panda habitat in the late 1970s, Hu has also overseen the successful breeding of pandas in captivity in Chinese zoos. Author of many published works on the giant panda, notably (in Chinese) *Trailing the Giant Panda* (*Xunzong Guobao*), he has also studied Asiatic leopards (*bao*) and Asian black bears (*hei xiong*), as well as the red panda (*xiao xiongmao*). Preservation of the giant panda, according to Hu, depends heavily on the preservation of old-growth forest in China and an ample supply of arrow and square bamboo (*zhuzi*), on which the animals depend as a major food supply.

HU JINTAO (1942–). Serving as general secretary of the **Chinese Communist Party (CCP)** and president of the People's Republic of China (PRC) from 2002/2003–2012, Hu Jintao joined with Premier **Wen Jiabao** in shifting national policy away from singled-minded economic growth to include environmental protection and restoration, even as degradation continued during their 10-year tenure, including higher levels of **water pollution** and several major **ecological disasters**. A native of Anhui Province, one of China's poorest regions, Hu graduated from Tsinghua University with a degree in water conservancy and engineering, followed by early work experience at the Liujia Gorge Engineering Bureau under the Ministry of Water Conservancy. Working in remote and relatively poor Gansu and Guizhou provinces, as well as the **Tibet** Autonomous Region, from 1968 to 1982, he became acquainted with the major imbalances in China between economic growth and environmental conditions, along with increasing social inequality.

Coining the term "ecological civilization" (*shengtai wenming*) at the 17th National Congress of the CCP (October 2017), Hu declared that with the PRC becoming the world's largest emitter of greenhouse gases (GHG) in 2007, environmental protection had become of "vital interest" to the Chinese people. Endorsing the use of the **green gross domestic product (GDDP)** as a criterion in making personnel decisions within the party and in government, he advocated a model of "scientific development" (*kexue fazhan*) for addressing the increasing problems of **climate change**, which emerged as a major international issue during his tenure. Stressing the importance of developing such "green" technologies as **solar power** and **wind power**, Hu announced the goal of **renewable energy** reaching 15 percent of the country's total **energy** production by 2020, along with plans of expanding **forest** cover by 40 million hectares (98 million acres).

For all the official rhetoric, progress on the environmental front in the PRC under Hu's leadership was limited, as rapid economic growth, especially in unsustainable sectors of the **economy**, continued unabated with concomitant environmental damage, including a rise in carbon dioxide (CO_2) and other GHG emissions from 2001 to 2006, along with the depletion of drinking **water resources**. Participating in the United Nations Conference on the Human Environment (December 2009/Copenhagen, Denmark), the Chinese delegation was subject to criticism for allegedly torpedoing an **international agreement** on emission controls and reductions. Disasters like the major **chemical** spill on the **Songhua River** in November 2005, demonstrated the lack of emergency response capability in the country, while such major **hydropower** projects as the Pubugou Dam on the Dadu River in **Sichuan province** were pursued under Hu's notoriously vague concept of "scientific development," with the Sinohydro Company, which involved Hu, participated in these major construction projects. While neither Hu nor Premier Wen Jiabao attended the opening ceremonies of the **Three Gorges Dam**, Hu's hardline policy on speech and publication proved inimical to open debate and grassroots action that real progress on environmental issues required. Yet, with an emphasis on **sustainable development** and a more "holistic approach" (*zhengti fenxi*) to the environment, Hu set the stage for the more vigorous pursuit of environmental issues by the subsequent administration of President **Xi Jinping** (2013–).

HUAI RIVER. Located midway between the **Yellow River** and **Yangzi River**, and running 1,110 kilometers (670 miles) in length, the Huai River flows from west to east through **Henan**, Hubei, Anhui, and Jiangsu provinces, with a total basin area of 174,000 km^2 (67,182 mi^2). Central to the economic development of the People's Republic of China (PRC), plans for "harnessing the Huai" (*liyong Huai he*) involved major policy initiatives for both **industry** and **agriculture**, which led to major environmental degradation and several **ecological disasters**. Most notable was the collapse of a series of dams on major Huai tributaries, including the Ru and Hong rivers, during a typhoon in August 1975, killing as many as 170,000 people. Major pollution crises have also afflicted the river on several occasions, the first in 1974, which led the central government to create the Leading Group on Environmental Protection under the State Council. Similar crises occurred in 1989, 1992, 1994, 1998, and 2001, when billions of gallons of heavily contaminated water from thousands of factories built along the **river** and agricultural runoff flowed downstream through China's breadbasket.

A major site of industrial development, including highly polluting factories producing asbestos (*shimian*), **chemicals**, dyes (*ranliao*), electroplating (*diandu*), and **paper** pulp, as well as **tanneries**, along with generally unregulated township and village enterprises (TVEs), the Huai has sustained immense damage to **fisheries** with enormously deleterious effects on **water quality**. Ranked as the fourth most polluted river system in the PRC, water extracted from the Huai has produced serious **health** problems among the local population, including dysentery (*li*), diarrhea (*fuxie*), and vomiting (*qutu*). Targeted for major cleanups with the allocation of substantial funds by the central government in the 1980s and 1990s, thousands of polluting factories, for example, 999 paper mills, were shuttered, while wastewater treatment plants were constructed, leading to a sufficient reduction of pollution and improvement in water quality, disputed by local environmental critics. A national grade III rank, suitable for drinking and fishing, was set for the Huai in 1998.

With the inauguration of Operation Zero Hour in 1998, accompanied by substantial **media** attention, the Huai was dubbed a "national key-point area" (*guojia guanjiao dian lingyu*) for complete environmental amelioration in 2000. Yet, disaster struck again in July 2001, when torrential rains brought a **flood** of polluted water stored upstream to the lower reaches of the river and **lakes** (Hongze and Gaoyou) through which the Huai flows, resulting in major fish die-offs and threats to local drinking water supplies. Despite renewed financial and bureaucratic commitments to clean up the Huai, pollution from industry and agriculture remains a serious problem, with one study indicating heightened cancer rates between 1973 and 2006, among basin residents in

Henan province, including some of the notorious **cancer villages**. Serious flooding has also occurred along the Huai and major tributaries, with significant damage in 1954, 1968, 1974, 1991, 1998, 2003, 2007, 2010, and 2016.

HUANG MING (1958–). A researcher of **solar power** and founder of Himin Solar Research Group, the largest maker of solar water heaters in the People's Republic of China (PRC), Huang Ming holds 600 patents and has been dubbed the "solar king of China" (*taiyang zhi wang Zhongguo*). Originally trained as a petroleum engineer with a degree from the Chinese University of Petroleum, Huang, upon realizing **oil** was a finite commodity, shifted to solar in 1987, establishing Himin in 1995, in **Dezhou**, Shandong province, the world's first solar city. Headquartered in Sun-Moon Mansion, the largest solar-powered office building in the world, Himin is financed in part by Goldman Sachs Investment Bank of the United States and is a prominent presence in Dezhou's solar valley. A member of the National People's Congress (NPC), Huang helped draft China's Renewable Energy Law (2005) and has spoken at many **international conferences** and gatherings on **sustainable development**, notably to the United Nations. Committed to a goal of "solar everything" (*taiyang neng yiqie*), Huang has received many honors, one of which is the Right Livelihood Honorary Award (2011), bestowed by Sweden, also known as the "Alternative Nobel Prize."

HUANG WANLI (1911–2001). An engineer and hydrologist, Huang Wanli became an outspoken critic of **water-control** policies pursued by both the Nationalist government (1927–1949) and the People's Republic of China (PRC). As a young man, Huang witnessed massive **floods** in 1931, and a major dam failure on the **Yellow River** in 1933, and then trained as a meteorologist in the United States, where he received a Ph.D. Having participated in the construction of the Norris Dam in Tennessee, he returned to China in 1937, where he continued work on hydraulic engineering for the Nationalist government and headed the Water Resources Bureau in Gansu province. Following the establishment of the PRC in 1949, Huang became chief advisor to the Northeast China General Bureau of Water Resources and, beginning in 1953, served as a professor at Tsinghua University, while also conducting intensive research surveying more than 3,000 kilometers of the country's **rivers**. Once the Chinese government, with the assistance of experts from the **Soviet Union**, decided to construct the Sanmenxia Dam on the Yellow River on the border of Shanxi and **Henan** provinces, Huang openly expressed his concerns. The potential buildup of Yellow River sediment behind the dam, he argued, would rapidly fill up the storage area, which indeed has occurred on a large scale, as 10 billion m^3 of silt are added to the reservoir annually, dramatically reducing the dam's electrical generating capacity.

During the relatively liberal period of the Hundred Flowers Campaign (1956–1957), Huang launched criticisms of the **Chinese Communist Party (CCP)** and, cryptically, Chairman **Mao Zedong**, for what he considered the abusive treatment of intellectuals. Once the political tables were turned in the Anti-Rightist Campaign (1957–1958), Huang became a primary target of the CCP's political and ideological ire, notably an explicit personal attack on him by Mao Zedong in a June 1958 *People's Daily* article. Banned from further teaching, Huang was sent to a labor camp, where he was forced to engage in manual labor on dam construction and persecuted once again during the **Cultural Revolution (1966–1976)**. Surviving this political maelstrom, he emerged as a major critic of the proposed **Three Gorges Dam** project, going so far as to write a personal letter to U.S. President Bill Clinton requesting that the United States cut off any support for its construction. For many in China, Huang remains an example of a scientist who was both intelligent and direct despite the threat of political persecution.

HUNTING AND POACHING (*SHOULIE, TOULIE*). Following the establishment of the People's Republic of China (PRC) in 1949, **Chinese Communist Party (CCP)** chairman **Mao Zedong** issued a proclamation banning shooting and hunting that remains in force today and ended centuries-old traditions, especially among **minorities** living in areas with plentiful **wildlife**. Unlike their counterparts in the West, where animals are often bagged as trophies, Chinese hunters engage in their craft largely to acquire food and/or ingredients for **traditional Chinese medicine (TCM)**, particularly horn (*jiao*), bone (*gutou*), fin (*qi*), scale (*guimu*), and other body parts from a host of animals, including tigers (*hu*), rhinoceros (*xiniu*), black bears (*hei xiong*), and musk deer (*she*), most now listed by China as **endangered species**. Tigers, most notably the South China tiger (*Huanan hu*), which numbered several thousand in the 1950s, also became a target following a declaration by Mao Zedong in 1959, that the animal was an "enemy of man" and should be eradicated, leading to a reduction in their numbers from 80 to 30, with none left in the wild, as an outright ban on tiger hunting was not issued until 1979, following Mao's passing.

Other animals frequently targeted by illegal hunters included the **Tibetan antelope (*chiru*)**; pangolin anteaters (*chuanshanjia*), for their tasty meat and scales, thought to contain medicinal benefits; and the rich variety of **birdlife**, for example, the yellow-breasted bunting (*huangxiong caiqi*), prized as a delicacy in Chinese cuisine. As a major transit point for large-scale bird migrations, especially along the East Asian–Australasian Flyway in spring and summer, China is beset by large numbers of bird poachers employing tree nets and even Carbofuran **pesticide** to trap an estimated 7 to 10 million birds a year, with large numbers ending up as prized dishes in local restaurants or ensconced in breeding farms and some species, for instance, the spoon-billed sandpiper (*shaozuiyu*), driven to near-extinction. Offenders face stiff fines, jail time, and property confiscation, with execution as the ultimate penalty for killing a giant **panda bear**, with a commensurate crackdown on TCM pharmacies. While limited hunting of nonthreatened species is allowed on regulated hunting grounds, established in the 1980s, for example, the Taoshan International Game Land, with guns still banned in China and urban residents generally unattracted to the sport, legal hunting is largely the province of the nouveau riche, often at the expense of wildlife in Africa and other less-regulated lands.

HYDROLOGY (*SHUI WENXUE*). Defined as the study of global **water resources** and the movement of water in relation to land, along with its distribution, allocation, and quality, hydrology, or hydrological science, is a major field in the People's Republic of China (PRC). With approximately 22,000 **rivers**, water resources in the country are unequally divided, plentiful in the south and southwest but scarce in the north and northwest. Among the major topics of study and research that feed into important national **policy-making** affecting the handling and distribution of water resources in China are **floods**, water shortages, **drought**, and **water pollution**, with hydrological sciences playing a major role in determining the country's **sustainable development**, in which stable water supplies are crucially important. Drawing on nationwide data on water resources collected by national surveys, central components in hydrological analysis include the following: the study of natural causes, especially the all-important hydrological cycle (*shuiwen xunhuan*) conditions through which water passes from vapor in the atmosphere to precipitation on the Earth; and human influences, including the construction of **hydropower** stations and heavy water use for **industry** and **agriculture**.

Major hydrological issues in China include **Yellow River** dry-up, **Yangzi River** floods, and the potential effects on water resources of impending **climate change**. Remote sensing, geographical information systems, and computer modeling are employed in hydrological analysis, with such major national institutions as the United Center for Water Research, Chinese Academy of Sciences (CAS), and the Chinese Academy of Engineering conducting studies, with universities like Tsinghua, Hohai, and Wuhan University of Hydraulic and Electrical Engineering offering academic programs.

HYDROPOWER (*SHOUDIAN*). With approximately 22,000 **rivers** and three major river systems consisting of the **Yellow**, **Yangzi**, and **Pearl**, China has perhaps the world's greatest potential for hydropower, with installed capacity of 319 gigawatts (GW) in 2012, and plans for expansion to 350 GW by 2020. Following the establishment of the People's Republic of China (PRC) in 1949, the **Chinese Communist Party (CCP)** pushed immediately for the construction of a series of dams on major rivers and their tributaries, notably the massive **Three Gorges Dam** on the Yangzi, which had also been pursued by the previous Nationalist government (1927–1949).

Altogether 25 large-, 130 medium-, and approximately 90,000 small-sized hydropower stations were built from 1949 to 1986, many constructed out of clay and masonry, with annual electrical power output of 86 billion kilowatt hours (kWhs) in 1983. Major dams and hydropower projects include the Gezhouba, Banqiao, Shimantan, Xiaolangdi, Sanmenxia, and Three Gorges. Located on the Yangzi River in Yichang, in Hubei province, Gezhouba was built in the 1970s and 1980s, with an installed electrical capacity of 2,715 megawatts (MW). The facility is 2,595 meters long, with a maximum height of 47 meters and a reservoir capacity of 1.58 km^3, with a navigation lock on the third channel among the 100 largest in the world. The Banqiao and Shimantan dams were built with the assistance of consultants from the **Soviet Union** in **Henan province** on the Ru and Hong rivers, major tributaries of the **Huai River**, largely in response to severe **flooding** throughout the Huai basin in 1949 and 1950, and completed in June 1952.

Lacking basic **hydrology** data and with a lower design standard for the dams, following major flooding on the Huai in 1954, upstream reservoirs, with the Banqiao dam being an example, were extended, including an increase of three meters in dam height. Total capacity of the reservoir was 492 million m^3 (398,000 acre-feet), with 375 million m^3 (304,000 acre-feet) reserved for flood water storage and a maximum discharge at 1,742 m^3/s. Following the sudden appearance of cracks in the dam and sluice gates, repairs were made on the advice of Soviet engineers, which, along with a new design, led to dubbing the facility an "iron dam" (*tieba*) considered unbreakable, while similar features and safety problems beset the Shimantan dam on the Hong River. In the midst of a massive typhoon in August 1975, both dams collapsed, producing one of the greatest **ecological disasters** in the PRC, with more than 170,000 fatalities and enormous damage to agricultural lands and infrastructure.

The Xiaolangdi dam was, for a time, the second-largest dam in China, behind the Three Gorges Dam, with a similar construction design and also shrouded in controversy, with critics predicting a silting up of the reservoir within 30 years. Located near Jiyuan, Henan province, construction of the dam on the Yellow River took six years (1994–2000). The dam is 154 meters (505 feet) high and 1,317 meters (4.321 feet) in base width, with 1,836 MW installed capacity. The Sanmenxia Dam is a concrete gravity dam on the middle reaches of the Yellow River on the border of Shanxi and Henan provinces, with a 400 MW installed capacity. Originally proposed during the Nationalist era and constructed with assistance from Soviet engineers, the dam was constructed from 1957 to 1960, for flood and ice control, along with **irrigation**, electrical power production, and navigation. The first major **wa-**

ter-control project on the Yellow River, Sanmenxia was viewed as a major engineering achievement, with an image of the dam printed on Chinese currency. Due to serious sediment accumulation in the reservoir, the dam was later renovated to include 12 deep sluices and alterations in the release of water, but with continued sediment buildup that threatened upstream navigation and inundation of farmland, along with the necessary relocation of an additional 1 million people.

Responding to the potential crisis, a high-level meeting was convened in December 1964, attended by Premier **Zhou Enlai**, during which a decision was made to reconstruct the Sanmenxia's outlet works for improved discharge and silt control. With sediment buildup causing severe flooding on the upstream Wei River, Chinese engineer Zhang Guangdou would later describe the project as a "mistake" and remarked how hydrologist **Huang Wanli** had been sent to hard labor for opposing the project. In August 2010, Chinese **journalist** Xie Chaoping was detained for writing *The Great Migration*, a book that criticized the Sanmenxia dam and the government's handling of the project.

Recent hydropower projects include the Jinping-I arch dam, at 305 meters (1,000 feet) the tallest dam in the world, on the Yalong River, a tributary of the Yangzi, in **Sichuan province**, with 3,600 MW installed capacity, completed in 2015; the Nuzhadu on the **Lancang River** in **Yunnan Province**, with 5,850 MW installed capacity, completed in 2014; the Pubugou on the Dadu River in Sichuan, with 3,300 MW installed capacity, completed in 2010; the Yangqu on the Yellow River in Qinghai province, with 1,200 MW installed capacity, completed in 2016; the Changheba on the Dadu River, Sichuan, with 2,600 MW installed capacity, completed in 2016; the Xiluoduo, an arch dam on the Jinsha River in Yunnan province, with 13,860 MW installed capacity, completed in 2014; and the Xiaowan arch dam on the Lancang in Yunnan, with 4,200 MW installed capacity, completed in 2010.

Also planned are two massive dams, Lianghekou on the Yalong River and Baihetan on the Jinsha, both in Sichuan, with the latter slated to generate 16,000 MWs, the second largest in the world, after Three Gorges. Altogether 23 dams are slated for the Yalong and 25 for the Jinsha, but only nine have been completed as of 2017, as project locations are increasingly in remote areas. In addition to inundating such historical sites as temples and monasteries, many sacred to Buddhism, dams contribute to **soil erosion**, flooding from overflowing reservoirs, and the destruction of **habitat** for **wildlife** and **birdlife**. Planned construction of dams on the **Nu River**, China's last wild river, and in Tiger Leaping Gorge (*Laohu Kuaye Xiagu*), on the Jinsha River, have been canceled due in part to protests from environmental nongovernmental organizations (ENGOs) and widespread local opposition. The planned Xiaonanhai, on the Yangzi River, near the city of **Chongqing**, was similarly terminated in the midst of construction as a threat to the Yangzi Nature Fish Reservoir, the last stretch of the river supporting hundreds of different fish, with many species endemic to China.

Considered a primary form of "green" technologies and **renewable energy**, with strong political support from giant energy firms, major hydropower projects are often followed by such heavily polluting industries as **steel**, **coal**, and **chemical** smelters, as occurred in Sichuan and Yunnan provinces. Modern hydropower in China is also compared unfavorably to more benign traditional forms of water-control, for instance, the ancient Dujiangyan project on the **Min River** in Sichuan, which created an elaborate system of canals by splitting the river into two channels with the construction of artificial islands, providing irrigation to the entire Red Basin, with few deleterious side effects. More ecologically friendly small and micro hydropower projects have been constructed since the 1980s (especially in rural and heavily mountainous areas) to replace dirty and inefficient diesel generators while providing electrical power to 300 million people. Consisting primarily of small "run of the river" facilities with generating capacity between 0.1 and 13 MW, these projects currently produce 27 percent of China's total annual hydropower with regional concentrations in south and southwestern provinces, primarily Guangdong, Sichuan, Yunnan, Fujian, and Hunan. Under the guise of the Green Small Hydropower Project (GSHP) inaugurated in 2016, many of these facilities, some so simple as water flowing through penstocks over natural height differences, prevent unnecessary flooding of **forests** and **grasslands**. China is also supporting development of new technologies to completely replace dams, with one example being the hydrodynamic energy conversion device, which takes a mere three months to install.

I

INDUSTRY. A major source of **air pollution** and **water pollution**, along with serious **soil** contamination, the industrial sector in the **economy** of the People's Republic of China (PRC) is a major target of environmental amelioration, particularly state-owned enterprises (SOEs), with their archaic equipment, notably antiquated **coal**-burning power plants. Operating at relatively low fuel conversion rates, especially by industrial boilers that have functioned at a mere 50 percent level of efficiency, industry is a major contributor to national pollution levels. Involved in such heavy-polluting production sectors as cement (*shuini*), **chemicals** and petrochemicals (*shiyou huaxue*), electric power (*dianli*), **mining**, **paper**, **pharmaceuticals**, and **steel**, SOEs accounted for 645 million tons of waste in 1995. No less important are the relatively small-scale and generally unregulated township and village enterprises (TVEs), numbering more than 500,000, which have also contributed to dramatically elevated pollution levels by producing 410 million tons of solid waste in 1996, largely in medium and small **cities** and the countryside. With many industrial facilities employing antiquated equipment installed before 1970, and often too small to afford relatively expensive treatment equipment, a survey conducted in 2009, by the **Ministry of Environmental Protection (MEP)**, indicated that industry accounted for 63.7 trillion m^3 (2,249 trillion ft^3) of waste gas, 3.8 billion tons of solid waste, and 209.8 billion tons of wastewater.

The most damaging industrial pollutants to air, water, and soils include the following: fly ash; sulfur dioxide (SO_2), half of which in the state sector of the economy is produced by power plants, along with more than 400,000 largely antiquated and inefficient industrial boilers; industrial dust (*gongye chenfu*); chemical oxygen demand (COD); **oil**; cyanide (*qing huawu*); arsenic (*shen*); mercury (*gong*); lead (*qian*); cadmium (*ge*); and hexavalent chromium (*liujia ge*), a major carcinogen. Of the reported 9 million pollution sources existing in the PRC (2017), 7.4 million are from industrial sources, an increase of 50 percent since 2010. Studies of industries in Northern China conducted in 2017 also show that in 28 cities, a total of 14,000 facilities failed to meet environmental standards, with large percentages operating in unauthorized locations and lacking proper certification. One thousand factories reportedly discharge wastewater directly into the **Yellow River**, while many facilities store their waste at production sites with potential threats to groundwater.

Despite the inauguration of mitigation efforts in the industrial sector beginning in 1978, environmental experts warn that the Chinese environment is currently "nearing a point of no return," with President **Xi Jinping** warning in May 2018, that the country was failing to achieve major environmental protection goals. Yet, with increasingly aggressive government policy targeting industry, Chinese domestic industries confront significant increased costs for new and cleaner production processes mandated by the Cleaner Production Promotion Law (2002), along with treatment equipment ordered by **environmental inspections**. Substantial fines and penalties for polluting factories are collected by powerful local **taxation** bureaus, with the most serious offenders confronting closure, usually temporary, but permanent in extreme cases, as tens of thousands of such facilities were shuttered in October 2017. With an estimated 300,000 premature deaths annually attributed to pollution, with cancer the number-one cause of fatalities, public pressure for improved environmental performance by Chinese industry is unrelenting.

INSECTS. *See* BEES AND BUTTERFLIES (*MIFENG/HUDIE*).

INSTITUTE OF PUBLIC AND ENVIRONMENTAL AFFAIRS (IPEA/ *GONGGONG HE HUANJING SHIWU YANJIUSUO*). Founded in 2006, by **Ma Jun**, an environmental **journalist** and author of *China's Water Crisis*, IPEA is one of many environmental nongovernmental organizations (ENGOs) in the People's Republic of China (PRC) devoted to conveying important information to the Chinese public on environmental conditions in the country, especially **air pollution** and **water pollution**. Headquartered in **Beijing** and with important connections to domestic and international environmental groups, most notably **Friends of Nature (FON)** in the PRC and the National Resources Defense Council (NRDC) in the United States, IPEA collects and collates **environmental statistics** from the national and local governments in China, including all 32 provinces/central municipalities and 338 **cities**, along with domestic enterprises and China-based multinationals.

Core projects include the Blue Map website and app available on smartphones, which presents real-time data (*shishi shuju*) on **air quality** and **water quality**, as well as industrial emissions, using dynamic maps, while also providing a social **media** channel for Chinese citizens to report polluting offenders to local officials. Also developed in conjunction with NRDC is the Green Supply Chain Map, which links multinational corporations to their Chinese suppliers' environmental performance using information drawn from the institute's database, developed since 2006, on more than 15,000 industrial facilities in the PRC, with links to another half-million, as export-oriented industries in China produce 25 percent of carbon dioxide (CO_2) emissions. Other IPEA environmental projects include the Air Quality Transparency Index, Corporate Information Transparency Index, Pollution Risk Index for corporations, and Professional Inspection Training Institute assessment for 113 cities in the PRC.

INSTITUTES AND RESEARCH ORGANIZATIONS. With commitment to a policy of environmental protection and restoration in the People's Republic of China (PRC) beginning in the 1980s and 1990s, major institutes and research organizations were established for addressing the multiple issues posed by the country's damaged environment and the national goals of achieving **sustainable development** and **renewable energy**. Notable institutes, along with major functions, include the following: Beijing Environmental and Development Institute (applied research on environmental issues using market-based approaches); China National Cleaner Production Center, set up under the former State Environmental Protection Administration (assisting businesses in achieving more environmentally sound production methods); Environmental Research Institute (renewable energy); Global Environmental Institute (marine conservation, **ecosystem** conservation, **energy** and **climate change**, renewable energy, **organic farming**, and **ecotourism**); Institute for Environment and Development (teacher training and information development, headed by noted environmentalist **Qu Geping**); Institute of Environment and Sustainable Development in Agriculture, Chinese Academy of Sciences (agroenvironment, sustainable agricultural development, and impact of climate change on **agriculture**); Institute of Geography, Chinese Academy of Sciences (environmental training program); **Institute of Public and Environmental Affairs (IPEA)** (collecting and conveying information on **air pollution** and **water pollution**, and environmental performance of Chinese suppliers to major multinationals); China Research Academy of Environmental Sciences (scientific research on environmental protection and sustainable development, and technological support to national environmental strategy); Environmental Law Research Institute, Wuhan University (drafting of major environmental **laws and regulations**); and South–North Institute for Sustainable Development (pilot demonstrations of biogas production and usage).

INTERNATIONAL AGREEMENTS. Shifting from an oppositional regime to embracing global efforts at addressing major environmental issues, the People's Republic of China (PRC) is a signatory to more than 50 multilateral environmental agreements, including the following, with dates of ratification: Convention on the Continental Shelf (1970); Protocol Concerning Establishing a Research Center to Protect Pandas (1979); International Convention for Regulation of Whaling (1980); Convention on International Trade in Endangered Species of Fauna and Flora (1981); Agreement Concerning the Protection of Migratory Birds and Their Habitat (1981); Antarctica Treaty (1983); Convention on Prevention of Ocean Pollution by Dumping Waste and Other Matter (1985); Amendment to the International Convention for the Prevention of Pollution from Ships (1986); International Convention Relating to Intervention on the High Seas in Case of Oil Pollution (1990); Montreal Protocol on Ozone Depletion by Substance (1991); United Nations Convention on Biological Diversity (1992); Convention on Wetlands of International Importance, Especially as Waterfowl Habitat/Ramsar Convention (1982); Vienna Convention for the Protection of the Ozone Layer (1989); Protocol on Environmental Protection for the Antarctica Treaty (1994); Basel Convention on the Transboundary Movement of Hazardous Waste and Their Disposal (1995); United Nations Convention on the Law of the Sea (1996); United Nations Convention to Combat Desertification (1997); International Natural Rubber Agreement (1997); International Tropical Timber Agreement (1997); United Nations Framework Convention on Climate Change and Kyoto Protocol (1998); International Convention on Oil Pollution Preparedness, Response, and Cooperation (1998); Stockholm Convention on Persistent Organic Pollutants (2001); and Paris Accord on Climate Change (2015).

INTERNATIONAL CONFERENCES. Following the ascension of the People's Republic of China (PRC) to the United Nations in 1971, after years of international isolation, China participated in its first international environmental conference, sending a delegation at the behest of then-premier **Zhou Enlai** to the United Nations Conference on Human Environment in Stockholm, Sweden, in 1972. While assuming an overall low profile, Chinese delegates at the conference participated in drafting a Declaration on the Human Environment but failed to sign the final agreement, although China pledged to implement the conference resolutions. Following years of internal political turmoil, especially during and immediately following the **Cultural Revolution (1966–1976)**, with little involvement in international environmental gatherings, China participated in a meeting of signatory nations to the Montreal Protocol on Substances That Deplete the Ozone Layer in Helsinki, Finland, in 1989, where PRC delegates facilitated an agreement between developed and developing nations on funding the protocol. China then convened a Ministerial Conference of Developing Countries on Environment and Development in **Beijing** in 1991, where the relationship of environmental protection to global economic development was discussed, as participating nations issued the Beijing Declaration (*Beijing Xuanyan*), calling for international cooperation on global environmental issues.

Attending the United Nations Conference on Environment and Development, known as "Earth Summit I," in Rio de Janeiro, Brazil, in 1992, Chinese premier **Li Peng** presented several proposals for managing the global environment, while also noting the importance of economic development and national sovereignty. Reluctant to accept the concept of **sustainable development** highlighted at the conference, seeing it as a restriction on developing countries, China quickly switched position, laying out a plan for sustainable development with the adoption of **Agenda 21** in 1994. Support for sustainable development by China continued at the World Summit on Sustainable Development in Johannesburg, South Africa, in 2002, while at the United Nations Climate Change Conference in Copenhagen, Denmark, in 2009, China was faulted for failure to meet targeted goals for global pollution amelioration aimed at halting impending **climate change**. A major player at the United Nations Conference on Sustainable Development, held in Rio de Janeiro, known as "Rio+20/2012," China reaffirmed its support for sustainable development, while at the Paris Climate Change Conference in 2015, China joined the United States and other nations in agreeing to major goals for combatting climate change, assuming global leadership on the issue following the U.S. withdrawal from the agreement in January 2017. China hosted its first international environmental conference on the United Nations Convention to Combat Desertification in Ordos, Inner Mongolia, in 2017.

INTERNATIONAL ORGANIZATIONS. With the emergence of a nascent environmental movement in the People's Republic of China (PRC) and the formation of more than 7,800 environmental nongovernmental organizations (ENGOs), prominent international environmental organizations have become increasingly involved in the country's environmental affairs, beginning in 1994. Relying on broad experience and expertise in global environmental matters, international groups working in the PRC have helped shape the national dialogue on environmental issues, including in the **media** and among Chinese **journalists**, bringing considerable resources (including finances) and promoting greater openness on environmental challenges in China's growing "green public sphere" (*lüse gonggong lingyu*). Sharing a common mission with their Chinese counterparts, international groups find the Chinese government generally receptive to international standards, even as relations with political authorities can be somewhat precarious.

Among the most prominent international groups are **Greenpeace** East Asia, the World Wildlife Fund (WWF), Conservation International (**wetlands** protection), Earthwatch Institute (**panda** protection), Friends of the Earth (**environmental awareness**), the International Fund for China's Environment (**biodiversity**), the International Union for Conservation of Nature (**endangered species**), the National Resources Defense Council (**climate change**), the Nature Conservancy (**wildlife** and **habitat**), and the World Resources Institute (**sustainable development** and climate change), while groups like the International Rivers Network and the Blacksmith Institute have provided crucial financial support to Chinese **nongovernmental organizations (NGOs)** and environmental **activists**, including opponents of constructing **hydropower** projects on such scenic waterways as the **Nu River** in **Yunnan province**.

INTERNET (*HULIANWANG*). From a mere 10,000 users in 1994, when the People's Republic of China (PRC) was first linked to the international internet, and growing to 111 million in 2005 and more than 730 million users in 2017, the internet has played a major role in expanding the public sphere of the emerging environmental movement and creating a "virtual community" (*xuni shequ*) of increasingly informed "net citizens" (*wangmin*). Through major microblogging (Weibo) sites like Sina and Tencent, environmental issues have been brought to the forefront with information, photos, and videos, for example, the highly influential *Under the Dome*, with 155 million downloads, effectively circumventing the traditional channels of newspapers and other **media** more easily susceptible to government blockage or interference. Along with online publishing and discussion forums, internet users are offered the opportunity to sponsor environmental actions, for instance, the protection of migratory **birds**, and express, within limits, personal opinions, while providing text message links to government and media websites.

Adding to the supervisory role over government actions or inactions, the internet is a major tool for environmental nongovernmental organizations (ENGOs), almost all of which maintain websites like Greener Beijing and Wetlands of China for conveying mission statements, information on upcoming events, and links to domestic and international environmental groups. Facilitating the creation of an environmentally oriented **civil society** in the PRC, the internet is also the vehicle for multinationals the likes of IBM to develop a capacity via the Internet of Things to predict and convey to the public **air pollution** levels in real time. Although environmental issues and organizations have not been targeted by Chinese government censors, with the country's highly sophisticated "firewall" (*huoqiang*), topics ancillary to the environmental movement, for example, citizen **journalism** and civil society activity, have been periodically subject to state interference. One of the most significant negative impacts of the internet on the environment has been the explosion in online retailing, with the profusion of packaging materials amounting to 14 billion parcels in 2014, with these materials requiring proper disposal and waste management.

INVASIVE SPECIES (*RUQIN WUZHONG*). Defined as an organism that causes ecological and economic harm in a new environment where it is not native, bringing about serious threats or even extinction to plant, animal, and other indigenous life forms, invasive or nonnative species are an increasingly serious global problem that has intensified in China since 1850. Rich in **biodiversity**, China is ripe for invasive species, with 488 detected, composed of 171 animals, 265 plants, 12 viruses, 11 prokaryotes (single cell without nucleus), and 3 protists (single cell with nucleus). With trade as a major conduit for their transmission as the People's Republic of China (PRC) has engaged **globalization**, nonnative species are distributed primarily in the **coastal** provinces of Guangdong, Jiangsu, Zhejiang, Fujian, and **Shanghai** municipality, along with **Yunnan province**, while being virtually nonexistent in the more insular western and northwestern provinces.

Regional sources of invasive species into the PRC include primarily North America and South America (51 percent), and Asia (17 percent), with a substantially smaller percentage from Europe and Africa, as most species, including 50 rated as the world's worst, were introduced unintentionally, with half inhabiting the country's many **nature reserves and national parks**, where they have wreaked havoc on many native species and required major costs for control and removal.

Prominent examples of highly destructive invasive species include the water hyacinth (*shui hulu*), originating from South America and blocking **rivers and waterways**, while threatening local **birdlife**; the red-eared slider turtle, spreading diseases harmful to the native turtle population; smooth cord grass, introduced intentionally to improve **soil quality** but with harmful effects on **mangrove forests**; the pine-wood nematode (roundworms), responsible for so-called pine wood cancer; German cockroaches; the oriental wood borer; the tropical fire ant; common pokeweed; the tropical ageratum plant; and bullfrogs. The most critically affected areas include the giant **Three Gorges Dam** reservoir in Hubei Province, where 42 nonnative aquatic and other species have been detected, notably the highly destructive water hyacinth and the mosquitofish. Concerns have also been raised by international environmentalists that the planned One Road, One Belt Initiative (OBOR) promoted by the PRC, with plans to facilitate links to four continents and 70 countries, will become a conduit for the spread of invasive species from China, with one notable example being the river-bound boa constrictor.

IRRIGATION (*GUANGAI*). With the largest agricultural land area under irrigation in the world, at 66 million hectares (163 million acres), including 75 percent of the country's vital grain-growing regions, the People's Republic of China (PRC) confronts multiple environmental problems brought about by this elaborate system of **water control**. Bringing irrigation to Chinese **agriculture** was a major state function throughout the imperial period, exemplified by the massive Dujiangyan water-control project on the **Min River** in **Sichuan province**, built during the 3rd century BCE and still in use today. With the widespread destruction brought on by the Second Sino–Japanese War (1937–1945) and the Civil War between the Nationalists and Communists (1946–1949), reconstructing and expanding irrigation networks was a top priority of the PRC, especially in the arid regions of the north.

Beginning in 1955, with the construction of reservoirs and **river** diversion projects, water control devoted to expanding irrigation was a prominent feature of the **Great Leap Forward (1958–1960)**, with such major projects as the Red Flag Canal (1960–1969), which, built entirely by hand through rough **mountain** terrain, brought water to poor and infertile areas in **Henan province**, although many other projects were hastily built, with poor design and shoddy construction. The installation of pump wells in the late 1960s through the mid-1970s in the countryside further increased the percentage of irrigated land, with only modest increases coming in the 1990s, reaching current levels.

The primary environmental impact of irrigation is the consumption of an estimated 55 percent of the country's **water resources**, 340 billion m^3 in 2013, with projections of 372 billion m^3 by 2020. Operating at a low rate of efficiency of 0.532, substantially lower than levels recorded for developed countries like the United States, irrigation has led to the substantial waste of scarce water resources, especially in North China, where high pollution levels in surface rivers and **lakes** has led to increasing reliance on groundwater (100 billion m^3 annually in 2012), with water tables declining an average of two meters annually as **soil** salinization (*yanhua*) increases. Relying on mechanical pumps to access ever deeper underground aquifers in the **drought**-prone north, many composed of small-scale operations difficult to regulate, emissions of carbon dioxide (CO_2) are estimated at 33 million tons annually.

Amelioration measures include government programs, including the National Irrigation Agricultural Water-Saving Program (1996), which have encouraged a shift from highly wasteful "shallow flooding irrigation," with rice paddies retaining water layers throughout most of the growing cycle, to "water-saving irrigation" (WSI). WSI has been adopted by 48 percent of Chinese farmers, with a goal of reaching 64 percent by 2020. In addition to reducing water consumption and the **energy** needed to meet excess demand, WSI leaves rice paddies without water layers for 75 to 85 percent of the growing cycle, reducing **water pollution** and soil pollution from **fertilizer** runoff, improving both soil aeration and field microclimatic conditions by decreasing air humidity and reducing pests, requiring less application of harmful **pesticides**. Other measures include improving irrigation canal lining, employing drip irrigation, and engaging in intermittent irrigation and other agronomic practices, with water consumption for irrigation reduced by 33 percent, even as agricultural production continues to increase.

J

JOURNALISM AND JOURNALISTS (*XINWEN XUE, JIZHE*). With the emergence of environmental protection and restoration as a "core priority" (*hexin youxian*) of the People's Republic of China (PRC) from the late 1990s onward, the field of environmental journalism has expanded dramatically with the task of environmental journalists substantially improved. Whereas one of the first major and critical works on the environment, *Yangtze! Yangtze!* by Ms. **Dai Qing**, dealing with the controversial **Three Gorges Dam**, was banned, with the author imprisoned in 1989, Chinese reporters and **media** have been increasingly able to investigate and produce stories in newspapers and especially on websites and microblogs on a variety of environmental issues and scandals, often resulting in government action led by the **Ministry of Environmental Protection (MEP)**. Aided by the passage of a Chinese version of the Freedom of Information Law (2008), professional journalists and so-called "citizen journalists" (*gongmin jizhe*) have brought to light cases of serious threats of pollution, which, when given attention by national television coverage or high-profile stories on the **internet**, force the hand of the Chinese government to take such ameliorative action as the dismissal and punishment of local officials for cover-ups and other obstructionist tactics.

Even as the Chinese press has become more commercialized and sensationalist, investigative journalism on the environment continues to thrive, with journalists more than eager to cover such major environmental events as the deadly explosion at a **chemical** storage facility in **Tianjin** in August 2015, with follow-up coverage of similarly dangerous conditions involving the storage of toxic chemicals in other **cities**. While aggressive reporters still often confront intimidation and harassment, and even death threats, from local government officials and company thugs, once a story has reached the public through coverage in local and/or national newspapers, like the frequently disputatious *Caixin Weekly* and *Southern Weekend*, along with more conventional and state-controlled press, mainly *Xinhua, China Youth Daily*, and such city and provincial newspapers as *Beijing News* and *Sichuan Daily*, pressure for action escalates. This is particularly the case when video and/or dramatic photos by photojournalists like **Chen Jie** of polluted sites with people suffering become an integral part of the story.

Among the major newspapers focusing almost exclusively on the environment is *China Environment News* (inaugurated in 1984), while events like "Access China Environmental Protection Action" (1993) have led to a reported 6,000 journalists covering environmental issues. Among the domestic organizations for environmental journalists is the China Forum of Environmental Journalists, while international organizations like Oxpecker, which tracks international trade in **endangered species**, have included Chinese journalists in their work.

L

LAKES (*HU/CHI*). Numbering 24,800, of which 2,800 are larger than 1 km^2 (0.386 mi^2), lakes are a major **water resource** in the People's Republic of China (PRC), offering rich **fisheries** and **habitat** for **birdlife** and **plant species**, as well as water supply for major **cities**, but with serious problems of pollution, shrinkage, and dry-up. Among the most important are Qinghai Lake, the largest saltwater lake in the country, at 4,583 km^2 (1,769 mi^2), located in Qinghai province; the transboundary Xinghai, or Khanka Lake, in Heilongjiang province, with 72 percent of the surface area in **Russia**; **Poyang Lake**, the largest freshwater lake, at 3,500 km^2 (1,351 mi^2), and a major bird habitat for white cranes (*baihe*) and black storks (*heiguan*) in Jiangxi province; **Dongting Lake**, 2,800 km^2 (1,080 mi^2), in Hunan and Hubei provinces; Hulun Lake (*Hulun Buir* in Mongolian), 2,339 km^2 (902 mi^2), located in Inner Mongolia; and Namtso Lake in **Tibet**, the second-largest saltwater lake in the country, with pristine, turquoise-colored water. With more than 1,100 saltwater and salt lakes in the country primarily in major saline areas such as the intermontane Qaidam Basin, salt production is a major industry with Chaka Lake (Qinghai), Dingbian (Shaanxi), and Qarhar (Qinghai) containing major deposits of natural salts, solid and liquid. Qarhar is also a major source of lithium used in electric batteries.

Other prominent lakes and their provincial locations include Bosten (**Xinjiang**); Chao (Anhui); **Dian**, Erhai, and Lugu (**Yunnan**); Xuanwu and Hongze (Jiangsu, the latter on the **Huai River**); Qiandao (Zhejiang); Qingshan and Sha (Hubei); Sẽrling (Tibet); Shahu (Ningxia); **Tai** (Jiangsu/Zhejiang); Tianchi (Jilin), a crater lake and deepest in the country, and Yinghu (Ankang, Shaanxi), which has 1,000 islands. Of the 51 largest lakes in the PRC, 14 have suffered permanent dry-up, notably Lop Nur Salt Lake in Xinjiang. Other important lakes, for example, Baiyang, near **Beijing**, once known as the "kidney of north China" (*Zhongguo beifang de shenzang*), are now replete with **garbage** and **pesticide** residue, incurring periodic dry-up or, as in the case of Poyang Lake, substantial shrinkage. Poyang Lake and Hongjiannao Lake, the latter located on the edge of the Mu Us Desert (Shanxi), lost one-third of its water volume to destructive **coal** mining and the effects of

climate change. Large-scale fish and crustacean die-offs periodically afflict lakes, as occurred from unknown causes in Hongze Lake in August 2018, while fish stocks in many lakes have undergone significant declines, with examples being the renowned white shrimp (*baixia*) and white fish (*baiyu*), along with the hairy crab (*dazhaxie*) in Tai Lake, the traditional center of China's "land of fish and rice" (*yu mi zhi xiang*).

A major cause of lake and **river** pollution in the PRC is industrial waste, especially from **chemical** plants, which, requiring substantial water supply for the production of dyes (*ranliao*), solvents (*rongji*), adhesives (*nianheji*), and other such chemical agents, are located near lakes, which also frequently serve as dumping sites for industrial wastewater. Other major sources of lake pollution include agricultural runoff composed of **fertilizer** and pesticide residue; urban and rural **sewage**; and animal waste, especially from large-scale pig and livestock operations, which contribute to algae plumes (*zaolei yumao*), for example, cyanobacteria algae blooms ("pond scum") on lake surfaces. The water circulation and waste-filtering functions of many lakes have also been disrupted by the construction of dams and weirs (*yan*) built for **flood** control and **irrigation**, along with major land **reclamation** projects of lakes and nearby **wetlands** to expand **agriculture**, especially during the 1950s and 1960s. While disruptive and unattractive, some algae plumes consisting of phytoplankton are used as feedstock for aquatic life like silver carp (*lianyu*), as well as other **birds** and animals.

Major cleanup campaigns begun in the 2000s produced substantial improvements, with the number of highly polluted lakes dropping by two-thirds from 2006 to 2014, and major declines in phosphorous (*lin*) residue, a major cause of algae plumes and eutrophication (*shuiti fu yingyanghua*). Improvements in sanitation with the construction of waste pipeline and waste treatment plants has also reduced direct sewage dumps, although **water quality** in many lakes continues to deteriorate, especially in **Sichuan**, Shanxi, and Inner Mongolia, the last the site of highly polluting and radioactively toxic **rare earth** production near the city of Baotou. While lake dry-ups continue from the effects of **desertification** and changing patterns of precipitation, some dried-up lakes have reappeared including Jutinghai Lake (Inner Mongolia) and Har Lake in the Gobi Desert (Gansu) which after 50 years of dry-up now measures 24 km^2 (9 mi^2).

LANCANG RIVER. The longest north-to-south river in the People's Republic of China (PRC), at 2,179 kilometers (1,329 miles) and constituting the upper reaches of the Mekong River flowing through Southeast Asia, the Lancang River is faced with a number of environmental issues, most notably the deleterious effects of numerous **hydropower** projects on river **hydrology**, **fisheries**, and sedimentation flows. With headwaters in the Tanggula Mountain range in Qinghai province, the Lancang flows through **Tibet** and

Yunnan province, where a cascade of eight dams began with the construction of the Manwan embankment dam (1995), followed by Dachaoshan (2003), Xiaowen (2010), and Nuozhadu (2012) dams, the last with an installed capacity of 5,850 megawatts (MWs). Five more hydropower projects are in various stages of construction, and another 11 are in the planning stages, with one dam cancelled based on concerns about its impact on fish migration. As multipurpose structures, the dams were designed to enhance China's low-carbon **economy** by providing pollution-free electricity, along with **flood** control, increased freshwater supply, and enhanced shipping through the additional construction of navigation channels. One of the most **biodiverse** rivers in the world, second only to the Amazon in Brazil, the Lancang/Mekong contains 1,700 varieties of fish, with 62 species endemic to the river, notably the Mekong Giant Catfish, the largest freshwater fish in the world, with several species considered **endangered**.

Among the major threats to river fisheries are changes in water flow with increases in the dry season and decreases in the wet season, lower water temperature, and blockages of fish migration, all of which have altered fish composition, reducing the ratio of large fish to medium and small fish, with a drop in the number of species from 139 to 80. Riverbank **soil erosion** has also increased, along with trapped sedimentation in reservoirs, reducing nutrient flow into the flood plains with their highly productive rice paddies in Cambodia, Laos, Thailand, and Vietnam, while also incurring potential saltwater intrusion into the Mekong River Delta. Agricultural runoff containing toxic **fertilizer** residue, **sewage**, and chemicals like ammonia (*an*) from nearby **mining** operations is also an issue, as dams reduce the self-flushing capacity of the river, with concomitant increases in algae (*zaolei*) and zooplankton (*fuyou dongwu*).

Nearby mining operations have also resulted in riverside tailing ponds, waste piles, and wastewater pollution along with destructive effects of sand mining on the river bottom in Yunnan province. Periodic dry-ups in the headwaters have also occurred from regional reductions in precipitation and **glacier** melt. Distinctive geological formations on the river include the Melli Grand Canyon located in Yunnan with deep and long valleys, 150 kilometers (93 miles) in length and considered sacred by Tibetans. The Lancang–Mekong Environmental Protection Center was founded in 2018, to enhance cooperation between China and the five other nations in the river basin.

LANDSLIDES (*SHANTI HUAPO*). The second most destructive natural hazard after **earthquakes**, landslides are a major threat in the People's Republic of China (PRC), especially in the **mountain** regions in and around the **Qinghai–Tibetan Plateau** and in areas of heavy rainfall in the southwest, especially **Sichuan** and **Yunnan** provinces, with major losses in lives and property. Defined as a downward and outward movement of slope-forming materials, including rock, **soil**, and artificial filling, landslides and mudslides in China are triggered by such natural forces as topographical (steep mountain slopes) and geological features (fault lines and active **earthquake** zones), along with **climate change** (increased El Niño activity). Human-induced actions include underground **mining**, slope cutting for railway and road construction, **deforestation**, and the impoundment of reservoir water by such large projects as the **Three Gorges Dam**. While 200 catastrophic landslides leading to significant human fatalities and property destruction have occurred in China since 1900, the frequency of landslides has increased since the 1980s, largely due to accelerating climate change and inappropriate construction projects in fragile areas, particularly in mountainous areas, which constitute 69 percent of the country's land mass and contain 50 percent of the population.

With the country divided into 12 zones, four of them listed as "highly susceptible," with only one zone listed as "low susceptibility," located in the northeast, regions like the Qinghai–Tibetan Plateau, with an average elevation of more than 4,500 meters (14,800 feet), are most vulnerable to landslides, primarily of snow and ice. Other areas prone to landslides include the monsoon-affected areas of Central and Southeast China, especially where the lithology (*yanxing*) consists of clayey soils, subvolcanic (hypabyssal) rock, mudstone, and/or sand slate with mountain roads and entire village in remote areas virtually swept away. Scientific studies of landslides by such organizations as the China Society for Rock Mechanics and Engineering (CSRME) did not begin until the establishment of the PRC in 1949, with more recent surveys and mapping employing high-resolution remote sensing equipment carried out by the Ministry of Land and Resources (now the **Ministry of Natural Resources**).

A national warning system has been installed in 1,640 of the most vulnerable counties, with information input provided in part by the local populations. Compulsory geohazard risk assessments are now also required for proposed construction in high susceptibility zones, especially for such potentially hazardous projects as natural gas pipelines. Recent landslides occurred in Sichuan province following a massive earthquake in 2008, and mountainside collapses stemming from heavy rainfall also took place in Sichuan in June 2017 and August 2017.

LANZHOU. The capital of Gansu province, located in the northwest of the People's Republic of China (PRC), and once rated as one of the 10 most polluted **cities** in the world, Lanzhou carried out a major assault on **air pollution** that has dramatically improved **air quality** with substantial reductions in fine particulate matter (PM 2.5) concentrations. A center of highly polluting heavy and light **industry**, including brick kilns (*zhuanyao*), petrochemicals (*shiyou huaxue chanpin*), locomotives (*jiche*) and rolling stock (*cheliang*), industrial **chemicals**, rubber (*xiangjiao*), and **fertilizers**, along with **coal**-fired power plants, Lanzhou sits in a narrow valley on the southern banks of the **Yellow River** with **mountains** to the north and south that inhibit air circulation. Adding to low air quality were the increasingly frequent sandstorms from the nearby Gobi Desert and precipitous drops in annual precipitation in the surrounding semiarid region, along with a rise in average temperatures. Dramatic increases in air pollution were also brought about by the growth in the number of **motor vehicles**, totaling 232,000 in 2007, in a city of 2.8 million people, leading the municipal government to encourage government employees to walk to work.

City programs aimed at improving air quality began in the 1980s and accelerated dramatically in the 2010s with radical measures aimed at reducing pollution that was so bad the city was shrouded from view by overhead satellites. Measures included shutting down the most heavily polluting industrial plants, many with excessive capacity; shuttering other polluting facilities during the cold winter months, moving yet others to an industrial park located on the city's outskirts; and relying on clean-coal technology, while shifting power plant fuels to natural gas, as coal consumption dropped from 10 million tons to 6 million tons annually.

Employing as many as 10,000 pollution supervisors roaming the city in search of violators, antipollution regulations were strictly enforced, turning Lanzhou into a model for other similarly blighted cities, for example, Zhengzhou (**Henan province**) and **Shijiazhuang** (**Hebei province**). Recording the fastest improvement in air quality of any Chinese city, Lanzhou was awarded the country's Climate Progress Title in 2015. **Water quality** in the Lanzhou basin of the Yellow River also improved, with local sightings reported of the **endangered** giant salamander, although the downstream "resource-depleted" (*ziyuan kujie*) industrial and **mining** city of Baiyin, also in Gansu, remains a major polluter of the river.

LAWS AND REGULATIONS. Legal remedies to environmental problems in the People's Republic of China (PRC) began with the enactment of a Trial Environmental Protection Law (1979), followed by a second Environmental Protection Law (1989), with new provisions added in 2015. By 2010, the PRC had adopted and enacted 30 major environmental laws involving a variety of issues, including, among others, **air pollution** and **water pollution**, management of solid waste and **chemical** substances, and **noise pollution**, along with 90 administrative regulations and numerous environmental standards.

Drawing on initial experiences with environmental amelioration, the 1979 Trial Law did the following: defined the goals and tasks of environmental protection; established the legal basis for institutions of environmental protection, beginning with the State Environmental Protection Agency (1987–1998); outlined legal liability; and established a pollution-charging system. In addition, it called for **environmental impact assessment (EIA)** and mandating environmental standards for new construction projects, for instance, pollution control facilities. The more comprehensive 1989 law contained 47 articles, with specific measures for supervision and management of environmental protection, improvement of environment protection, prevention and control of environmental pollution and other hazards, and clarification of legal liability.

More uniform national and local standards for environmental protection were created by authorized environmental units for pollutant discharges, along with an environmental monitoring and assessment system for on-site **environmental inspections**. According to the law, commercial enterprises were held responsible for eliminating and controlling pollution, while special attention was given to mitigating environmental damage to **agriculture**, including **soil** pollution, salinization (*yanhua*), and alkalization (*jianhua*), along with **desertification**. Additions in the 2015 Environmental Protection Law increased the number of articles to 70, with such major provisions as increased penalties for offending parties, along with regulations for public-interest litigation and **lawsuits**. The authority to create regulations at the subnational level was granted to **environmental protection bureaus (EPBs)**.

LAWSUITS AND LEGAL RULINGS (*SUSONG/FALU CAIJIE*). The transformation of the legal and court system, along with the emergence of environmental nongovernmental organizations (ENGOs), in the People's Republic of China (PRC) has led to an increase in lawsuits and numerous legal rulings with major implications for environmental policy and practice. Most notable are the revisions to the Civil Protection Law (2013) and the Environmental Protection Law (2015), which empowered established **nongovernmental organizations (NGOs)** to bring "public interest lawsuits" (*gongyi susong*), while Chinese prosecutors are encouraged by legal authorities to be more proactive in pursuing suits against environmental offenders by enterprises and lax government enforcement agencies, with the power to file cases nationwide granted in 2017.

With 700 ENGOs qualified to file public-interest lawsuits and a pilot program in 13 provinces instituted for local prosecutors to litigate on behalf of "protecting the **ecosystem** and natural resources," major types of cases include ENGOs filing separate suits against individual polluters or local enterprises involving such large state-owned enterprises (SOEs) and multinationals as Hyundai and Volkswagen, or multiple suits in cases of a major pollution incident. Public prosecutors, including the Supreme People's Procurator, are also encouraged to file suits against government agencies, largely for failing to enforce existing environmental statutes, with prosecutors often "nudging" potential plaintiffs to take formal legal action.

While most environmental disputes are settled out of court, major cases have brought important legal rulings, including imposing large fines on offending parties, overruling lower court decisions barring suits by some ENGOs, and allowing suits between government agencies in different jurisdictions but with overlapping environmental issues. The establishment of specialized Environmental Courts beginning in 2007, in Guizhou province, numbering 130 such tribunals and collegiate panels in 16 provinces with specially trained judges, has also advanced the legal process, although the time required for resolving cases is generally quite substantial.

Major problems confronting environmental litigants include the following: the high cost of court filings, totaling as much as RMB 400,000 ($60,000), and the bureaucratic difficulties of bringing such filings; the burden of proof placed on plaintiffs with the technical difficulty of establishing causation and calculating monetary costs to the environment; the limitation on suits by private entities against government authorities; and the possibility of the imposition of substantial legal fees on ENGOs in failed public-interest suits. From July 2016 to July 2017, 850 public-interest lawsuits were filed, primarily by government prosecutors, with the largest percentage of such suits involving **air pollution**. In the first public lawsuit involving **endangered** animal life, authorities in Guangxi province were sued by ENGOs in May 2019 for failing to save a group of smuggled pangolins that died while in custody.

"LEARN FROM DAZHAI". *See* AGRICULTURE (*NONGYE*).

LEGAL ASSISTANCE TO POLLUTION VICTIMS. Founded by Professor Wang Cangfa of the Chinese University of Political Science and Law in **Beijing** in 1998, the Legal Assistance to Pollution Victims is an environmental nongovernmental organization (ENGO) devoted to providing legal assistance to victims of pollution in the People's Republic of China (PRC), ensuring legal measures enacted by the state are fully enforced. Composed largely of environmental lawyers and scholars, along with technical experts and even graduate students, most on a volunteer basis, the center runs a hotline offering free legal advice, while also bringing lawsuits against violators of the country's antipollution laws, including large public and private enterprises.

Registered with the Ministry of Justice, as required by Chinese law, the center has pursued more than 100 legal cases in court, winning approximately 50 percent, with major decisions including forcing a **steel** mill in **Hebei**

province to move and securing a $730,000 settlement against a **paper** mill in Shandong province. Other successful cases have involved settlements for fish and frog growers in Jiangsu, as well as fruit tree farmers in Fujian, whose land and **water supplies** were contaminated by local **chemical** facilities. The center also provides public-interest training for aspiring environmental lawyers and even judges, while educating the Chinese public on how to respond to infringement on their rights to a healthy environment. In addition to setting up a nationwide network of lawyers, many working for the Chinese government, Wang Cangfa and staff have assisted state authorities in drafting environmental protection **laws**.

LI JUNFENG. A specialist in **energy** environmental studies and **renewable energy**, Li Junfeng worked at the Energy Research Institute (ERI) from 1982 to 2011, serving as deputy director (1999–2011). An advisor to the **Ministry of Environmental Protection (MEP)**, Li helped draft the Renewable Energy Law of the People's Republic of China (PRC), enacted in 2013, along with the Medium- and Long-Term Development Plan for Renewable Energy Development. President of the Renewable Energy Industries Association, he has also served at the National Center for Climate Change Strategy and is coauthor of *Powering China: The Role of Renewable Energy*.

LI KEQIANG (1955–). Premier of the People's Republic of China (PRC), appointed in 2013, Li Keqiang is most noted for his declaration of a "war on pollution" (*dui wuran de zhanzheng*) during a meeting of the National People's Congress (NPC) in March 2014, when he also warned that the heavy smog engulfing many Chinese **cities** constituted "nature's red-light warning against inefficient and blind development." Trained as an economist with a Ph.D. from Peking University, Li served as governor of Liaoning province from 2004 to 2007, a center of heavy industrial production with such cities as **Shenyang**, followed by a stint as a vice premier, closely allied with President **Hu Jintao**.

While economic policy and finance constituted Li's primary policy focus, he also dealt with important environmental issues, including **sustainable development** and green **energy**, along with overseeing major infrastructure projects, including the **Three Gorges Dam** and **South-to-North Water Diversion Project (SNWDP)**, opposed by many environmentalists both inside and outside the PRC. Committed to improving ambient **air quality** in Chinese cities confronting the serious "Airpocalypse," Li has called for reducing the high levels of fine particulate matter (PM 2.5) and PM 10 by eliminating outdated energy production facilities, including shutting down 50,000 small **coal**-fired furnaces and shifting to **nuclear power** and **renewable energy** as centerpieces of a low-carbon **economy**.

LI PENG (1928–2019). Premier of the People's Republic of China (PRC) from 1987 to 1998, Li Peng influenced environmental policy, most notably by his strong support for the construction of the giant **Three Gorges Dam** on the **Yangzi River**, the largest **hydropower** project in the world, along with China's nascent **nuclear power** industry. Trained as a hydraulic engineer in the **Soviet Union** at the Moscow Power Engineering Institute, Li headed the Beijing Electricity Power Administration (1966–1980) and served as a vice minister and minister of power (1979–1983). As premier, he played a central role in authorizing the military crackdown against prodemocracy demonstrations in June 1989, after which the Three Gorges project received government approval in 1992.

Leading the PRC delegation to the United Nations Conference on Environment and Development in Rio de Janeiro ("Earth Summit I") in June 1992, Li stressed the obligation of developed countries to find solutions and develop the technology and financing for confronting global environmental problems, while defending the sovereignty and independence of all countries, especially in the developing world. While members of Li's family gained prominence in China's power sector through their control of a major **energy** company, upon his final retirement from government in 2002, the entire power sector was reorganized.

LIANG CONGJIE (1932–2010). The founder of **Friends of Nature (FON)** in 1994, the first environmental nongovernmental organization (ENGO) in the People's Republic of China (PRC), Liang Congjie was also one of the country's first environmental **activists** to pursue a nonconfrontational approach to ameliorating China's severe environmental problems. The grandson of Liang Qichao, a reformer during the last days of the Qing Dynasty (1644–1911), and son of Liang Sicheng, the noted architect who argued, unsuccessfully, for the preservation of ancient buildings and walls in **Beijing** during the 1950s, Liang Congjie was trained as a historian at Peking University and served for years as editor of *Encyclopedic Knowledge*, where he became aware of the country's mounting environmental problems stemming largely from rapid and at times uncontrolled economic growth.

An early admirer of **Greenpeace** and an associate of such early Chinese environmental activists as Ms. **Dai Qing** and **Ma Jun**, Liang used his position at the privately run Academy for Chinese Culture to establish FON as a legally recognized **nongovernmental organization (NGO)**, pursuing such nonconfrontational activities as tree planting and **bird** watching, while also concentrating on **environmental education**, especially for teachers and young people in the countryside. A member of the advisory Chinese People's Political Consultative Conference (CPPCC), with access to the national **media** and political leaders, and a master at organizing and impacting official policy, he pursued campaigns against the illegal felling of old-growth trees that ultimately led to a nationwide ban on **logging** and the preservation of **endangered species**, most prominently the **Tibetan antelope (*chiru*)** and **golden snub-nosed monkey**.

Recognized as an "environmental ambassador" (*huanjing dashi*) by the State Environmental Protection Administration (SEPA), which had once refused to sponsor FON, as required by Chinese law, Liang was awarded the Ramon Magsaysay Award for Public Service in 2000. He also served as an environmental consultant to the Beijing Olympic Committee in 2008, and was named one of China's 50 most influential people by the frequently disputatious *Southern Weekend.* Advocating greater government transparency on environmental matters, Liang helped transform the environmental movement in China into a mainstream activity led by independent **civil society** groups and environmental nongovernmental organizations (ENGOs), which now number more than 3,500.

LITERATURE. The emergence of **environmental awareness** among the general public in the People's Republic of China (PRC) has included a boom in writings on nature in the form of fiction, poetry, and drama, aimed at cultivating an aesthetic appreciation on the relationship of human kind to natural surroundings. Stressing the coexistence of humanity and nature, literary works with an environmental subtext encourage humans to adapt to the surrounding environment, avoiding any attempt to make an enemy of it or conduct a **"war on nature"**—as was pursued by the **Chinese Communist Party (CCP)**, especially under the leadership of Chairman **Mao Zedong** in the 1950s and 1960s. Calling for a transcendence of the dominant anthropocentrism that China's approach to modernization has promoted at the expense of the country's natural environment, especially **forests** and **grasslands**, contemporary literary works reassert traditional Chinese values and philosophical perspectives, captured in such classical phrases as "nature is most beautiful" (*ziran damei*) and "heaven and humanity in harmony" (*tian ren heyi*).

In addition to translating Western literary classics on the value of **wilderness**, for example, *Walden* by Henry David Thoreau, Chinese novels have captured a similar ethos, particularly in the case of *Wolf Totem* (*Lang Tuteng*) by Jiang Rong (pen name of Lü Jiamin), a novel, also made into a film, chronicling the experiences of a Chinese student "sent down" (*xia xiang*) during the **Cultural Revolution (1966–1976)** to the grasslands of Inner Mongolia. Confronting the wildness and freedom of wolves, and appreciating the role of local nomads in maintaining an ecological balance between human activity and the natural world, *Wolf Totem* contrasts the fiercely independent wolves with the "sheepish" (*xiuqie*) Chinese people. The novel is also an indictment of the Han Chinese approach to "civilization" involving the transformation of wild grasslands into a "tame" (*xunfu*) land of modern technology with resultant **desertification**, while wiping out the wolf population, which had kept such environmentally destructive animals as squirrels and marmots in check. Returning to the area in Inner Mongolia 30 years later, the protagonist discovers that the wolves have disappeared, a victim to extermination campaigns promoted by the state, while the once-verdant but fragile grasslands have turned to yellow dust, a monument to destructive effects of rapid "modernization." Foreign films depicting destruction of ecosystems and man's separation from the natural world have generally done poorly in China as occurred with the 2017 dystopian *Blade Runner 2049*, this despite scenes of environmental degradation and uncontrolled urban sprawl suggestive of present-day **Guangzhou**.

LIU CHANGMING. A hydrologist with the Chinese Academy of Sciences (CAS), Liu Changming is an advisor to the government of the People's Republic of China (PRC) on the **South-to-North Water Diversion Project (SNWDP)**, undergoing construction since 2002. Commenting on the controversial western channel (one of three), which will draw water from the headwaters of the **Yangzi River** in **Tibet** and cross the **Qinghai–Tibetan Plateau** and the highlands of western **Yunnan province** through a series of long tunnels and dams, Liu has acknowledged that the project presents major "technical challenges" to the country's scientists and engineers.

LIU JIANGQIANG (1969–). A research associate at the Center for Nature and Society, Peking University, Liu Jianqiang was, for several years, an investigative reporter at *Southern Weekend* (*Nanfang Zhoumo*), one of the most aggressive investigative newspapers in the People's Republic of China (PRC). Trained as a **journalist** at Tsinghua University, Liu worked around government censors and warded off threats from agents of powerful construction firms to publish several provocative articles on environmental issues. Most notable was a series published in September 2006, on the proposed building of **hydropower** projects on the **Nu River**, the country's last wild river, which, when read by Premier **Wen Jiabao**, led to a suspension and investigation of the project. Other influential articles included the issue of **genetically modified** rice; the role of newly formed environmental nongovernmental organizations (ENGOs); and the proposed reconstruction of the Imperial Summer Palace Lake outside **Beijing**, which led the State Environmental Protection Administration (SEPA) to conduct the first-ever state-level public hearing on an environmental issue.

Liu has also published work on environmentalists in **Tibet** and is involved in training China's next generation of environmental journalists. An ardent opponent of hydropower projects, notably the **Three Gorges Dam**, viewing them as particularly destructive to the environment, especially to **fisheries** and **birdlife**, he believes the greatest need of the Chinese public is for information on environmental matters, which, in his view, is enormously enhanced by the wide availability in the PRC of the **internet**.

LIUZHOU. *See* "FOREST CITIES" (*SENLIN CHENGSHI*).

LOESS PLATEAU (*HUANGTU GAOYUAN*). Covering an area of 640,000 km^2 (250,000 mi^2), encompassing seven provinces in the northwest and centered in Shanxi and Shaanxi, the semiarid Loess Plateau, rendered as "yellow earth plateau" in Chinese, is one of the most environmentally afflicted areas in the People's Republic of China (PRC). Once the center of highly productive **agriculture** during the imperial era, with its thick, fertile yellow **soils** composed of calcium carbonate, the plateau gradually became seriously denuded throughout the centuries, primarily by poor agricultural practices and overgrazing by large herds of sheep and goats, leading to serious **deforestation** and the destruction of 90 percent of the surface vegetation, including extensive **grasslands**. Described as the "most highly erodible" soil on Earth, annual **soil erosion** amounted to 5,000 to 30,000 tons per km^2, creating a landscape dominated by deep gullies (*shuigou*), tunnel inlets (*suidao rukou*), and barren **mountains** subject to frequent **landslides** and encroaching **desertification**. Spread by ferocious wind storms, plateau soils, while contributing to the enormous fertility of the neighboring North China Plain, became the primary source of sedimentation, estimated at 90 percent, into the **flood**-prone **Yellow River**.

Supported by the **World Bank**, the plateau was singled out for major ecological restoration by the Loess Plateau Watershed Rehabilitation Project (1994–2002), with goals of creating sustainable agricultural practices, increasing **forest** cover and rainwater retention, and reducing sediment inflows into the Yellow, Wei, and other local **rivers**. Major project provisions included restricting agricultural cultivation, primarily corn and millet, to flatlands, ending the traditional practice of tilling steep hillsides with an incline of greater than 25 degrees; replacing free-range grazing by sheep and goats with net pens and fodder feed instead of plateau grasses; and a ban on highly destructive **logging** and the collection of firewood.

Restoration measures involved the construction of sediment storage dams, halting soil loss and filling in gullies with productive agricultural flatlands, regenerating grasslands, and transplanting trees on steep, fish-scale terraces carved out by local laborers hired as project employees. With 920,000 hectares (2.27 million acres) restored, the significant reduction of annual sedimentation inflows of 100 million tons into rivers and waterways was achieved, along with water retention, soil stabilization, and the reappearance of local **wildlife**. While the project is considered a model for **ecosystem** restoration through the creation of a macrolevel management structure, large sections of the Loess Plateau remain denuded, contributing to a major flood on the Wei River afflicting the city of Xi'an in 2011.

LOGGING (*JILU*). Suffering substantial **deforestation** during the 1950s and 1960s, especially during the enormously destructive **Great Leap Forward (1958–1960)** and the broader **"war on nature"** declared by **Chinese Communist Party (CCP)** chairman **Mao Zedong**, the People's Republic of China (PRC) has, in recent years, imposed strict controls and regulations on **logging** and the **timber industry**. While the expansion of the country's depleted **forests** emerged as a major policy goal inaugurated during the 1980s, following serious and highly destructive **flooding** on the **Yangzi River** in 1998, bans on commercial logging in natural forests were imposed in 13 provinces, especially in areas bordering major **river** basins. Similar bans were imposed in 2014, 2015, and 2017, largely in Northeast China, including forested regions of Heilongjiang, Inner Mongolia, and Jilin provinces, where mature and usable forests are in short supply, even in plantation forests designed for commercial use.

While 49 million m^3 of natural forest is logged annually, China's goal is to end logging in natural forests by 2020, although illegal logging, especially in remote areas, remains a persistent problem. With demand for wood in the Chinese **economy** increasing dramatically for infrastructure projects, the booming real estate and furniture industries, and charcoal production, China's imports of timber from Southeast Asia and other forest-rich regions have increased exponentially, often at the expense of native reserves.

M

MA JUN (1968–). Founder and director of the **Institute of Public and Environmental Affairs (IPEA)**, one of the first and most **activist** environmental nongovernmental organizations (ENGOs) in the People's Republic of China (PRC), Ma Jun is author of *China's Water Crisis* (*Zhongguo shui weiji*), the first major book documenting the country's enormous problems of **water resources**, water scarcity, and **water pollution**. Working as an investigate **journalist** for the *South China Morning Post* (1993–2000) out of **Beijing**, Ma confronted the stark reality of China's environmental woes when reporting from the highly polluted Fen River area in Shanxi province, a center of heavy industrial and **coal** production, while also learning of **Yellow River** dry-up. Collecting data gathered by local and national government environmental protection authorities, for example, the **Ministry of Environmental Protection (MEP)**, Ma and members of IPEA put together a Water Pollution Map, the first public database of water pollution in China.

A consultant to the Sinosphere Corporation, which provides sustainable solutions for international businesses operating in the PRC, Ma is the recipient of the Goldman Environmental Prize (2012) and appeared in *Before the Flood*, a documentary film on global water pollution and **climate change** produced by the National Geographic Society. Ma is a strong advocate of public participation in addressing China's environmental problems, believing that the country is at a crucial "tipping point" (*yinbao dian*) in achieving a balance between environmental protection and economic development, while advocating stronger enforcement of China's environmental **laws**, especially by local officials. He was named one of the world's 100 most influential people by *Time* magazine in 2006, and received the Skoll award in 2015, for innovative use of technology.

MA SHIJUN (1915–1991). A specialist in insect bionomy and a member of the Harmful Insects Expert Committee of the United Nations, Ma Shijun was a strong supporter of **sustainable development**, with additional specializations in biology and ecological control. As a delegate to the United Nations, Ma participated in the drafting of the *Brundtland Report: Our Common*

Future (1987), which advocated for an international approach to sustainable development. He is also known as the "father of **ecological engineering**" in China for his pathbreaking work in combining base and applied sciences in employing natural systems to address problems of environmental protection.

MA YINCHU (1882–1982). A prominent economist and strong advocate of family planning in the People's Republic of China (PRC), Ma Yinchu predicted a dramatic increase in China's population based on the first national census, conducted in 1953, and in accord with Ma's own "New Population Theory" (*Xinde Renkou Lilun*). Arguing that an accelerated rate of population growth in the newly established PRC would disrupt the country's development and put enormous pressure on environmental resources, he called for the adoption of population control policies, which were roundly denounced in the official **media** as "Malthusian" (*Maersaisi*) and an attempt to "discredit socialism" (*huaiyi shehuizhuyi*). Subject to personal attack and humiliation by **Chinese Communist Party (CCP)** chairman **Mao Zedong** in 1957, Ma was dismissed from his position as president of Peking University until his rehabilitation in 1979, following Mao's passing (1976) and the emergence of **Deng Xiaoping** (1978–1979) as China's paramount leader.

With the country's population growing from 600 million in 1953 to 700 million in 1964, and 950 million in 1976, a One-Child Policy was finally instituted in 1979, based in part on Ma's theory, which, throughout the years, has prevented an estimated 400 million additional births. Reaching 1.3 billion in January 2015, four years later than previously predicted, China's huge population continues to have enormous consequences and effects on the country's environment, notably threatening **sustainable development**.

MANGROVES (*HONG SHULIN*). Found in subtropical and tropical **coastal** areas between 25 degrees north and south latitudes in the saline and brackish waters of intertidal mudflats, mangrove swamps and **forests** are characterized by trees, shrubs, and other deep-rooted **plant species**. Decreasing from 42,000 hectares (103,000 acres) in the 1950s to approximately 20,000 hectares (49,400 acres) in the 2010s, mangroves are located along the coastline of the People's Republic of China (PRC) in the four provinces of Fujian, Guangdong, Guangxi, and Hainan. Concentrated in 12 **nature reserves**, six devoted primarily to mangroves and constituting 0.1 percent of the global total, China's mangroves are composed of 37 separate species, two of which are endemic to the country.

Commonly known as the "nursery of the sea," mangrove forests, with their tangled network of spike-like and partially submerged roots, provide crucial feeding and spawning **habitat** for **fisheries**, including shrimp and crab, while serving as posts for migratory **birds**. Other major benefits include maintaining **biodiversity**; protecting coastal areas from erosion, especially **beaches** and banks from the destructive effects of storm surges and tidal intrusion; purifying coastal waters; and storing more "blue carbon" (carbon captured by the oceans) than any other plant species.

Maintaining a symbiotic relationship with **coral reefs**, mangroves also retain nutrients from **rivers** and estuaries. Mangroves provide materials for dyes and medicines, for instance, extracts for secondary metabolites used in **traditional Chinese medicine (TCM)**. Since 1949, 73 percent of mangroves in the PRC have been lost to **aquaculture** and rice farming, widespread cuttings for firewood, and rapid, uncontrolled urban coastal development, with the greatest decline in Guangdong province, from 21,200 hectares to 3,813 hectares (51,870 acres to 9,418 acres), while the largest remaining mangrove is in Guangxi province, at 5,640 hectares (12,930 acres), a decrease from 10,000 hectares (24,700 acres).

Major efforts at mangrove restoration began in the 1990s, with the planting of 150,000 new trees, involving such environmental groups as the China Mangrove Protection Project, headed by Liu Yi. Persistent threats to mangroves include a decline in species richness, with a decrease from 35 to only nine, along with outbreaks of algae plumes (*zaolei yumao*) caused by industrial pollution, as occurred in the Haibin Natural Wetland Park, Guangxi province. Scientific research on mangroves relies heavily on remote-sensing, employing Landsat earth-observing satellites to monitor changes in mangrove canopy and density, considered major indicators of ongoing **climate change**. Mangrove nature reserves in the PRC include Qinglan and Dongzhai (Hainan), Futian (Guangdong), Shanko (Guangxi), Jiulongjiang (Fujian), and Shenzhen (Special Economic Zone).

MAO ZEDONG (1893–1976). Chairman of the ruling **Chinese Communist Party (CCP)** and generally the undisputed leader of the People's Republic of China (PRC) from 1949 until his passing in 1976, Mao Zedong was committed to policies of national military and economic power, along with ideological orthodoxy. During the early years of central economic planning modeled on the **Soviet Union** and beginning in 1953, Mao evidently advocated a program for "greening the motherland" (*lühua zuguo*) in 1956, to deal with the enormous problems of **deforestation**. While the program was slated to run for 12 years, Mao quickly shifted gears and declared a **"war on nature"**—rejecting the traditional concept of "heaven and humanity in harmony" (*tian ren heyi*), which the chairman considered part of the "old culture" in need of destruction. Voicing slogans like "make the high mountain bow its head, make the river yield the way" (*rang gaoshan ditou, shi heliu shunqi*) and invoking the ancient mythology of the **"Foolish Old Man Who Re-**

moved the Mountains," Mao called on the Chinese people to carry out a mission of conquering nature that would cause enormous environmental damage in the PRC.

Among the most disastrous of the Maoist policies was the **"Four Pests"** campaign, in which millions of sparrows, described by Mao as grain-consuming "pests," were killed, leading to a massive infestation of locusts that contributed to the Great Famine (1959–1961). Equally, if not more, destructive was the **Great Leap Forward (1958–1960)**, in which the promotion of **backyard furnaces** for the production of **steel**, largely in rural areas, led to massive deforestation, along with innumerable **water-control projects**, which, combined with other harmful practices, for example, deep plowing and close planting, afflicted agricultural lands. Followed by the "Grain First Campaign" (*Liangshi Diyi Yundong*) in the early 1960s, with the slogan "taking grain as the key link" (*yi liang wei gang*), an overemphasis on grain production was pursued throughout the country, especially in marginal areas, for instance, **Tibet**, with attendant deleterious effects on **grasslands**, **wetlands**, and **lakes** subject to **reclamation**. Finally, came the **Cultural Revolution (1966–1976)**, when Mao ordered youth "sent down" (*xia xiang*) to the countryside, where, in such areas as **Xishuangbanna (Yunnan province)**, highly destructive land reclamation projects were carried out, leading to a major shrinkage of **lakes** and destruction of old-growth forests, which were replaced by water-guzzling rubber trees.

Mao also promoted political assaults on intellectuals and scientists, particularly during the Anti-Rightist Campaign (1957–1958), which stifled informative discussion and the debate of policies potentially harmful to the environment, for example, the construction of the Sanmenxia **hydropower** station on the **Yellow River**. Exhibiting a penchant for "gigantomania," apparently inherited from Soviet leader Josef Stalin (1927–1953), Mao supported massive infrastructure projects like the **Three Gorges Dam** project and the **South-to-North Water Diversion Project (SNWDP)**, which, while not inaugurated until after his passing, have resulted in substantial environmental damage. Not until the administration of President **Hu Jintao** (2003–2013) was the general Maoist distain for the environment replaced by policies of environmental protection and restoration, which have continued with the presidency of **Xi Jinping** (2013–).

MARINE CONSERVATION (*HAIYANG BAOHU*). With extensive damage to **coastal** areas, islands, and waters along the 14,500 kilometers (9,000 miles) of coastline of the People's Republic of China (PRC), major efforts at remediation and restoration of **biodiversity** and **fisheries** have been adopted, particularly since the establishment of "ecological red lines" (*shengtai hongxian*) of protection in 2016. Major targets of restoration policies include boosting the conservation of 16 heavily polluted inland bays and repairing the **habitat** of 50 other coastal areas suffering from significant erosion, with limits imposed on land **reclamation**, while also pursuing large-scale reforestation and renewal of **mangroves** and reed and **seagrass** beds. Rebuilding ocean fish stocks involving 270 **endangered species** has also been carried out, along with concomitant constraints on unregulated overfishing by the country's giant seagoing fishing fleet. Endangered ocean **wildlife**, including dolphins and several turtle (*gui*) species, are also being afforded preservation and protection, many in national **nature reserves**, with assistance from marine conservative organizations and student groups, both domestic and international (such as the China Sea Turtle Conservation Alliance), that have become involved in marine conservation in the PRC since 1979, especially in Hainan and Guangdong provinces, and in **Beijing**.

Organizational initiatives include creation of Aquatic Wildlife **nature reserves** (200), Special Marine Protection Areas (SMPAs/111), and Marine Natural Resource Areas (MNRAs/35) beginning in 2005 and covering 124,000 km^2 (4.1 percent) of coastal areas, low by international standards. Multiple uses of marine areas combine restoration with limited economic activity, mainly fishing and exploiting deep-sea resources, in an attempt to achieve **sustainable development** as coastal regions contribute 9 percent to the country's gross domestic product (GDP). While impressive on paper, many of these plans suffer from major problems, including weak implementation, inadequate funding, lack of baseline data, and poor public awareness of the multiple threats to coastal and marine **ecosystems**, along with a lack of major Chinese environmental nongovernmental organizations (ENGOs) engaging the issue. International complications are also created by the inclusion of the internationally disputed Nansha Islands in the **South China Sea** in the overall plan for marine conservation pursued by the PRC.

MARINE RANCHING AND FARMING (*HAIYANG YANGZHI*). The depletion of natural **fisheries** and scarcity of land in the People's Republic of China (PRC) has led to a dramatic increase in the creation of marine ranches for increasing fish stocks in **coastal** waters, with similar operations for growing agricultural crops, both involving major risks to the ocean environment. Developed in the 1970s, marine ranching, also known as mariculture (*haishui yangzhi*), involves the cultivation of marine organisms for food and other products by placing artificial reefs on seabeds or enclosed sections of the ocean. While the most simple and original method involved sinking old boats, which, by changing sea currents, would encourage algae (*zaolei*) to grow, in turn attracting other marine species, more advanced, large-scale technologies entail building large cages and steel enclosures for releasing into the waters juvenile fish that reproduce and develop into mature species for capture. The fastest-growing form of ocean **aquaculture** in China's rapidly growing fishing industry, marine ranches number about 200, with plans to extend their coverage from 850 km^2 to 2,700 km^2 (328 mi^2 to 1,042 mi^2) by 2025, with locations primarily in the **Bohai** and **Yellow** seas but not in the tropics because China now breeds more fish than is consumed.

While touted as a minimally disruptive approach for restoring stocks of such commercially viable and popular fish species as tilapia (*luofeiyu*) and the increasingly rare yellow croaker (*huangyu*), the practice can involve considerable environmental risks, including causing damage to fragile seabeds and **coral reefs** from large-scale marine infrastructure, including the world's largest salmon cage, now off the coast of Shandong province. Attendant problems include the disruption of natural food chains; pollution from the release of untreated water back into the sea containing fish waste and antibiotics (*kangshengsu*), which kill bacteria and reduce the diversity of microbes in the environment, and continued overfishing. Environmentally proper procedures involve avoiding damage to natural **habitat**, with coral reefs or beds of aquatic grass (*shuicao*) off limits to artificial reefs and prohibitions on the cultivation of single-fish species. Regulatory provisions installed by the Chinese government include banning carcinogenic antibiotics (*zhi'ai kangshengsu*) and providing enhanced protections of **mangroves**, oyster beds *(muli chuang*), and coral reefs, although many Chinese practitioners of marine ranching show little or no concern for disrupting the **ecosystem**. Marine farming consists primarily of cultivating vegetation, primarily kelp, on artificially floating rafts, which began in 1952, but with a gradual takeover by the more lucrative field of marine ranching.

MEDIA (*MEITI*). Consisting of conventional television, radio, magazines, **journalism**, films, and the **internet**, media in the People's Republic of China (PRC) has become increasingly influential on issues involving the environment, especially the role of text messaging by "social media" (*shejiao meiti*). Under tight state control from 1949 to the 1980s, and dominated by such official outlets as *People's Daily*, the newspaper of the **Chinese Communist Party (CCP)**, Xinhua News Agency, of the Chinese government, and Central China Television (CCTV) and Central People's Broadcasting, both government owned, any and all coverage of environmental issues adhered to national policy with little, if any, independent investigative coverage.

Beginning with the economic reforms (1978–1979) and cutbacks in government subsidies to news organizations, media outlets became increasingly commercialized, with environmental issues more open for discussion and debate with relatively objective coverage of events like as the massive **chemical** explosion in **Tianjin** in August 2015, when reporters for CCTV and other outlets rushed to the scene to provide live coverage, with major follow-up stories on the impact and possible causes of the **ecological disaster**, which lasted for several months.

Along with **publications** explicitly devoted to the environment and with national policy favoring the vigorous pursuit of environmental protection and restoration, especially under the presidencies of **Hu Jintao** (2003–2012) and **Xi Jinping** (2013–), a range of environment issues and stories are covered in the Chinese media, with text messaging via such outlets as WeChat acting as major channels of communication for engaging in environmental **social protests**. Some topics and stories in the environmental arena remain off limits, for example, the country's continued commitment to **hydropower**, especially involving the controversial **Three Gorges Dam** and planned projects in politically sensitive areas like **Tibet**. Censorship of the media has also occurred with documentary films on the environment, for instance, *Under the Dome*, which, portraying the deleterious effects of **air pollution**, with millions of downloads within 24 hours, was quickly pulled off the Chinese internet. A similar fate befell other investigative documentaries, for example, *Plastic China* by filmmaker Wang Jiuliang, although other films, like *Behemoth*, documenting the environmental and human costs of the **coal** industry, have been left untouched. Films like *Damming the Angry River* (2005) have also influenced government policy, in this case contributing to protests, evidently successful, against hydropower construction on the **Nu** [Angry] **River**.

Wanting to raise public **environmental awareness** and encourage "green" habits among the Chinese public, especially aimed at undermining the pervasive **"waste culture"** through the media, the Chinese government nevertheless is still willing to wield the censorship cudgel whenever a story appears to touch on sensitive political issues. Media attention to acceptable environmental issues remains a major tool for generating mitigating actions, especially by local government authorities who fear the public spotlight and make possible public-interest **lawsuits** on environmental issues.

MEDICINE. *See* TRADITIONAL CHINESE MEDICINE (TCM/*ZHONGYI*).

MIN RIVER (*MIN JIANG*). Located in central **Sichuan province** and 735 kilometers (457 miles) in length, the Min River is a major headwater and source of downstream **water pollution** into the **Yangzi River**, the largest inland waterway in the People's Republic of China (PRC). Revered in Chinese history as the site of the ancient Dujiangyan **irrigation** system, built in the 3rd century BCE and still in use today, the Min has served as a subject of scientific studies by Chinese hydrologists on the characteristics and deleterious effects of **river** water pollution. As the central Sichuan basin underwent rapid economic development in **agriculture** and **industry** beginning in the 1970s, **water quality** in the Min gradually deteriorated, with 30 percent of the water volume rated unsuitable for domestic use (grade IV, V, V+) in the late 1990s, to a high of 41 percent rated unsuitable in 2006, but with some improvement, reduced to 29 percent unsuitable in recent years, although still at a relatively high volume.

The impact on river **fisheries** in the Min has also been dramatic, with only 16 of the 40 fish species found in the 1950s surviving to the present day. Prominent **hydropower** projects on the river total 27 dams built, undergoing construction, or planned, notably the Zipingpu Dam in Sichuan, with 760 megawatts (MW) of installed capacity. The dam developed dangerous fissures and cracks in the wake of the 2008 **earthquake**, which some seismologists suggest was caused by the dam's large reservoir. Research on pollution in the Min notes the importance of tailoring remediation efforts to the specific form and source of contamination.

Another Min River (same spelling in English but different Chinese character) is located in Fujian province, which, at 577 kilometers (359 miles), is the largest waterway in the province, known to locals as the "mother river" (*muqin he*). The site of a devastating fish die-off in 2011, controversy arose as to whether the catastrophe, in which millions of fish perished, derived from oxygen deprivation, the official view, or the consequence of upstream industrial pollution, the view of local fishermen and environmentalists.

MINING (*KUANGYE*). A major **industry** in the **economy** of the People's Republic of China (PRC), providing 40 percent of the world's supply of key minerals, including more than 90 percent of **rare earths**, mining, both underground and open-pit (*lutian*), has wrought considerable environmental damage on the country. With approximately 10,000 mines, mostly in the **coal** sector and many small-scale, China's mines are located primarily in the northern provinces of Heilongjiang, Jilin, Liaoning, Shandong, and **Hebei**, the last ranked as the most polluted province in the country. Major environmental effects include serious **water pollution** by wastewater, contamination of **soils**, and destruction of **habitat** for **plant species** and **wildlife**.

Undergoing rapid expansion since the introduction of economic reforms (1978–1979) with consistently high prices for such minerals as lead (*qian*) and lithium (*li*), the latter used in batteries for **electric vehicles**, mining operations are producing more than 20,000 hectares (49,4000 acres) of derelict land annually, with serious contamination from highly toxic minerals. Included are, among others, boron (*peng*), cadmium (*ge*), coal, copper (*tong*), gold (*jin*), gypsum (*shigao*), iron ore (*tie kuang*), lead (*qian*), lithium (*li*), manganese (*meng*), mercury (*gong*), molybdenum (*mu*), nickel (*nie*), platinum (*bo*), pyrites (*huangtie kuang*), tin (*xi*), tungsten (*wu*), and vanadium (*fan*), with the mining of minerals like graphite (*shimo*) producing all-encompassing dusts destructive to tree life. Involved in the sector are some of the largest mining **companies** in the world, including Shenhua Energy Co. and Zijin Mining Group.

After many years of poor regulations, weak enforcement, especially of the many illegal, highly polluting mining operations, and short-lived cleanup efforts, the mining sector, with the exception of separately regulated coal, became subject to new rules and tighter regulations. Included are requirements for the treatment of 85 percent of previously raw, untreated wastewater; remediation of soil contamination; and use of all tailings and solid waste, with lead and cadmium production obligated to employ treatment by biological and **chemical** technologies. Confronting substantial financial penalties under the newly enacted Environmental Protection Tax Law (2018), mining companies are also required to clean up abandoned mines and attendant contamination of surroundings, including **water resources** and 16 percent of national soils. With more than 1,000 coal mines closed nationwide in 2016, authorities in the **Xinjiang**-Uighur Autonomous Region also closed mines, mostly small-scale, from operating in the Altun **nature reserve**. Prospectors seeking gold and other valuable minerals have a history of causing enormous environmental damage, especially to water resources, as occurred in the Kekexili region on the **Qinghai–Tibetan Plateau**.

MINISTRY OF ECOLOGY AND ENVIRONMENT. *See* MINISTRY OF ENVIRONMENTAL PROTECTION (MEP/*HUANJING BAOHU BU*).

MINISTRY OF ENVIRONMENTAL PROTECTION (MEP/*HUANJING BAOHU BU*). Established in 2008, under the authority of the State Council in the government of the People's Republic of China (PRC), the Ministry of Environmental Protection (MEP) was charged with enforcing environmental **laws and regulations**, while also managing the funding and organization of research and development in the environmental field, along with several "key laboratories," including the Urban Air Particles Pollution Prevention and Wetland Ecology and Vegetation Recovery. Preceded by the State Environmental Protection Agency (1987–1998), followed by the State Environmental Protection Administration (SEPA/1998–2008), and succeeded by the Ministry of Ecology and Environment (MEE) in 2018, MEP oversaw **water quality**, ambient **air quality**, solid waste disposal, soil pollution and **noise pollution**, and radioactivity (*fangshexing*) releases.

Major offices and departments in MEP included Policies, Laws, and Regulations; Science and Technology Standards; Pollution Control; Natural Ecosystem Protection; Environment Impact Assessments; International Cooperation; Nuclear Safety; and Environmental Inspection. These were in addition to appropriate administrative units, including a branch of the **Chinese Com-**

munist Party (CCP). Regional divisions (with the location of primary offices indicated in parenthesis) included eastern (Nanjing), southern (**Guangzhou**), northwestern (Xi'an), southwestern (Chengdu), northeastern (**Shenyang**), and headquarters (**Beijing**). Previous MEP ministers include Zhou Shengxian (2008–2015), Chen Jining (2015–2017), and Li Ganji (2017–2018), the last becoming minister of the newly created MEE, one of two super-ministries established in 2018, to end the bureaucratic fragmentation afflicting environmental protection by bringing about a consolidation of authority and power.

With MEP frequently competing with other ministries involving overlapping agendas of environmental protection, for example, the Ministry of Land and Resources and the Ministry of Water Resources, MEE inherited the responsibilities and enforcement areas of MEP, while being granted authority over several new areas, including the following (with previous responsible administrative units in parenthesis): **climate change** emissions (National Development and Reform Commission [NDRC]); groundwater pollution (Ministry of Land and Resources); watershed environments (Ministry of Water Resources); agricultural pollution (Ministry of Agriculture); and marine conservation (State Oceanic Administration [SOA]).

See also MINISTRY OF NATURAL RESOURCES (MNR/*ZIRAN ZIYUAN BU*).

MINISTRY OF NATURAL RESOURCES (MNR/*ZIRAN ZIYUAN BU*). Established as part of a larger institutional reform of the State Council of the People's Republic of China (PRC), enacted in March 2018, the Ministry of Natural Resources (MNR) is one of two super-ministries, including the Ministry of Ecology and Environment (MEE) created to bring greater consolidation and bureaucratic efficiency in **policy-making** and **policy implementation** in environmental protection and conservation. Granted official ownership and management of China's vast trove of natural resources, MNR replaced the now-defunct Ministry of Land and Resources, State Oceanic Administration (SOA), and National Administration of Surveying, while assuming responsibilities in the environmental sector from the National Development and Reform Commission (NDRC), Ministry of Agriculture, Ministry of Housing and Urban-Rural Development, and Ministry of Water Resources, and the reorganized State Forestry and Wetlands Administration (SFWA), which MNR now oversees.

One of the major tasks of the MNR is establishing a spatial planning system, along with measures for enforcing user payments of natural resources, in line with the Master Plan for Ecological Civilization (2015), by ending previous practices of plundering and poaching the country's valuable but limited natural resources. The minister of natural resources is Lu Hao, former governor of Heilongjiang province (2013–2018).

MINORITIES (*SHAOSHU MINZU*). Numbering 55 officially recognized ethnic groups in the People's Republic of China (PRC), many residing in some of the country's most environmentally rich but vulnerable regions, for example, **Yunnan province**, minorities have a mixed relationship to the natural world, in some cases living in apparent harmony with their physical surroundings and in others portrayed as serial abusers. Constituting about 7 percent (108 million) of the total population in the PRC, minorities are concentrated in the so-named autonomous regions (*zizhiqu*) where they collectively constitute a majority (Guangxi-Zhuang, Inner Mongolia, Ningxia-Hui, **Tibet**, and **Xinjiang**-Uighur), along with significant populations in Guizhou, **Sichuan**, and Yunnan provinces.

Praised for revering in religious terms surrounding **forests**, **rivers**, **lakes**, and even **wildlife**, especially in Tibet and among other groups concentrated in mountainous areas like the Naxi in Yunnan and Sichuan provinces, minorities in Inner Mongolia have been singled out for having allegedly caused significant degradation of **grasslands** through the overgrazing of livestock, while defenders of **nomadic pastoralism** attribute the deleterious effects to widespread and often illegal **mining**.

Social protests by minorities against alleged environmental abuse have been carried out, for example, by resident Tujia and Miao minorities in the **mountain** areas of South Central China (2008) against local manganese (*meng*) mines for having caused serious shortages of **water resources**, including vulnerable groundwater sources, with wastewaters containing ammonia (*an*) and hexavalent chromium (*liujia ge*), the latter a known carcinogen. Along with major reductions in rice production, highly disruptive dynamiting prompted then-president **Hu Jintao** to launch an official investigation. Ethnic conflicts over state sovereignty involving such independent minded minorities as the largely Muslim Uighurs (*Weiwu'er*) have also been married to environmental issues, as in the case of terrorist attacks on a **coal** mine in Xinjiang (2015). Granting minorities an important role in the country's **nature reserves** has been promoted given their history of opposition to excessive **logging** and other environmentally destructive practices often carried out in the name of rapid growth in the national **economy**.

MOTOR VEHICLES. Numbering 310 million in 2018, with two-thirds, 200-plus million, consisting of automobiles, most privately owned, along with buses and trucks, motor vehicles in the People's Republic of China (PRC) have become the primary source of increased **air pollution** in recent years, especially nitrogen oxide (NO_x), as contamination by industrial gases and **coal**-burning power plants has remained flat or declined in many urban areas. Even as the country adopted relatively stringent **fuel and emission standards** for a variety of motor vehicles, environmental pollution and damage, especially in urban areas with large numbers of automobiles, more than

1, 2, and 3 million units in 49, 19, and 6 **cities**, respectively, the last comprised of **Beijing**, **Chongqing**, Chengdu, **Shanghai**, **Shijiazhuang**, and Suzhou, is significant.

With 10 of the world's 25 most congested cities, where monumental traffic jams occur largely from poor road design, air contaminated with vehicle exhaust contributes between 12 and 52 percent of total air pollution in several of these cities, a figure likely to increase as the number of automobiles continues to grow, despite limits on the allocation of licenses in such heavily congested cities as Beijing and Shanghai. Increasing the number of **electric vehicles (EVs)**, including buses, deployed in major cities like Shenzhen, numbering 16,700, is helping to mitigate the overall environmental effects of an ever-expanding vehicle market. An action plan for dealing with highly polluting diesel-powered trucks essential for the **transportation** of goods between and within cities, and used in large urban construction projects, is also being considered. An offshoot of motor vehicle expansion is the growing problem of scrap tires, 15 million tons annually, ending up in local dumps and in **rivers** and **lakes**. Major remediation efforts include construction of an innovative scrap rubber recycling plant in **Henan province** for rinsing, sorting, drying, pyrolysis, and converting rubber into carbon black, fuel oil, steel wire, and synthetic gas, the last used to power the plant, resulting in **zero waste**.

MOUNTAINS (*SHAN*). Constituting two-thirds of the 9.6 million km^2 (3.2 million mi^2) land mass of the People's Republic of China (PRC), mountains are a prominent topographical feature, especially in the south and southwestern regions of the country south of the **Yangzi River**. With more than 30 major ranges and seven of the world's top 12 peaks higher than 8,000 meters (26,000 feet), including the two highest in the world (Mt. Everest/*Qomolangma* and K-2), mountains in China also pose such serious environmental challenges as frequent **landslides** and major **flood** events, brought on by substantial **deforestation** in headwater areas of major **rivers** and accelerated shrinkage of the country's more than 48,000 **glaciers**. Home to 20 percent of the nation's population, numbering 280 million people, many of them **minorities**, mountain regions provide substantial economic benefits, including rich mineral deposits, **forests** and **timber** resources, **hydropower** sites, and tourist attractions, especially the many **nature reserves**.

Major ranges include the following: Himalayas (*Ximalaya*), (2,400 km along the PRC–India–Nepal–Bhutan border); Kunlun, 2,500 km (1,552 miles) from the Pamir Plateau to Tibet and Qinghai provinces; Tianshan (traversing **Xinjiang** province); Tanggula (the **Qinghai–Tibetan Plateau** and headwaters of the Yangzi River); Qinling (southern Shaanxi province, demarcating the north and south of the country); Greater Khingan (1,000 km in Heilongjiang province); Taihang (400 km on the eastern fringes of the

Loess Plateau); Qilian, 800 km (496 mi) marking the border between Qinghai and Gansu provinces; Changbai, 320 km (198 mi) ranging across Heilongjiang, Jilin, and Liaoning provinces to Democratic People's Republic of Korea; and Hengduan 900 km (560 mi) at the juncture of **Tibet**, **Sichuan**, and **Yunnan** provinces. Others (with primary locations in parentheses) include: Altai and Altun (Xinjiang); Amne Machin, Bayan-Har, and Beishan (Qinghai); Daba (Sichuan, Shaanxi, Hubei); Dabie (Hubei); Daxue (Sichuan); Dayao (Guangxi); Dayunwu (Guangdong); Gangtise (Tibet); Helan (Ningxia); Huangshan (Anhui); Jinping (Sichuan); Jundu (Beijing, **Hebei**); Karakoram (Xinjiang, Tibet); Kuaiji (Zhejiang); Min (Sichuan, Gansu); Nanling (Guangdong); Nyainqêntanglha (Tibet); Pamir (Xinjiang); Qian (Jilin); Shaluli (Tibet, Sichuan); Thistle (Guangxi); Wanda (Heilongjiang); Wudang (Hubei); Xiao (**Henan**); Yan (Hebei); Yandang (Zhejiang); Yin (Inner Mongolia, Hebei);Yuecheng (Hunan, Guangxi); Zhongnan (Shaanxi); and Zhongtiao (Shanxi).

Surrounding such major agricultural centers as the Sichuan basin and encircling major **cities** located in mountain valleys, for instance, **Lanzhou** (Gansu province), mountain peaks have become the prime targets of removal (700 in the case of Lanzhou) in land-creating operations aimed at extending urban landscapes and allowing better air flow and circulation. An example of the penchant in China for gigantic transformations of the natural landscape, these earth-moving projects are also taking place around major cities like **Chongqing**, Shiyan and Yichang (Hubei province), and Yan'an (Shaanxi province), provoking criticism by Chinese environmentalists for inadequate evaluation, with potential negative environmental effects, including landslides, flooding, **air pollution**, **water pollution**, and the disruption of river flow.

N

NATURE RESERVES AND NATIONAL PARKS (*ZIRAN BAOHU QU, GUOJIA FENGJING MINGSHENGQU*). Modeled on a system developed in the **Soviet Union** during the reign of Josef Stalin (1927–1953) and introduced into the People's Republic of China (PRC) in the 1950s, nature reserves, national parks, and other forms of protected areas are designed to preserve the **biodiversity** of **wildlife** and **plant species** against encroachments by human settlement and disruptive economic activity like **mining** and **logging**. Following the establishment of the first nature reserve at Dinghu Mountain, Guangdong province, in 1956, the number grew to 19 (1965), 481 (1987), 2,000 (2004), and 2,750 (2017), covering 147 million hectares (363 million acres), 15 percent of the country's total land area. Established at the national, provincial, prefecture, county, and municipal levels, national nature reserves, numbering 474 in 2018, are generally the most well-funded and competently staffed, while many at the provincial level and below (2,274) suffer from inadequate funding, poorly trained personnel, and ineffective management, and are frequently referred to as "paper parks" (*zhigong yuan*). Concentrated in the country's southwest and northwest, with the largest protected land area in **Tibet** (34 percent of the territory), followed by Qinghai (29 percent) and Gansu (15 percent) provinces, the operation of nature reserves, especially at the national level, is guided by the Nature Reserve Law (1994), with priority given to the protection of **forests**, **wetlands**, steppe, desert, and **oceans**, along with rare and **endangered** animal wildlife and plant species.

Four types of nature reserves have been established in China based on their primary function: scientific research, for example, Changbai Mountain (Jilin), **Wolong (Sichuan)**, Wuyi Mountain (Fujian), Fanjing Mountain (Guizhou), and **Xishuangbanna (Yunnan)**, all considered national showcases; historic, scenic, and tourist value, for example, Jiuzhai Valley (Sichuan), a network of valleys with cascading waterfalls and multi-colored, pristine lakes; the protection of endangered and rare wildlife and plant species, for example, Gangkou National Sea Turtle Reserve (Guangdong), Snake Island (Liaoning), and Ganjia Lake Saxoul [Tree] (**Xinjiang**); and the preservation of special geological (*dizhi*) and geomorphological (*dimao*) features, for example, Stone Forest (Yunnan) and Zhangye Danxia Landform National Geopark (Gansu).

The largest national nature reserve in the PRC, also the highest and second-largest in the world, is Qiang Tang Nature Reserve, in northern Tibet, at 29.8 million hectares (73 million acres). Located at an altitude of more than 5,000 meters (16,400 feet), Qiang Tang lacks virtually any permanent human habitation and serves as a refuge for a rich variety of wildlife, including brown bears (*zong xiong*), wild asses (*ye lü*), black-necked cranes (*heijing he*), **Tibetan antelopes (*chiru*)**, and snow finches (*xue que*), with many pristine **lakes** and a small staff charged with warding off **hunting and poaching**, especially of the *chiru*, with its valuable fur. The top five provinces with the greatest number of nature reserves include Heilongjiang (52), Inner Mongolia (38), Sichuan (32), Shaanxi (31), and Yunnan (26).

Other well-known and internationally recognized national nature reserves and national parks, with involvement by the United Nations and the World Wildlife Fund (WWF), are as follows: Puducao National Park in western Yunnan, the first national park in China with white water stone terraces (*Bashuita*); Mt. Everest (*Qomolangma*), with peacocks (*kongque*), gibbons (*changbiyuan*), Tibetan bears (*Xizang xiong*), and snow leopards (*xue bao*); Hoh Xil/Kekexili, in an isolated region of the **Qinghai–Tibetan Plateau** near the headwaters of the **Yangzi River**, with wild yaks (*ye maoniu*), Tibetan antelopes, and brown bears, and cited as a World Heritage Site, also featured in the film *Mountain Patrol: Kekexili*; Shennongjia (Hubei), with some of the wildest **mountains** and forests in the PRC, and reputed home of China's "bigfoot" (*yeren*); Wolong (southwest Sichuan), the earliest and largest protected area for the giant **panda bear**; Yancheng Littoral Mudflats and Rare Birds (Jiangsu), with 377 protected bird species, 65 annelids (*huanjie dongwu*), and 310 insects (*kuanchong*); Kanas (Xinjiang), with **glaciers** and forests; Fanjing Mountain, with 80 percent virgin forest cover, and Bali Azalea, with the largest rhododendron forest (*dujian shulin*) in the world (both in Guizhou); West Lake near the desert city of Dunhuang (Gansu) designed to halt encroaching desertification with yardang land forms; and Xishuangbanna (Yunnan), with rainforests (*redai yulin*), 5,000 wild plant species, tropical gem plasma (*baoshi dengliziti*), along with China's few remaining Asian elephants.

Divided into core, scientific and experimental, and economic development spatial zones, nature reserves are designed to minimize disruption by prohibiting humans from the core with heavy restrictions on logging, livestock grazing, and cultivation of crops. While these regulations are generally followed by national-level nature reserves, this is not true for reserves at the provincial level and below, where more than 1 million people live in core areas and extensive economic activity takes place within reserve boundaries, including logging, mining, farming, and industrial construction (e.g., pipelines), all highly destructive to **habitat**, which in many of these reserves is increasingly fragmented and degraded. Other problems include the overexploitation of wildlife and excessive clearing of vegetation, as occurred in the Zhalong Nature Reserve (Heilongjiang), where the habitat of local **birdlife** like the white crane (*bai he*) was systematically destroyed.

Responses to these endemic problems include the creation of more highly regulated national parks, numbering 244 in 2017, and the inauguration of the Green Shield program in 2018, aimed at cracking down on illegal activities in nature reserves, notably the illicit sale of wildlife and plant species. Other organizational forms of land and resource protection areas include geoparks (224); national wood or flower parks (17); national key parks (63); national wetland reserves and parks (262); urban wetland parks (58); national sandy land protection areas (92); fossil conservation areas (59); national mining parks (87); and an unspecified number of national forests and marine and aquatic parks, along with botanical gardens.

Areas lacking sufficient protection in China include steppe, meadows, **coastal** areas and oceans, and lands with wild plants, as economic more than ecological factors determine the creation of protected areas under the authority of the State Council at the national level. Nature reserves established in areas of considerable tension between **nomadic** herders and Chinese government officials committed to gaining complete control of the territory include, most notably, Three Rivers National Nature Reserve on the Qinghai–Tibetan Plateau (Qinghai). Plans exist to create a national park system modeled on the United States.

NOISE POLLUTION (*ZAOYIN WURAN*). With little or no effect on the physical environment, noise pollution harms human **health**, causing tinnitus and, in more severe cases, hearing loss. According to data supplied by the China National Environmental Monitoring Center, one-fourth of the **cities** in the People's Republic of China (PRC) suffer from high levels of noise pollution, with **Guangzhou** graded as the worst city in the world for this type of pollution, with concomitant high rates of hearing loss, followed by **Beijing** (graded sixth) and **Shanghai** (graded 12th). Sources of excessive noise include construction work (50 percent), industrial operations (17 percent), human activity (12 percent), and **transportation**, including **motor vehicles** (12 percent), with excessive noise occurring at night, as well as during daytime hours, heightened by narrow streets and poorly constructed buildings. Factory noise is particularly severe, with estimates that as many as 20 million workers in the 1980s were subject to excessive noise levels, largely from antiquated machinery. Measures at amelioration introduced in the late 1980s included stiff fines imposed by municipal governments on violators and the inclusion of planned controls of noise pollution in **environmental impact assessments (EIAs)** for proposed construction and other projects.

NOMADIC PASTORALISM (*YOUMU XUMU YE*). A long-standing feature of the rural **economy** in the far western regions of the People's Republic of China (PRC), nomadic pastoralism involves as many as 16 million herders of sheep (*yang*), wild yaks (*ye maoniu*), and goats (*shanyang*), largely by Tibetans and Mongolians, on the extensive **grasslands** of the Mongolian, **Qinghai-Tibetan**, and **Loess** plateaus. Blaming the herders and their livestock for the degradation of large swaths of grasslands (estimates up to 90 percent by overgrazing), the Chinese government banned the free-range grazing of goats and sheep, limiting herds to enclosed net pens with fodder feed. The government has been engaged in a 15-year effort to relocate herders, officially referred to as "**climate change** refugees." Although the herders have traditional lifestyles and livelihoods, they have been forced to sell their animals and move to permanent housing near towns and **cities**, where they often confront joblessness, indebtedness, and a dependence on shrinking government subsidies while conditions on the grasslands continue to substantially deteriorate.

Chinese government efforts to "tame" (*xun*) the highly mobile and independent herders began soon after the establishment of the PRC in 1949, with **Chinese Communist Party (CCP)** chairman **Mao Zedong** showing an indifference to nomadic lifestyle and outright hostility and disdain, especially for the highly religious Tibetans. Pressured to increase herd size beyond sustainable levels to supply the growing population of in-migrants engaged in the development of petroleum, **mining**, and other forms of mineral extraction in both Tibet and Inner Mongolia, herders were deprived of ownership of livestock during the **Great Leap Forward (1958–1960)** and forced into the large People's Communes by government officials, who saw the locals less as able stewards of environmental balance and more as ignorant and "backward" (*luohuo*) people.

Following the inauguration of agricultural reforms in the 1980s, herders regained ownership of their animals but not the land and immediately cut back on the size of bloated herds, while the degradation of grasslands continued with an explosion in the number of such rodents as pikas (*shutu*) and other destructive pests, along with unregulated **logging**, the cultivation of crops on marginal lands, and the exploitation of the watershed for water-guzzling industries like **mining**. Campaigns like the Great Western Development Strategy (*Xibu da kaifa*) and Closing Pastures to Restore Grasslands (*Tuimu huancao*), with grazing severely limited, including outright bans in particularly fragile areas, introduced from the late 1990s onward ignored the fact that grasslands benefit from moderate grazing, with little consumption of

water and sound ecological management by herders, based on generations of pastoralist experience and trial-by-error practices. Equally poor policies were adopted in the 1990s, for instance, the Four-Way Scheme (*Sipei Tao*), which deprived herders of lease-land rights, while mandating widespread fencing off of grazing lands and permanent sheltering of animals and herders, as toxic weeds invaded dormant grasslands, with wild weather tearing up weakened **soils**, creating conditions described as "black beach" (*heise shatan*). Environmental nongovernmental organizations (ENGOs) operating in the region and working with herders include the Snowland Great River Environmental Protection Association.

NONGOVERNMENTAL ORGANIZATION (NGO/*FEI ZHENGFU ZUZHI*). Beginning with the establishment of **Friends of Nature (FON)** in 1994, and quickly followed by Beijing Global Village and Green Home, environmental nongovernmental organizations (ENGOs) in the People's Republic of China (PRC) numbered more than 7,800 in 2012. Representing a major structural change in **policy-making** in the PRC from top-down command and control led by the **Chinese Communist Party (CCP)** to a greater reliance on public participation and market-based incentives, ENGOs have become an integral part of the country's approach to environmental protection, providing government authorities with scientific expertise and models for the effective monitoring and enforcement of environmental **laws and regulations**.

Drawing on foreign examples of organization, with their specialized focus on a particular environmental issue, for example, the International Rivers Network and such well-financed operations as the World Wildlife Fund (WWF), Chinese ENGOs concentrate on three areas of activity, generally tolerated by political authorities, which include the following: educating and guiding the Chinese public on major environmental issues and problems; promoting citizen involvement in matters of environmental protection and advising government authorities on policies and programs; and monitoring environmental conditions in both urban and rural areas, while assisting domestic enterprises and multinational corporations in developing greater concern for environmental protection.

Laws governing the organization and operation of ENGOs include the Civil Procedure Law (amended 2012), Environmental Protection Law (revised 2015), and Charity Law (2016), the last setting regulations for domestic NGOs officially referred to as "social organizations" (*shehui zuzhi*). Foreign NGOs operating in China, for instance, WWF, are governed by the Foreign Non-Government Organization Law (2017), which requires registration with public security authorities in the PRC and restricts both the range of activities and financing arrangements.

Major restrictions on domestic NGOs include the prohibition of the formation of regional branches and the prevention of an individual serving as the legal representative of more than one NGO, along with requiring NGOs to maintain a branch of the CCP. Registration with government authorities requires a sponsoring agency, a process that can take considerable time and is often difficult to obtain, leading many NGOs to register as private companies, which entails paying taxes on received donations, the primary form of financing and one of the major difficulties confronting most ENGOs, as they often rely on foreign sources.

While the 2016 Charity Law relaxed registration and funding requirements, recruitment of volunteers into NGOs remains a problem, as large segments of the Chinese public remain suspicious of privately run organizations. Major ENGO activities include reclaiming **wastelands**, observing **birdlife**, planting trees, protecting **endangered species**, and establishing "green" communities, while compensating for declining central government influence in the environmental arena. Public-interest **lawsuits** against environmental violators have also been pursued by such ENGOs as **Legal Assistance to Pollution Victims**, with the first victory in 2015, against a quarry accused of illegal dumping of waste in Fujian province. Government-organized nongovernmental organizations (GONGOs) closely affiliated with state agencies include such organizations as the China Environmental Science Association.

NU RIVER (*NU JIANG*). With headwaters in the melting **glaciers** on the **Qinghai–Tibetan Plateau** and 2,815 kilometers (1,700 miles) in length, and flowing through the **mountainous** regions of **Yunnan province** into Myanmar (formerly Burma) and Thailand as the Salween, the Nu River basin, at 325,000 km^2 (125,000 mi^2), is one of the largest in Southeast Asia and a hotspot of **biodiversity**. The last wild river in the People's Republic of China (PRC), the *Nu* ("angry") is home to 6,000 **plant species** and 50 percent of the **wildlife** species in China, notably snow leopards (*xue bao*), black snub-nosed monkeys (*heisi hou*), and red pandas (*xiao xiongmao*). Described by the United Nations Educational, Scientific, and Cultural Organization (UNESCO) as the most biologically diverse temperate zone on earth, the Nu also flows through the western section of the Three Parallel Rivers World Heritage Site and is distinguished by the 370-mile "Grand Canyon of the East" (*dongbian da xiagu*), also known as "China's Grand Canyon" (*Zhongguo da xiagu*), with gorges as deep as 4,000 meters (13,000 feet), located in northwest Yunnan.

One of the two remaining undammed rivers in the PRC, the Nu was targeted in 2003, for the construction of five to 13 **hydropower** projects, suspended by Premier **Wen Jiabao** in 2004, following **social protests** from environmentalists and local residents slated for mass relocation. A topic of major **media** attention in the film *Damming the Angry River* (2005), plans were renewed by the State Council in 2014, only to be discarded once again by the Yunnan provincial government in 2016. Damage to the river comes from extensive **logging** in the basin with attendant **soil erosion**, while domestic environmental nongovernmental organizations (ENGOs) committed to the river's preservation include Green Watershed of Yunnan, which has pushed for government transparency, especially involving the unreleased **environmental impact assessment (EIA)** of proposed dams, with support from **Friends of Nature (FON)**, Green Earth Volunteers, Beijing Global Village, and the international **nongovernmental organization (NGO)** International Rivers Network.

NUCLEAR POWER AND RADIOACTIVE WASTE. Beginning in the late 1970s, the People's Republic of China (PRC) inaugurated the construction of nuclear power plants (*hedian chang*), which, by 2017, numbered 44 active light-water pressured reactors, with 13 more under construction, and a plan to produce 6 percent of total electric power (70 gigawatts) by 2020. Considered an integral part of China's commitment to nonfossil fuel **renewable energy**, nuclear power is touted as an alternative to the continued burning of **coal**, boosting the production of electric power (*dianli*) in densely populated areas, while keeping abreast of advanced technology. Like most countries pursuing nuclear power for electrical generation, China began its program with little or no consideration of how to handle the spent fuel, most of which remains on plant sites in water-filled pools until a permanent underground depository is made available sometime in 2040.

Assisted by the French nuclear power corporation Areva, which also aided in the construction of several nuclear power plants, for example, Daya Bay near Hong Kong, China pursued the path of reprocessing spent fuel for producing reusable plutonium (*bu*) and uranium (*you*). Intense **social protests** against a proposed reprocessing facility in the city of Lianyugang, Jiangsu province, just north of **Shanghai**, led to the 2016 cancellation of the project, slated for completion in 2030, while other pilot reprocessing projects have failed to meet production goals.

The disposal of high-level waste (HLW) is currently planned for a deep geological site of tunnels, selected in 2006, dug into solid granite in the Gobi Desert at Beishan, Gansu province. Despite the promise of clean electrical power, the nuclear power **industry** in China, favored by such prominent **Chinese Communist Party (CCP)** leaders as **Zhou Enlai**, **Li Peng**, and **Wen Jiabao**, has confronted serious popular resistance, including demonstrations in Heshan, Guangdong province, which led to the cancellation of a proposed uranium processing facility in 2013. Control of nuclear waste is administered by the China National Nuclear Cooperation (CNNC), with nuclear safety handled by the National Nuclear Safety Administration (NNSA) in the former **Ministry of Environmental Protection (MEP)**, now the Ministry of Ecology and Environment (MEE). Considerable investment and research into the development of the Thorium Molten Salt Reactor (TMSR) has been pursued in China since the 1970s because these fourth-generation nuclear power plants with their closed nuclear fuel cycle produce substantially less waste and require considerably less cooling water. Operating at low atmospheric pressure and high temperature, TMSR can be built in modular fashion relatively cheaply and installed underground throughout the country—especially in the arid northwestern and western areas of the country.

O

OCEAN POLLUTION AND TRASH. The largest contributor to pollution of the Pacific Ocean, primarily by toxic effluents from major **rivers** emptying into the sea, the People's Republic of China (PRC) has launched several programs to ameliorate the situation. Of the 18 ecological zones in the ocean waters off the PRC, 14 were determined to have unhealthy levels of pollution in 2011, with major pollutants including organic nitrogen (*youji dan*), reactive phosphates (*fanying xing linsuan yan*), **plastics**, **oil** from agricultural runoff, and industrial wastewater, which have contributed to the increasing proliferation of green, brown, and red tides (83 incidents annually), as well as algae plumes (*zaofei yumao*), especially during the last decade, with a deleterious impact on local **fisheries**.

Of particular importance is pollution from plastics, with China at 8.8 million tons in 2010, the largest contributor of plastic bottles, bags, and other plastic rubbish, clogging global sea lanes, with some patches floating in the Pacific larger than entire countries. Predictions are that total accumulated plastics in the oceans will reach 155 million tons by 2025, with major inflows from Asian rivers, four of which flow from the PRC—the **Yangzi**, **Yellow**, Hai, and **Pearl**—with the Yangzi, the largest in the region, pumping in an estimated total of 3.53 million tons of plastics annually. Research indicates that one piece of microplastic (plastic fragments less than five millimeters in length) exists for every 3.5 square meters of seawater in the **Bohai**, **South China**, and **East China** seas, while fragments smaller than 150 micrometers can end up in the circulatory systems of marine life and ultimately be consumed by humans.

In addition to rules by the Chinese government issued against the production and use of ultra-thin plastic bags (2008) and new rules on the use of plastics (2018), China is also engaged in developing plastic material that will dissolve upon contact with seawater, although a purely technological fix to the problem is considered unfeasible. China also issued regulations to reduce pollution from oceangoing shipping (2009), with accidents involving ships (753 between 1998 and 2008) adding to the spike in ocean contamination. While reducing pollution of nearby ocean waters remains a high priority of the PRC, with some marginal improvements in **water quality** in nearby seas in 2017, problems persist, including a lack of sufficient long-term monitoring networks and the perennial problem of uncoordinated approaches among competing and overlapping bureaucratic units and agencies preventing effective action. China also remains outside the United Nations Environmental Program on Clean Seas, with scientists predicting that current pollution trends will mean more plastics in the oceans than fish by 2050, notably in some of the most remote and deepest parts of the five oceans.

OIL INDUSTRY, SPILLS, AND NATURAL GAS EXPLOSIONS. The fourth-largest producer of oil and other petroleum products in the world and a major consumer of natural gas (*tianranqi*), the People's Republic of China (PRC) has experienced a number of major oil spills and serious explosions of natural gas with significant impacts on the natural environment and human **health**. Producing 3.7 million barrels a day and 191 million tons of crude oil (*yuan you*) annually from both onshore and offshore wells, and 112 billion m^3 of natural gas, with substantial imports of liquified natural gas (LNG/*yehua tianranqi*), China has constructed a vast system of pipelines, storage tanks, and LNG terminals for storing and transporting vital **energy** supplies.

In the case of oil, with considerable production located in northwestern fields of the Junggar and Tarim basins, **Xinjiang** province, with consumers concentrated on the east coast, heavy reliance is on pipelines that are subject to leakage and periodic accidents. The most serious occurred at the Xingang port near the northeastern city of Dalian (Liaoning province) in June 2010, when an explosion in two pipelines led to the spillage of 1,500 tons (400,000 gallons) of crude oil, which entered the nearby harbor. Creating a 430 km^2 (160 mi^2) oil slick threatening the **Yellow Sea**, the slick burst into flames, with the fire lasting 15 hours and seriously impacting nearby animal life and sea **birds**. No less devastating was an explosion from the rupture of an underground oil pipeline in Qingdao, Shandong province, in November 2013, which killed 35 people and contaminated local waters, although without long-term environmental effects.

At sea, oil spills have occurred at offshore drilling rigs operated by Chinese oil firms, while the worst oil spill in decades occurred in the **East China Sea** from the collision of an Iranian oil tanker with a freighter, leading to the release of condensate mixture of oil and natural gas covering 100 km^2 (39 mi^2), killing millions of fish, with the tanker burning for an entire week. China's lack of monitoring buoys in affected areas contributed to a tardy government and company response, with **media** coverage also lacking. Plans have been announced to establish a scientific laboratory for dealing with the many environmental effects of oil spills, with satellites and robots assisting in detecting and dealing with underwater ruptures. Breaks in a diesel fuel pipeline in Shaanxi province in January 2010, contaminated tributaries of the

Yellow River, while unregulated and privately run oil fields remain a major threat of spillage and contamination. For the interior city of **Chongqing**, an explosion at a natural gas well in December 2003, led to the death of 198 people in one of the country's most serious **ecological disasters**.

ORGANIC FARMING (*YOUJI GENGZUO*). Adopted based on a concern for food safety and the excessive use of inorganic **fertilizers** and **pesticides** in conventional agricultural production, organic farming has undergone rapid growth in the People's Republic of China (PRC) for both international and domestic food markets. Beginning in the early 1990s with the first export from China of organically grown tea, organic farming and management is practiced on 1.39 million hectares (3.4 million acres), constituting 1 percent of agricultural land in the PRC (2010), as 3 million tons of organic product consisting of beans, vegetables, mushrooms, and meat are grown, primarily in the northeast and southwest.

With an estimated loss of topsoil in rich black-soil regions in the northeast totaling one millimeter annually, with 6 million hectares (14.8 million acres) suffering from the excessive use of the 32 different pesticides available in the country, organic farming and the similar practice of **ecofarming** have been adopted. Employing natural products like hot peppers (*lajiao*) to ward off pests and organic fertilizers composed of rapeseed and animal manure, organic farmers in China also avoid toxic herbicides, removing weeds by hand. The third-largest producer of organic products in the world, primarily for export, domestic demand is centered in such wealthier, heavily middle-class **cities** as **Beijing**, **Guangzhou**, and **Shanghai**, where organic farmers' markets now operate, with Shanghai opening up the country's first organic supermarket in 2005.

While not providing direct government financial support to organic farmers, China has promoted organic markets with "green food" certification since the 1990s, and the certification is applied to 13,000 products nationwide, although many Chinese consumers continue to exhibit considerable mistrust, even as national organic product standards were issued in 2005, along with an organic logo. Sustainable agricultural practices have also been promoted by organizations like China Ecological Agriculture (CEA), along with backing by the Ministry of Agriculture. Community Supported Agriculture (CSA) is an international group involved in promoting organic farming, directly linking consumers to organic producers, although overall the PRC remains one of the worst offenders of food safety in the world, primarily because of the excessive use of pesticides, with residue ending up in **rivers** and **lakes**. Prominent organic product companies include YMT Organic Farm, Shanghai, which offers online purchasing and home delivery of organic products and local land-adoption farms, where consumers grow their own crops following organic guidelines outside of the city.

OVERPOPULATION (*RENKOU GUOSHENG*). A perennial problem since the imperial era, overpopulation and rapid growth in the **economy**, especially from the 1950s onward, have been the two most important factors putting pressure on the environment in the People's Republic of China (PRC), including **air pollution**, **water pollution**, and severe degradation of agricultural **soils**. Reaching 120 million people during the Song Dynasty (960–1279) and 436 million in 1850, with the establishment of the PRC the country's population was 552 million in 1950, 700 million in 1964, more than 1 billion in 1982, and 1.3 billion in 2017, with expectations of topping out at 1.4 billion sometime in the 2020s. Failing to institute family planning policies, as recommended by demographer **Ma Yinchu**, population growth surged throughout the period of rule by **Chinese Communist Party (CCP)** chairman **Mao Zedong** (1949–1976), slowing down with the adoption of the One-Child Policy in 1979.

Relying throughout history on extensive **agriculture**, whereby major increases in food production required opening up new cropland, virtually all arable and semiarable land in the country has been brought under the plough, including **grasslands** and lands reclaimed from **lakes** and waterways, **coastal** areas, and in **mountains** through **deforestation** and other environmentally deleterious practices. The increasingly large population has also led to massive urbanization, with 65 **cities** surpassing 1 million people, including such megacities (in the millions) as **Chongqing** (32), **Shanghai** (24), and **Beijing** (20), with concomitant pressures on **energy** and water supplies, along with increases in municipal **garbage** and waste that append to large, densely populated urban areas, which have also experienced increases in income and consumption, especially among the expanding middle class.

OZONE DEPLETION AND POLLUTION (*CHOUYANG HAOJIE, WURAN*). A 1991 signatory to the Montreal Protocol (1987), banning production of **chemicals**, primarily trichlorofluoromethane (freon-11 or CFC-11), the People's Republic of China (PRC) joined the international effort to halt the depletion of the ozone layer, which protects the Earth from ultraviolet radiation (*ziwaixian fushe*) emitted by the sun. A powerful greenhouse gas (GHG) with a greater impact than carbon dioxide (CO_2) in trapping heat from the sun, China announced a reduction in the production of 100,000 tons of freon (*fuli'ang*), with concomitant reductions in consumption in 2009, with a total ban on its use, especially in new appliances, in 2010.

In July 2018, China was revealed as the source of the illicit production of CFC-11, which represented a potential setback to the gradual closing of the massive hole in the ozone layer over Antarctica discovered in 1985. Traced to 18 construction firms operating in 10 Chinese provinces, production of the banned chemical was effectively concealed from regulatory authorities, with apparent collusion of local officials. Both low in price and high in quality compared to nondepleting substitutes, CFC-11 was applied to the production of polyurethane foam used for home insulation in the country's booming housing **industry**. Informed of the revelation by an international environmental nongovernmental organization (ENGO) operating in the United Kingdom, Chinese authorities immediately committed to taking legal action against the offending enterprises with fines and other enforcement tools. Ozone pollution, with the potential for lung damage to humans, also increased in the PRC by 8 percent in 2017, hitting record-high levels in **Beijing** during the summer months.

P

PAN YUE (1960–). A longtime official in environmental agencies of the People's Republic of China (PRC) and one of the first environmental **journalists** and **activists**, Pan Yue has been a strong advocate of taking on violations of environmental **laws**, particularly by commercial enterprises, seeing environmental protection as a central element of the county's national rejuvenation. Working as a reporter for *Economic Daily* and the *China Environmental Journal* in the 1980s, Pan entered government service, becoming a deputy director of the State Environmental Protection Administration (SEPA) and a vice minister in the **Ministry of Environmental Protection (MEP)** from 2003 to 2015, with an important position in the **Chinese Communist Party (CCP)** apparatus and within the government's environmental bureaucracy. Nicknamed "Hurricane Pan" for his aggressive pursuit of environmental protection and tough enforcement of environmental statutes, he oversaw several high-profile campaigns against polluting industries, while also halting 30 projects that were in apparent breach of environmental regulations.

An advocate of the **green gross domestic product (GGDP)**, while noting the relatively high level of pollution in the PRC as a developing country, Pan emerged as something of an "environmental nationalist" (*huanjing minzuzhuyizhe*), believing that overcoming serious environmental problems is essential for the country to become a great international power. Arguing for holding such political jurisdictions as **Hebei province** responsible for polluting neighboring areas by bearing the costs of environmental cleanup, Pan also endorsed the importance of **environmental impact assessments (EIAs)** and enhanced government transparency in addressing the multitude of environmental problems confronting the PRC. While invoking traditional Confucian principles in his general environmental philosophy, he has interpreted environmental degradation as a product of the capitalist system and free markets with the potential to undermine the country's economic foundation, which must be addressed, especially by the younger generation. In 2010, Pan received the Ramon Magsaysay Award for environmental public service.

PANDA BEAR (*DA XIONGMAO*). Literally the "big bear cat" and native to South Central China in the **mountain** regions of **Sichuan**, Gansu, and Shaanxi provinces, the giant panda bear is the iconic representative of **wildlife** in the People's Republic of China (PRC). Generally found in broadleaf (*kuoye*) and coniferous (*zhenye*) **forests** rich in the primary food source of arrow and square bamboo (*jianzhu*, *fangzhu*), pandas in China number approximately 2,060, with 1,864 adults in 2017, as the species was brought back from near-extinction in the 1990s, when total numbers dropped to 1,200. Achieved primarily by the restoration of **habitat**, especially in the 40 panda **nature reserves** located in Southwestern China, most prominently the **Wolong National Nature Reserve**, 66 percent of the entire panda population now inhabits these protected areas but with future survival in the wild still questionable, although the species was upgraded from "**endangered**" to "vulnerable" by the International Union for Conservation of Nature (IUCN) in 2016.

Once widespread throughout the southern and eastern regions of China, pandas are now largely confined to six mountainous areas at an altitude between 1,525 and 3,050 meters (5,000 and 10,000 feet), often enshrouded in misty fog. Concentrated in the Minshan (Sichuan) and Qinling (Shaanxi) Mountain ranges, the latter is home to the same-named Qinling subspecies, which sports a brown fur unlike the black and white fur of the dominant species. Requiring the consumption of 12 kilograms (26 pounds) or more of bamboo daily, pandas are equal in size to the American black bear and weigh between 200 and 260 pounds for females and males, respectively. Suffering from a low birthrate and the loss and fragmentation of historical habitat due to **deforestation** and the relentless expansion of farmland, pandas have also faced threats from **hunting and poaching**, with their highly prized skins supplied to black markets, especially in Hong Kong and Japan.

Once subject to such backward practices as long-term caging to ensure their survival, pandas now benefit from more scientific conservation and preservation practices overseen by the State Forestry Administration (SFA), with assistance from such international wildlife groups as Earthwatch Institute and the World Wildlife Fund (WWF), the latter heavily involved in the original establishment of panda nature reserves. Four major breeding centers exist in China—Bifengxia, Chengdu, Dujiangyan, and Wolong—with 422 breeding pandas in 2015. Sichuan is also home to the red panda, a member of the racoon family and also known as the "lesser panda" (*xiao xiongmao*), native to southwestern China but with only a distant genetic relation to the giant panda. A rare, fully albino white panda was spotted in Sichuan in 2019.

PAPER INDUSTRY (*ZAOZHI YE*). A major sector in the **economy** of the People's Republic of China (PRC), with 10,000 operational mills and pulp factories at one time in operation, many of them small-scale township and village enterprises (TVEs) in rural areas, the paper **industry** has been a large contributor to **air pollution** and **water pollution**, and a primary target of Chinese government antipollution efforts, especially since the early 2000s. Undergoing double-digit growth since the introduction of economic reforms in 1978–1979, the paper industry is responsible for 10 percent of the total emission of wastewater and 25 percent of chemical oxygen demand (COD) in the country's **rivers and waterways**, including sulfur compounds, nitrogen oxide (NO_x), and chlorinated (*lühua*) and organic compounds and nutrients. One of the largest sources of industrial pollution in the Chinese countryside, offending paper and pulp production facilities have been hit with substantial financial levies, introduced in 1982, and in the most egregious cases the facilities have been subject to closure, numbering 1,000 since 1996.

While pollution control technologies are within the financial capacity of the country's large-scale paper mills, small- and medium-sized mills lack the capital for such investments, as pollution monitoring and enforcement of the many small operations has proved difficult for environmental protection authorities. Shutdowns reduced total paper production by 10 million tons between 2005 and 2009, with an additional reduction of 5 million tons, especially for nonwood pulp facilities, slated for 2020. Mills producing less than 5,000 tons annually are now banned from operation, while mills contributing significantly to water pollution have been targeted for government action. Major sources of pollution come from the bleaching process used to brighten paper for printing, along with the pervasive use in China of nonwood, low-grade straw fiber and effluents consisting of "black liquor" (*hei ye*) waste product and pulp residue. With the ban on unsorted and contaminated recycled paper imports imposed in July 2017, Chinese producers lacking foreign feedstock are relying more on domestic sources, which could result in increased **deforestation**, while the deinking process employed in paper **recycling** produces a variety of **chemical** wastes.

PEARL RIVER (*ZHU JIANG*). The third-longest **river**, at 2,400 kilometers (1,500 miles), and the second biggest in terms of water volume, after the **Yangzi River**, in the People's Republic of China (PRC), the Pearl River, with its major tributaries, including the Dong, Xi, and Bei rivers, was one of the country's most polluted waterways, making it a target for a major cleanup. With the Pearl River Delta, the fastest-growing economic region in the country, experiencing an average 13 percent expansion per annum, centered around such major **cities** as **Guangzhou**, Foshan, and Jiangmen, the Pearl River became heavily polluted from industrial effluents and agricultural runoff, along with high levels of domestic **sewage** from the 100 million people living in the densely populated delta.

Major pollutants into the Pearl include ammonia nitrogen (*an dan*), nitrate nitrogen (*xiao tai dan*), phosphorous (*lin*), and organic compounds (*youji huahewu*), with half of the contamination originating outside Guangdong province from upstream sources, primarily in Guangxi and Guizhou provinces, along with other areas in the vast river basin of 409,000 km^2 (158,000 mi^2). The biggest pollution hotspot in East Asia, with a substantial impact on **water quality** in the **South China Sea**, the Pearl River Delta has been a target of major amelioration since in 2004, when a regional organization of 11 provinces and administrative regions, notably Hong Kong and Macau, assisted by the **World Bank**, joined together in a coordinated effort to clean up the Pearl and its many even more polluted tributaries.

Measures aimed at reducing wastewater discharges into the river network included the construction of sludge treatment plants, the expansion of local sewer networks, and dredging the river's bottom, along with the expansion of water quality monitoring systems and the elevation of river banks to protect against **flooding** from the frequent typhoons (*taifeng*) in this subtropical zone. Efforts at halting saltwater intrusion into the river have also been pursued, as improvement in water quality of the Pearl has been achieved, with 95 percent rated grades I through III (suitable for human use) in 2015, with concomitant reductions of grade IV from 12 to 2 percent. Areas rated grade V and V+, unfit for any use, remain primarily in the Pearl River Delta, where 24 percent of the watershed receives these high rankings.

PESTICIDES (*NONGYAO*). With only 9 percent of the world's arable land, farmers in the People's Republic of China (PRC) use 30 percent of global pesticides accounting along with heavy application of **fertilizer** in substantial increases in agricultural yields per hectare but with serious problems of **soil** pollution, **water pollution**, and **air pollution**. Used primarily to ward off pests and control weeds, China consumed 1.8 million metric tons of pesticides in 2018, leading to the degradation of microbes in the soil, as only 35 percent of pesticides applied are functional, with 65 percent in excess, retained for long periods in the soil before degrading and contributing to toxic agricultural runoff, with some pesticide particles suspended in the air and carried away by combining with dust.

With food production as a paramount goal of agricultural policy in the PRC, particularly following the Great Famine (1959–1961), excessive pesticide use was of little to no concern until the 1980s, when China banned DDT (dichlorodiphenyltrichloroethane) in 1983, followed by commitments in the 2000s to eliminate the production and use of several of the 32 available pesticide variants in the country by 2020. Composed of sulfur, lime, nitrogen, and highly toxic chemicals, major categories of pesticides include insecticides (*shachongji*), herbicides (*chucaoji*), and fungicides and bactericides (*shajunji*), with natural pesticides composed largely of insecticidal plants and soil minerals used before 1970. From 1970 to 1980, inorganic synthetic pesticides were used, followed by organic synthetic pesticides from 1980 onward.

Major factors determining excessive pesticide use in the PRC include the small size of the average farm, of 0.6 hectares (1.67 acres), where pesticide use is most excessive and inefficiently applied, while larger plot farms employ more professional management practices, for example, integrated pest management programs that reduce overall pesticide use. In addition to the enactment of Pesticide Management Regulations in 2017, the Chinese government penalizes firms engaged in the production of illegal pesticides, although monitoring pesticide use and enforcing restrictions remains problematic in a country where local authorities often ignore or circumvent central state policies and regulations.

Phasing out of some of the most highly toxic chemicals in pesticides was announced in 2017. These include aldicarb, phorate, and isocarbophos (phased out in 2018), endosulfan and methyl bromide (due to be phased out in 2019), ethoprop, omethoate, methyl isothiocyanate, and aluminum phosphide (due to be phased out in 2020), and chloropicrin, cartoforan, and methomyl (due to be phased out in 2022). Supply problems have resulted in substantial price increases for some pesticides. Reliance on drones guided by satellite navigation to spray pesticides on fields reduces pesticide use by upwards to 30 percent and water consumption by 90 percent with crops retaining less pesticide residue.

PHARMACEUTICAL INDUSTRY (*ZHIYAO YE*). Composed of more than 6,000 manufacturers with annual growth rates between 15 and 25 percent and concentrated in the eastern provinces of Zhejiang, Guangdong, **Hebei**, and Jiangsu, and **Shanghai** municipality, pharmaceutical production has become a major sector in the **economy** of the People's Republic of China (PRC) and is a major source of **water pollution**. Producing 2 million tons of active pharmaceutical ingredients (API) annually, constituting 20 percent of the global total, pharmaceutical plants in China were constructed in the late 1990s, when enforcement of environmental protection **laws** and especially pollution controls were weak. Current Chinese government efforts to control the discharge of wastewater and **sewage** sludge containing such pharmaceutical pollutants as antibiotics (*kangshengsu*) has required many facilities to be shuttered, at the cost of robbing foreign drug companies of their major suppliers—in some cases their sole supplier.

Containing a greater diversity of pollutants than other conventional **chemicals** and with every new drug producing a new compound, pharmaceutical remediation is very costly, with small firms in the PRC unable to bear the costs, even as China remains the world's largest exporter of APIs—600 in all—especially to Europe and the United States. Deleterious environmental effects of pharmaceuticals, especially antibiotics, which can lead to the creation of antibiotic-resistant genes (ARGs), include damaging the reproductive capacity of **fisheries** and fostering environmental bacteria and algae. In addition to imposing tougher environmental regulations on pharmaceutical producers, China has encouraged producers to move into industrial zones and parks where pollution control is more effective, while encouraging Chinese firms to adhere to global standards of the production process. Among the major organizations in the PRC involved in this issue is the National Engineering Center for Pharmaceutical Pollution Water Discharge, located in Hebei province.

PHILOSOPHIES OF NATURE. While the dominant theme in the traditional Chinese philosophy of Confucius (551–479 BCE) and Mencius (372–289 BCE) concerned social relations and the content of human virtue, the natural world was also a topic of philosophical speculation and inquiry, especially by Laozi (?–533 BCE) and the Daoists. Conceived as a self-generating and complex arrangement of elements in a constant state of interaction and flux, nature (*xingzhi*) is dominated by the "Way" (*Dao*), a set of principles by which all things exist. Contrary to Western concepts of natural philosophy, the focus in the Chinese, especially Daoist, view of nature is not on causal or governing forces shaping nature, but on relationships between various elements, most notably the five elements (*wuxing*) of wood, fire, earth, metal, and water, which exist in a mutually productive and destructive relationship. Reflected in the concept of "heaven and humanity in harmony" (*tian ren heyi*), man is considered an integral part of the natural world and should follow the flow of nature's rhythms shaped by the complementary relationship in the entire universe of the yin (passive, dark, negative, weak, feminine, and cool) and the yang (active, bright, revealed, positive, masculine, and hot). Mimicking the forces of the "**wind and water**" (***fengshui***) in their exercise and breathing regime is for the Daoists man's way to comprehend the mysteries of nature, becoming part of nature's system of continuous cycles and achieving optimal moral and physical health.

Mountains and **forests** are of particular importance in Chinese natural philosophy as sacred places of refuge manifesting nature's vital "energy" (*qi*) and home to many ingredients for **traditional Chinese medicine (TCM)** and herbs, along with alchemist minerals, which hold the promise of longevity, one of the most prized values in Chinese culture. Natural refuges are sought out by men to purify their spirit, while Daoist and Buddhist holy men journeyed to sacred sites to build meditation huts and temples, with poets and artists celebrating the beauty of landscapes, which are recreated in elaborate Chinese gardens and featured in Chinese paintings.

While in constant flux, nature follows a regularity and order that humans should recreate in promoting "harmony" (*hexie*) both within society and with nature. Along with the five elements are the five animals (sheep, fowl, ox, dog, and pig) and the five classes of creatures (scaly, feathered, naked, hairy, and shell-covered). These and other traditional principles of natural philosophy were directly assaulted by the **Chinese Communist Party (CCP)** during the era of rule by Chairman **Mao Zedong** (1949–1976), especially in the proclaimed **"war on nature."**

PLANT SPECIES (*ZHIWU PINZHONG*). As the world's richest area of wild plant **biodiversity**, the People's Republic of China (PRC) is home to more than 32,000 species, constituting approximately 12 percent of the global total, including 10,000 species endemic to China. Almost all wild plant varieties that grow in frigid (*hanleng*), temperate (*shidude*), and subtropical (*yaredai*) zones of the northern hemisphere are found in China, where species wiped out by glaciation in North America managed to survive. Along with 2,000 species of edible plants, many used in spices for Chinese cooking, 11,000 medicinal plants are used in both modern medicine and **traditional Chinese medicine (TCM)**, with 7,000 species of woody plants, primarily trees.

Widely distributed throughout the country, 25 percent of plant species are concentrated in dense rainforests (*redai yulin*), mainly in **Yunnan**, Guizhou, and **Sichuan** provinces. Most notable among the plant varieties native to China are the camelia (*kameiliya*); Chinese wolfberry tree (*gouqi*), found primarily in Ningxia; forsythia (*lianqiao*); gardenia (*zhi*); ginseng (*can*); indigo (*dianqing*); jasmine (*moli*); magnolia (*yulan*); notoginseng (*sanqican*); primrose (*bao chunhua*); rhododendron (*dujuan*); safflower (*honghua*), found in **Tibet**; viburnum (*jia*); and peony (*mudan*), designated as one of the country's many national flowers.

During the last 50 years, 200 species have become extinct, with 5,000 more confronting extinction, mainly from the effects of rapid economic growth, loss of **habitat** to expanded cropland, and **overpopulation**. Most notable include a single rare cypress tree in northern Sichuan along with three remaining *baishanzu* fir, and a variety of diminishing angiosperms (flowered plants with seeds). The Chinese government has taken steps to protect wild plant life, especially species with small numbers, with such measures as the National Key Protected Wild Plants, a list of 419 species subject to legal protection against uncontrolled collection, issued by the State Council in 1999. This was followed in 2008, by the Strategy for Plant Conservation, as the PRC also remains a high priority of global biodiversity preservation. Along with the role of botanical gardens, China relies on small-scale reserves, also known as plant microreserves, as the most effective method for preserving threatened species, while continued surveys and research in large-scale **nature reserves** have led to the discovery of new species. A full catalog of plants in China is available in the 80-volume *Flora of China*, completed in 2004.

PLASTICS (*SULIAO*). The largest producer of plastics in the world since 2013, at 60 million tons a year (25 percent of the global total), and a major importer, the People's Republic of China (PRC) consumes 75 million tons of plastic annually. Cheap to produce but tough to **recycle**, plastic waste in China is disposed through landfills and incineration, contributing substantially to **air pollution** and **water pollution**, as well as the degradation of **soil quality**, with unregulated dumping into **rivers and waterways** and the **oceans**. Produced from **chemicals** sourced from fossil fuels, for example, **oil** and natural gas (*tianranqi*), plastics are used for packaging in bags, barrels, and bottles, along with such consumer goods as toys and in agriculture, primarily for heat- and moisture-preserving plastic sheeting. Per capita consumption of plastics in the PRC doubled from 22 kilograms (48 pounds) in 2005 to 46 kilograms (105 pounds) in 2010, with government plans to shift to other, less polluting materials.

Major factors contributing to the enormous growth of plastic consumption include rapidly expanding online food and delivery services, which used 14.7 billion plastic bags and 60 million plastic cartons in 2017. Reliance by food and other markets on single-use plastic bags and the 15 billion disposable plastic lunch boxes produced annually exacerbate the problem with inadequate waste management and infrastructure, particularly in rural areas, where piles of uncollected plastics accumulate. Especially hard hit are the country's major rivers, with the **Yangzi** listed as the world's top plastic-polluting waterway, along with the Huangpu in **Shanghai** and the Xi tributary of the **Pearl River**, both ranked in the world's top five. Dumping an estimated 3.5 million metric tons of plastics annually into the Pacific Ocean, 106 million metric tons since 1992, China is the world's most serious offender in mismanaging plastic waste (28 percent of the global total), while toxic emissions from plastic production in the form of chemical compounds like ethane equal the emissions of 200 **coal**-burning power plants, contributing to **climate change**.

Disposal of plastic waste in Chinese **cities** by an army of migrant scavengers, many children, adds to the problem, although recent price declines for such waste and stricter government regulations have reduced their role, with many small, family-run recycling workshops shut down. Yet, even as the State Council outlawed the provision of free plastic bags and containers by supermarkets and shopping centers, along with ultra-thin plastic bags, in 2008, so-called "white pollution" (*baise wuran*) by plastics persists. Reductions in 1.4 million tons of plastic waste were achieved in part by the introduction of a policy of "paid-use" (*youchang shiyong*) of plastic containers for customers, along with prohibitions on plastic foam, the latter lifted in 2013. Citizen efforts at addressing the issue include the Zero Plastic Challenge through WeChat on the Chinese **internet**, aimed at raising popular awareness of "white pollution," along with **nongovernmental organizations (NGOs)** like *Bichumei* (Closet Beauty) Studio, which monitors the frequent lax compliance of food markets to regulations on plastic bags.

With microplastics showing up in Chinese sea salt and absorbed by marine **wildlife**, for instance, whales, further government measures are being pursued, for example, enhanced monitoring of microplastics in sea waters and a ban on the import of **foreign waste**, including post–industrial plastics, enacted in 2018, although China has yet to sign on to the United Nations Agreement on Marine Waste. Remediation is also made difficult by the prevalence in contemporary China of a throwaway **"waste culture"** among significant segments of the increasingly consumer-oriented population in both rural and urban areas. As the largest plastic market in the world, China is ripe for such new technologies as chemical recycling, which turns polythyelene, the most common plastic, into cycloalkanes and aromatics suitable to be blended into cleaner-burning jet fuels or diesel, utilization of specific fungi to break down non-biodegradable polyurethane plastics, and production of biodegradable "green" plastics.

POLICY IMPLEMENTATION (*ZHENGCE SHISHI*). While national plans for environmental protection and ecological restoration have involved major targets for environmental amelioration embedded in the all-important five-year economic plans governing the **economy** of the People's Republic of China (PRC) since 1980, implementation of these ambitious environmental goals, have been governed by an "end-of-pipe" and "command-and-control" approaches relying primarily on regulatory and technological measures. Whereas national leaders the likes of President **Xi Jinping** and Premier **Li Keqiang** tout the vital importance of pursuing an "ecological civilization" (*shengtai wenming*) and the need for compliance with environmental protection policies and statutes, local officials at the provincial level and below struggle with the competing and often contradictory goals of improving the local economy versus pursuing costly and often disruptive environmental protection mandates issued in the country's heavily top-down, command-and-control political and administrative hierarchy. Relying on local industries, many of them often heavily polluting, for crucial revenues to fund their basic operations, as financial assistance from the central government to subnational authorities is often woefully inadequate, local officials granted enormous discretion by the highly decentralized authority structure instituted in the reform period beginning in 1978–1979 engage in "selective policy implementation," which amounts to enormous variation from the national goals of environmental protection. Despite statements by national leaders rejecting notions of economic growth at the expense of environmental protection, pursued by the PRC for decades, many local leaders, especially in the country's poorer areas, with some of the most backward, heavily polluting industries, for instance, the **mining** of **coal** and petrochemical (*shiyou huaxue*) production, continue to pursue planned rates of economic growth much higher than the national targets (15 percent versus 7 percent), while central environmental mandates that appear economically disruptive are downplayed or instituted at a snail's pace.

While specific environmental targets on **air quality**, **energy**, and carbon efficiency have been stated, the burning of nonfossil fuels, reduced water consumption, and expanded **forest** coverage, on the policy agendas of most local governments, enforcement authorities (involving **environment protection bureaus [EPBs]**), and other multiple government agencies, often lack coordination, as local governments frequently lack the necessary funds, technical staff, and resources needed to deliver on centrally imposed mandates. With local **Chinese Communist Party (CCP)** secretaries and urban and township officials subject to frequent job rotation every three to five years, politically driven projects with immediate short-term benefits are often pursued at the expense of more effective, long-term, large-scale environmental protection measures, even as cadre evaluation criteria now include meeting environmental targets, along with individual financial incentives.

While the emergence of environmental nongovernmental organizations (ENGOs) and greater government transparency with environmental policies have increased public input into the policy implementation process, which is often beneficial in achieving difficult and often disruptive environmental measures, the impact of the generally ill-informed Chinese public, especially among the less well-educated and those living in poorer regions, remains limited. Considering such barriers to policy implementation, national plans to bring about a shift to a resource-efficient and **sustainable development** model may require more fundamental changes in China's political structure to ensure more "accountability."

POLICY-MAKING (*ZHENGCE ZHIDING*). In the realm of environmental protection and ecological conservation in the government of the People's Republic of China (PRC), policy-making has evolved from an almost totally closed-door activity dominated by economic and bureaucratic interests, especially the powerful National Development and Reform Commission (NDRC), with little to no outside input from nongovernmental sources, to a more open and transparent process involving academics and **journalists**, along with **nongovernmental organizations (NGOs)**, both domestic and international. Beginning with the decision to address growing environmental issues following the PRC's participation in the United Nations Conference on the Human Environment in Stockholm, Sweden, in 1972, top leaders in the **Chinese Communist Party (CCP)**, primarily Premier **Zhou Enlai** and environmental specialist **Qu Geping**, dominated policy-making, as there were not yet the institutional actors that would gradually emerge with the creation of the Leading Group for Environmental Protection under the State Council in 1974, followed by the State Environmental Protection Agency (1987–1998), the State Environmental Protection Administration (SEPA/ 1998–2008), the **Ministry of Environmental Protection** (MEP) (2008–2018), and the newly created Ministry of Ecology and Environment (MEE/2018–).

With rapid economic growth the top priority of government policy beginning in 1978–1979, environmental protection policies included six major environmental statutes and 28 administrative regulations (*tiaoli*), as well as more than 427 environmental standards issued by SEPA, along with local **laws and regulations**. Along with the Environmental Protection and Resources Protection Committee of the National People's Congress (NPC), environmental policies throughout this period were a product of intense bargaining involving bureaucratic and economic interests scattered among many institutional actors and plagued by overlapping agendas and disproportionate power dynamics with little effective cross-territorial or cross-disciplinary thinking. A heavily informal structure and process, environmental policy-making is generally dominated by political appointees whose ill-informed preferences often trump the views of professionals and environmental specialists, as short-term perspectives focusing on economic concerns prevail.

The development of new environmental policy initiatives has been a notoriously slow and arduous process, requiring years before protection and restoration became a top priority in 2012–2013, with the elevation of **Xi Jinping** to the leadership of the CCP and the presidency. Replacement of MEP by the new Ministry of Ecology and Environment in 2018 was an attempt at consolidating and streamlining the policy-making process, eliminating major bureaucratic bottlenecks, although bargaining will undoubtedly continue with the competing **Ministry of Natural Resources (MNR)**.

See also POLICY IMPLEMENTATION (*ZHENGCE SHISHI*).

POPULAR ATTITUDES AND HABITS. *See* "WASTE CULTURE".

POVERTY. As a major factor in reducing environmental degradation, especially of **soil quality** and sedimentation buildup in **rivers**, poverty alleviation has been a top priority of the government of the People's Republic of China (PRC), as various policies have reduced the number of impoverished people by 500 million since 1949, down from 88 percent of the population in 1981, to 6.5 percent in 2012, with a further reduction of 86 million people from 2013 to 2017. Along with other major factors, primarily market integration pursued since the introduction of economic reforms in 1978–1979, and population control, decreases in poverty, particularly in the 592 poorest counties, largely in rural areas, has dramatically reduced such environmentally destructive practices as the following: slash-and-burn **agriculture**; inefficient and wasteful land use; and opening up marginal, sloping lands (*qingxie tudi*) to cultivation through widespread **deforestation**. Policies aimed at expanding off-farm economic opportunities have been pursued, especially in poorer, remote **mountainous** and desertified regions, mainly in western provinces, where environmental protection has been largely meaningless to people lacking adequate food, clothing, and shelter.

The expansion of local credit and off-farm labor markets, as well as the greater protection of land tenure and **property rights**, all of which suffered enormously during the period of central economic planning (1953–1978), along with the development of such local public goods as better education and **transportation**, have produced direct and indirect benefits for the environment. Efforts like the **Sloping-Lands Conversion Program (SLCP)**, introduced in 1999, have targeted destructive practices on almost 15 million hectares (37 million acres) of agricultural land, particularly in the upper reaches of such major **rivers** as the **Yangzi**, **Yellow**, **Pearl**, and Bai in Northern China near **Beijing**, with offers of compensation to farmers for returning marginal cropland to **forests** and **plant** life. With virtually all impoverished people living in rural areas, as urban poverty, which is defined in China as living on the equivalent of $1.90 a day, has been completely eradicated, plans to reduce the number of poor people by a further 10 million were announced by Premier **Li Keqiang** in March 2018, as poverty reduction and pollution control were listed as the top two national priorities.

POYANG LAKE (*POYANG HU*). The largest freshwater lake in the People's Republic of China (PRC), located in Jiangxi province, with an average size of 3,500 km^2 (1,400 mi^2), Poyang Lake is also a major **habitat** for migrating **birds**, numbering between 180,000 and 300,000 in the world's largest wintering **wetland**. Home to the **endangered** finless porpoise (*jiangtun*), whose numbers have dwindled to 1,400, one-half of the figure from 1997, Poyang is located on the south bank of the **Yangzi River** with five major feeder **rivers**. While normally undergoing enormous variation in size between the wet and dry seasons, since the early 2000s the lake has declined in average size, almost completely drying up in 2016.

Major causes for the lake's deterioration include the impoundment of water in the **Three Gorges Dam** reservoir, located 500 kilometers upstream, along with the impact of recent **drought** and the sizable sand dredging of the lake's bottom (236 million m^3 annually), which by muddying the waters has had a disruptive impact on **fisheries** and a concomitant denial of food to migrating birds. Suffering from substantial **soil erosion** and **desertification**, that has left ships stranded on the exposed lake bottom, Poyang Lake and nearby wetlands have been the target of major conservation and amelioration efforts, notably the establishment of a **nature reserve** (1983) on the northern lake shore as a winter home for such prized migrating birds as the endangered Siberian crane (*Xiboliya he*), white-naped crane (*baizhen he*), hooded crane (*lianmao he*), Oriental white stork (*Dongfang bai guan*), white swan (*bai tian'e*), and swan goose (*hong*), with international assistance by such organizations as the Paulson Institute of the United States.

Additional preservation activities involve the establishment of the Poyang Lake Eco-Economic Zone, committed to pursuing a strategy of **sustainable development** for the region, and the Poyang Lake Basin Town Water Environmental Management Program, for combatting the discharge of industrial effluents and other pollutants into lake waters, which threatened lake fish species, constituting 30 percent of all species found in the Yangzi River. Proposals for a Poyang Lake Water Conservancy Project, including a dam with sluice gates (*zhamen*) to restore water levels, has been criticized by domestic and international conservation groups, such as World Wildlife Fund (WWF), as potentially disruptive of lake **ecology** while poaching of endangered species persists.

PROPERTY RIGHTS (*CAICHANQUAN*). Under the system of property **law** governing farmland since the adoption of economic reforms introduced in 1978–1979, especially the Household Responsibility System (1983), land registration in the People's Republic of China (PRC) suffers from unclear ownership lines and disputed border claims with negative consequences for land use and rural environmental protection. With the state or the collective as the ultimate owner of land and land tenure use-rights allocated to individual households for as many as 30 years subject to expropriation, often at the whim of village leaders for profitable commercial purposes, Chinese farmers have little incentive to engage in long-term investments and sustainable agricultural practices, with attendant benefits to the local environment including **water quality** and **soil quality**, especially of marginal and sloping lands (*qingxie tudi*) subject to rapid degradation by overcultivation or overgrazing.

While statutes, for example, the Law on Rural Land Contracting (2003) and the Property Law (2007), are aimed at strengthening property rights protection, ongoing threats of expropriation remain as a major inhibition to long-term investments and land-use improvements. Rapid urbanization and urban sprawl into rural areas has also increased land takings, with concomitant negative effects on the surrounding environment. Similar problems afflict the use of **nomadic pastoral** lands, where plans for delegating use rights to individual households with expectation of improvements to rangeland sustainability have been undermined as property rights remain unclear, leading to overgrazing and the degradation of "common use" (*changyong*) **grasslands**.

PUBLICATIONS. Engaging environment protection and ecological restoration, especially in the late 1990s and early 2000s, policy-oriented and scientific publications on environmental issues in the People's Republic of China (PRC) have proliferated. Included are journals and books on general environmental issues and policies, along with specialized publications by various environmental institutes in the Chinses Academy of Sciences (CAS) and Chinese Academy of Social Sciences (CASS), in addition to such private research organizations as the **Institute of Public and Environmental Affairs (IPEA)** and environmental nongovernmental organizations (ENGOs) like **Friends of Nature (FON)**.

Q

QINGHAI–TIBETAN PLATEAU (*QING-ZANG GAOYUAN*). The highest plateau in the world, with an average altitude of 4,500 meters (14,800 feet), with more than 2,000 **mountain** peaks higher than 5,000 meters (16,000 feet), the Qinghai–Tibetan Plateau covers 2.5 million km^2 (970,000 mi^2) and is considered one of the cleanest places in the People's Republic of China (PRC), although with some outstanding environmental problems. Extending into western **Sichuan province**, along with Gansu and **Xinjiang** provinces, the plateau is the site of tens of thousands of **glaciers**, making it the planet's "third pole," as well as thousands of **lakes** considered sacred by the local population, many brackish, for example, Qinghai Lake, the largest saltwater lake in the country. Serving as a "water tower" (*shuita*) to Asia, the plateau contains headwaters of major **rivers**, including the **Yangzi**, **Yellow**, **Lancang**, and **Nu** (Indus in India), with a landscape composed of mountainous **grasslands**, alpine tundra (*gaoshan taiyuan*), monsoon-influenced shrubland (*guancong*), and **forests**, along with extensive regions of permafrost (*yongjiu dongtu*) stretching for 600 kilometers (375 miles).

High altitudes and hostile weather limit the size and variety of **wildlife** on the plateau, which consists of cranes (*he*), geese (*e*), gray wolves (*hui lang*), hawks (*ying*), snow leopards (*xue bao*), vultures (*tuying*), water buffalo (*shui niu*), wild donkeys (*ye lu*), and wild yaks (*ye maoniu*), among others. **Plant species** include 1,000 wild plants used in Tibetan and **traditional Chinese medicine (TCM)**, notably the highly prized and increasingly rare caterpillar fungus (*chongcao*), also known as "Himalayan Viagra," along with saffron (*zanghonghua*) and the tricholoma Matsutake mushroom (*songrong*). With **agriculture** largely restricted to barley, 40 percent of the plateau's population consists of herders involved in **nomadic pastoralism**, with yak manure and peatlands (*nitandi*) providing the primary sources of fuel.

Major environmental problems on the plateau include the degradation of grasslands and peatlands; loss of plant species and wildlife, especially the **Tibetan antelope (*chiru*)**; and with **water pollution** and **soil** contamination, especially by heavy metals. With rising temperatures on the plateau at a rate four times faster than at lower altitudes, glacier and permafrost melt are occurring, with the latter causing the massive release of methane (*jiawan*), a greenhouse gas (GHG), into the atmosphere. Culprits include the destructive effects of **mining** and widespread **logging**, the overgrazing of livestock, the excessive collection of valuable medicinal plants, and the disruptive intrusion into thinly populated areas like Kekexili by gold prospectors and animal poachers. Conservation and restoration efforts include the establishment of 155 **nature reserves** and other protected areas covering 775,000 km^2 (299,000 mi^2), constituting 31 percent of the plateau's land mass, part of which is the Qiang Tang National Nature Reserve in the northern part of the plateau, the largest in the country and the second-largest protected area in the world. Also shielded are the plateau's rivers, with 94 percent off limits to commercial development, along with the restoration of **wetlands** and the 1,000 lakes that disappeared during the 1990s, notably the expansion of Qinghai Lake.

Renewable energy on the plateau has been pursued, including **solar power**, **geothermal power**, and **hydropower** stations, with 11 planned projects on the **Yarlung Tsangpo River** (Brahmaputra in India), including a dam larger than the **Three Gorges Dam**. Construction of the Golmud (Qinghai)–Lhasa railway included tunnels and other measures to reduce threats to plateau wildlife, especially the Tibetan antelope, although critics point to the deleterious environmental impact of the **transportation** link, including discarded rubbish along the route and an increased influx of settlers, especially miners, and tourists into such pristine areas as the Kekexili area, long denuded of human habitation. Rising snow lines stemming from increased temperatures have driven female wild yaks to higher ground in search of much-needed snow for milk production, while Chinese officials claim improved environmental conditions on the plateau, as outlined in a white paper on ecological progress in 2018. Scientific research on the region is conducted by the Tibetan Plateau Research Institute in **Beijing**.

QIUGANG VILLAGE. Located near Bengbu city in south central Anhui province on a tributary of the **Huai River** in the People's Republic of China (PRC), with a population of 2,000 in 2010, mostly elderly and children, Qiugang village was the site of a major five-year (2003–2008) confrontation between local villagers and three highly polluting, privately owned **chemical** factories. One of China's notorious **cancer villages**, with 53 people dying of the disease, some of whom were small children, Qiugang underwent enormous environmental degradation, including loss of **biodiversity**, crop damage, **soil** contamination, and **air pollution** and **water pollution** of both surface and groundwater from pollutants generated by the production of **pesticides** and dyes (*ranliao*).

Assisted in their **social protests** and petitions to the local **environment protection bureau (EPB)** and higher authorities in **Beijing** by local **media** coverage and Green Anhui, an environmental nongovernmental organization (ENGO), village leaders, citing environmental **laws** against producing hazardous chemicals within 1,000 meters of residential property, ultimately prevailed in getting the factories shut down and moved to a nearby industrial park. The subject of the highly praised documentary film *The Warriors of Qiugang* (*Qiugang weishi*), produced in the United States, which chronicles the transformation of local villagers from abject resignation to greater organizational and **environmental awareness**, the village has had to deal with the remnants of the widespread pollution, including minimal government efforts at restoration and the problem of financing the expensive cleanup.

QU GEPING (1930–). A pioneer in the environmental protection movement in the People's Republic of China (PRC), Qu Geping served as director of the State Environmental Protection Administration (SEPA/1987–1993) and chairman of the Environmental Protection and Resources Committee of the National People's Congress (NPC/1993–2003), where he formulated some of the initial responses of the Chinese government to the country's mounting environmental crisis. A graduate in literature and the arts from Shandong University, Qu was serving as a division chief of the Chemical Industry and the State Planning Commission (SPC) when he was called upon by Premier **Zhou Enlai** in 1972, to investigate a massive die-off of shellfish by mercury poisoning (*gong zhongdu*) near the port city of Dalian, Liaoning province, followed by a convention of regional government leaders on the current state of environmental pollution throughout the country. Selected by Zhou to head the PRC delegation to the United Nations Conference on the Human Environment in Stockholm, Sweden, in 1972, followed by his appointment as PRC ambassador to the **United Nations Environmental Program (UNEP)**, Qu became convinced that China should avoid the mistakes of Western industrialized countries by avoiding the path of "pollute first, clean up later" (*xianwuran houzhili*), a position he promoted for several years against the interests of China's politically powerful industrial sector.

Supporting a major **media** campaign promoting environmental protection in the early 1990s after years of neglect, Qu, believing that government alone could not solve environmental problems, actively supported the creation of popular-based environmental nongovernmental organizations (ENGOs) in China, along with the enactment of the Environmental Impact Assessment Law (EIA/2003), just prior to his retirement. A professor at several universities in China and at Oxford University in the United Kingdom, he is recipient of numerous awards for his strong advocacy of environmental protection, notably the Sasakawa Prize by the United Nations, while he also served as a senior consultant to the World Environmental Forum.

A strong critic of China's "wild pursuit of economic growth," Qu also made the case for a close association between **overpopulation** and environmental degradation. His many authored works include the following: *Managing the Environment in China*; *Population and the Environment in China*; and *There Is Only One Earth*, published in 1972, when the dominant leftist ideology in China, promoted by Jiang Qing, wife of **Chinese Communist Party (CCP)** chairman **Mao Zedong**, held that pollution was an outgrowth of capitalism and did not exist in "socialist" China. Speaking in 2013, Qu berated the Chinese government for insufficient action in addressing environmental issues, despite the many **laws and regulations** enacted during the many years of his tenure that he assisted in drafting.

R

RADIOACTIVE WASTE. *See* NUCLEAR POWER AND RADIOACTIVE WASTE.

RARE EARTHS (*XITU*). Consisting of 17 chemically similar elements the likes of cerium (Ce), neodymium (Nd), gadolinium (Gd), and praseodymium (Pr), rare earths (RE) are primarily used to power magnates in such "green" products as wind turbines, smartphones, and **electric vehicles (EVs)**, along with such modern weapons systems as ballistic missiles. Relatively plentiful in the Earth's crust, rare earths are widely dispersed and exceedingly hard to mine, requiring a complex process of extraction employing acid baths (*suanyu*) and hydro-metallurgical techniques (*shuli yejin jishu*) that generate enormous amounts of highly contaminated and radioactive wastewater, along with toxic emissions of hydrofluoric (*qing fusuan*), hyposulfurous (*lianerya*) acids, and sulfonamide (*huang'an*).

The largest producer of rare earths in the world, China accounts for 85 percent of the global supply from only 30 percent of the total world deposits, as other countries with large rare earth deposits, for example, Brazil, India, South Africa, and the United States, avoid commercial production because of the immense environmental damage wrought by the **industry**. With half of the country's deposits in Inner Mongolia near the city of Baotou at the Bayan Obo mine, along with Shandong, **Sichuan**, Guangdong, Guangxi, Fujian, Hunan, and Jiangxi provinces, the last the site of ion-absorbed clay deposits, China began rare earth production in 1958, witnessing dramatic expansion in the 1980s and 1990s, primarily for export, with lax enforcement of environmental protections.

Attendant environmental damage is especially severe in the city of Ganzhou, Jiangxi province, known as the "kingdom of rare earths" (*xitu wangguo*), and in Baotou, where a 10 km^2 (3.86 mi^2) pond of rare earth tailings and slurry contains 70,000 tons of wastewater, also containing radioactive thorium (*fangshexing tu*), with discharges ending up in the **Yellow River** and bringing severe environmental degradation to surrounding villages. One ton of rare earth production generates 2,000 tons of wastewater, including such toxins as sulphates (*liusuanyan*), ammonia (*an*), and hydrochloric acid (*qinglusuan*). Adding to the problem are the many illegal and unregulated rare earth mines, which the Chinese government hopes to gain control of, with corporate consolidation and tighter regulatory authority of an industry that, for the first two decades of rare earth production (1958–1978), operated with little or no government oversight.

Waste material known as gangue (*maishi*) is enormously expensive to clean up, with cost estimates of amelioration in Inner Mongolia at RMB 4 billion ($600 million) in an industry suffering low prices and major revenue losses. Production sites with severely contaminated water, as in Guangxi province, have been completely shut down, as plans also call for the total reduction of rare earth wastewater by 20 percent, as laid out in the Five-Year Plan for Rare Earths (2016), along with national control of rare earth storage.

RECLAMATION (*KAIKEN*). Creating new land from **rivers** and **lake** beds, as well as the **ocean**, reclamation has occurred in China since the imperial era, continuing in the People's Republic of China (PRC) beginning in the 1950s, with major restrictions on the practice inaugurated in 2013. Substantial environmental damage from land reclamation includes loss of **wetlands**, 1.3 million hectares (3.2 million acres) between 2003 and 2013, and 16,878 km^2 (6,514 mi^2) on the coastlines; destruction of **mangrove** marshes (*hong shulin zhaoze*), 4,200 hectares (10,374 acres) in the 1950s and 14,600 hectares (36,062 acres) since 2013, also along the coasts; and shrinkage of tidal flats (*chao tan*), bays, and **coastal** waters, amounting to 2,469 km^2 (953 mi^2) since 2012. Moreover, there has been destruction of numerous islands and **beaches**, the latter shrinking by half. In addition to endangering the natural **habitat** of marine life, notably **endangered** white/pink dolphin and migratory **birdlife**, large-scale reclamation damages **coral reefs** (*shan hujiao*) and can result in blocked estuaries, contributing to **flooding**, along with **soil** liquefaction and **water pollution**.

While limited reclamation, especially of lakes, was carried out in the 1950s and 1960s, primarily to expand cropland, dramatic increases began with the introduction of economic reforms (1978–1979), as reclamation mitigated the effects of soaring land prices in densely populated and rapidly developing coastal regions, with 43 percent of the country's total population concentrated in the **Yangzi**, **Yellow**, and **Pearl** river deltas. While transforming existing land plots into more productive activities involved tearing down old buildings and untangling multiple legal claims, reclamation involves fewer complications and ultimately lower costs, with local authorities often facilitating the process in pursuit of higher tax revenues, despite national **laws** restricting unregulated reclamation projects. Examples include Jiaozhou Bay (Qingdao, Shandong province) on the Yellow Sea, which as a result of numerous reclamation projects has shrunk by 35 percent, and Beibuwan Bay

(Guangdong province), where substantial land reclamation was carried out before the requisite **environmental impact assessment (EIA)** was completed, making way for the construction of **chemical** and power plants.

With national land surveys revealing the extent of environmental problems brought on by excessive land reclamation, efforts by the Chinese government to bring about an absolute reduction in reclaimed land began in 2013, targeting such high-profile projects as the man-made Ocean Flower Island off Hainan province and the persistent damage to the **Bohai Sea**, the innermost gulf of the Yellow Sea off the **cities** of **Tianjin** and Dalian. Following the adoption of official reclamation control measures approved by President **Xi Jinping** in December 2016, the strictest standards ever were enacted, including a complete halt to almost all reclamation projects for 2018, save those for national defense, especially in the **South China Sea**, as local officials were stripped of their power to authorize such projects, with structures built on illegally reclaimed land subject to demolition. Major restoration work was also slated for 66 bays, 50 islands, and 2,000 kilometers (1,242 miles) of damaged coastline, some of which included 180 square miles of degraded wetlands where comprehensive restoration will take decades.

RECYCLING (*HUISHOU*). At 254 million tons a year and growing by 4 percent annually, with the average urban dweller generating a half-ton each year, the People's Republic of China (PRC) is the world's largest producer of **garbage** and rubbish, while recycling a mere 2 to 3 percent, with most waste ending up in landfills, estimated at 7 billion tons, or burned in highly unpopular incinerators. Major industries where recycling generally has yet to take hold include construction materials, where only 5 percent of waste is recycled, along with **steel** products, aluminum (*lü*), and highly polluting **rare earths**, where only 11, 21, and 1 percent, respectively, of wastes are recycled. In the booming **plastics** industry, polypropylene (*jubingxi*) and polyethylene (*juyixi*) are relatively easy to recycle in contrast to the more difficult polystyrene (*jubenyixi*).

Transitioning from a poor country with few packaged consumer goods to a relatively rich nation with modern supermarkets and a variety of packaged products, for example, food, household waste, especially in urban areas, has become a major problem requiring an army of scavengers numbering 160,000 in **Beijing** alone, many unlicensed and including children, to collect trash from households and businesses, often ending up in landfills or being illegally dumped into **rivers and waterways**. Seen primarily as a profit-producing enterprise employing more people in the country than any economic sector other than **agriculture**, recycling has been enhanced by the installation of reverse bottle vending machines in **cities** like Beijing, along with requirements in many urban areas to sort household waste prior to collection, increasingly by large-scale waste management **companies**. Along with a ban on the production of plastic bags in 2008, and a prohibition on imports of **foreign waste** in January 2018, current policy calls for recycling 35 percent of household waste by 2020, as greater standardization of recycling policies and domestic production of modern treatment and processing equipment is also being pursued.

Included is manufacture by both domestic and foreign companies of such crucial technologies as optical sorting machines employing intelligent imaging processing systems used in recycling and other agricultural and industrial sectors. Robotic recycling has also been introduced into China with the establishment of the first construction and demolition debris recycling facility in Jiangsu province, this in conjunction with Zen Robotics of Helsinki, Finland, in 2017. With China producing 15 million tons of construction waste annually, the largest in the world, and dumping 84 percent into landfills and unlicensed dumps and only 10 percent recycled, creating a viable system of reuse is vital with benefits, including reduced carbon dioxide (CO_2) from production of concrete.

REGIONS. While environmental pollution is a nationwide problem in the People's Republic of China (PRC), with few areas of the country spared significant damage, variations are apparent in the level of environmental degradation across major regions. Areas with high levels of **air pollution** and **water pollution** coincide with such factors as industrial development, higher gross domestic product (GDP), greater population density, and the number of **motor vehicles**, as urban and rural areas along the **coastal** manufacturing belt suffer from the highest levels of contamination, even as relatively well-off local governments have engaged in some of the strictest environmental alleviation policies, for instance, forcing highly polluting facilities to close or shift facilities to less densely populated areas.

With air pollution a prime target of government amelioration efforts, Northern China is afflicted by high emissions from motor vehicles and **coal**-burning power plants, conditions exacerbated by natural factors, mainly weather patterns and land relief. Serious air contamination is also problematic in the far west, especially in **Xinjiang**, where severe dust storms from increasing **desertification** contaminate the air for extended time periods. By contrast, Southern China experiences substantially lower levels of air pollution, with natural, meteorological factors playing a major role, along with substantially lower levels of heavy industrial production, especially **steel**, which is concentrated primarily in the northern and central regions, with **Hebei province** the country's largest producer, followed by Jiangsu, the country's richest province. Less developed regions in Central and Western China exhibit higher growth rates in air pollutants, notably higher fine particulate matter (PM 2.5) concentrations compared to northern and eastern regions, as local economies with lower levels of technology undergo initial stages of development, bringing spurts of air contamination to once-pristine areas only to level off later with increased wealth and tougher regulations, an inverted U-shaped correlation.

Equally serious problems of water pollution also exhibit regional variation, with the north suffering the highest percentage of contamination of local water supplies from **rivers**, **lakes**, and groundwater (*dixiashui*) sources. A total of 45 percent of **water resources** in the north are rated unfit for human use (grade IV and higher) versus only 10 percent in Southern China, with western water sources rated even better. While major **cities** like **Shanghai** confront serious water pollution and water scarcity problems involving significant declines in groundwater and aquifer (*hanshuiceng*) sources, rural areas suffer even more from poorly regulated **chemical**, **pharmaceutical**, and **paper** pulp facilities, along with highly contaminated agricultural runoff containing **fertilizer** and **pesticide** residue.

Serving as major conduits of pollution, of the 532 major rivers in China, 436 are afflicted by severe pollution, with the most significant impact on industrial and population centers in Eastern and Southern China, especially in the **Yellow**, **Yangzi**, and **Pearl** river deltas. Less developed areas like Guangxi and Hainan provinces, in contrast, rate high in terms of **water quality**, with 100 percent on Hainan rated grade I. Major national policy goals include the creation of unified **air quality** and water quality standards and less recrimination between regions in placing blame for increased pollution levels.

RELIGION (*ZONGJIAO*). Targeted for destruction, especially during the **Cultural Revolution (1966–1976)**, as a form of cultural "poison" (*du*), religions in the People's Republic of China (PRC), while experiencing a revival, with approximately 240 million believers in Buddhism, Daoism, Christianity, and Islam, have also become a major social force for environmental protection and ecological restoration. Long a haven for the protection of sacred **mountains** and **wildlife** in areas like Ganzi in western **Sichuan province**, traditional religious communities, including Buddhists and Daoists, with integrated ideas about nature and humanity, are being courted by the Chinese government in its quest to encourage the protection of natural **habitat** and preserve the country's threatened **biodiversity**. During the 1980s and early 1990s, when the issue of environmental protection was generally ignored in the Chinese **media**, religious groups, with the assistance of such international organizations as the Alliance of Religion and Conservation, became increasingly outspoken, with the China Daoist Association issuing a Statement on the Environment (1995), while conducting surveys of environmental conditions at sacred sites throughout the country.

With the China Buddhist Association taking similar positions stressing limits on human activity in protected areas, prominent Buddhists, particularly in **Tibet**, have joined with the Chinese government in addressing such major issues as **deforestation**, urban sprawl, and the emergence of a **"waste culture"** in the country's increasingly consumer-oriented, fast-paced **economy**. Invoking such traditional texts and concepts as karma (*yinguo baoying*) and sin (*zui*) to deride material excesses, religious leaders have established social service organizations as watchdogs against polluters, with religious figures like Li Yandong in Nanjing, Jiangsu province, advancing programs to clean up nearby polluted **lakes**.

Believing that "humans follow the earth" (*renlei gensui diqiu*) and "Daoism follows nature" (*daojiao zunxun ziran*), Daoists have crowned their founder, Laozi (?–533 BCE), as a "green god" (*lü shen*), with ecofriendly temples built on sacred mountains, as in Maoshon, Jiangsu province, stressing the traditional theme of "heaven and humanity in harmony" (*tian ren heyi*), with Daoist abbots pressuring government officials to set aside additional areas for **nature reserves and national parks**. While top political leaders like President **Xi Jinping** champion the study of China's religious traditions, converts to Buddhism and Daoism are urged to avoid looking to the government for leadership, relying instead on their own efforts and convictions to create a sustainable "ecological civilization" (*shengtai wenming*).

RENEWABLE ENERGY (*ZAISHENG NENGYUAN*). Consisting of **energy** produced from such nondepletable sources as flowing water (**hydropower**), **solar power**, **wind power**, **geothermal** heat, ocean wave energy, and biological processes (biogas), renewable energy is a top priority of the People's Republic of China (PRC), enshrined in the Renewable Energy Law (2005) and incorporated into the 12th Five-Year Economic Plan (2011–2015) with RMB 594 billion ($90 billion) invested in these sectors annually. Not limited by finite resources or dependent on imports, especially from unstable regions, for example, **oil** from the Middle East, renewable energy in China accounted for 36.6 percent of electrical power generation and 26 percent of total power production in 2017.

Current plans call for renewable energy sources to constitute 35 percent of total energy production in 2030, increased from an initial target of 20 percent, and become the country's main energy source by 2050, with concomitant reductions in highly polluting **coal** and other fossil fuel sources. Relying on the large-scale production of photovoltaic cells (*guangfu dianchi*), the largest in the world, and wind turbines, China is engaged in the development of large-scale solar power farms in such sun-intense provinces as Gansu, Ningxia, and Qinghai, and major wind power farms in Gansu, Inner Mongolia, and **Xinjiang**, with several major hydropower projects slated for construction. Major ocean wave energy experimental and modeling projects have been pursued since the 1980s. These primarily involved so-called oscillating water column (OWC) technology with electricity generated from the wave-driven rise and fall of water in cylindrical shafts or pipes powering air-driven turbines. Four OWC converters have been built in China, including a pilot semi-submersible floating platform named Sharp Eagle. Deployed in 2016, it has a production capacity of 10 kW, which, with low efficiency rates, makes commercial utilization unsustainable. Subsidies exist for renewable energy, notably the Golden Sun Programs, but with decreasing reliance on such incentives, as the cost of solar cells and other forms of subsidies for renewable energy continue to decline. Research organizations include the National Renewable Energy Center, the Renewable Energy Engineering Institute, and the Guangzhou Institute of Energy Conversion.

RESEARCHERS AND SCIENTISTS. *See* ENVIRONMENTAL SCIENCES (*HUANJING KEXUE*).

RESOURCE DEPLETION (*ZIYUAN KUJIE*). With proven reserves of almost every major mineral and commodity that has been consumed in large quantities by rapid economic growth and industrialization, especially since the introduction of economic reforms in 1978–1979, the People's Republic of China (PRC) is confronting increasingly serious problems of resource depletion. While rich in proven reserves of **coal**, at 114 billion tons of anthracite (*wuyanmei*) and lignite (*hemei*), petroleum reserves are relatively low, at an estimated 400 million barrels, 13th largest in the world, and natural gas (*tianranqi*), at 3,910 billion cubic meters, although with large reserves of recoverable shale deposits. With **oil** production peaking in 2018, China, at its current consumption levels, is confronted with a forecast exhaustion date in only 17 years, along with limited domestic natural gas production currently complemented by substantial imports. Overexploitation and widespread **energy** inefficiency have exacerbated the problem, as China records a high "burn rate," reserves-to-production, of most minerals and commodities, putting the country in the "red zone" for future supply shortfalls.

Relying on increased imports of commodities, especially oil and natural gas, the PRC is depleting global resources at a rate of 2.5 times sustainability. As coal production is slated to decline based on environmental considerations, so-called "resource-depleted **cities**" (*ziyuan kujie chengshi*) have emerged, particularly in the northeast provinces of Heilongjiang and Jilin, where relatively high production prices and depleting seams are leading to major mine closures. China also suffers from the classic "resource curse," where areas rich in resources and subject to greater environmental degradation, mainly in the west, combine with low levels of technology and education to produce a growing income gap with the more prosperous east, which the "Great Western Development Strategy," adopted in 2000, has failed to address.

RIVERS AND WATERWAYS. Vital sources of water for **agriculture** and human consumption, along with substantial freshwater **fisheries**, while also serving as essential **transportation** routes for goods and people, rivers and waterways in the People's Republic of China (PRC) have suffered enormously from pollution, particularly since the inauguration of economic reforms in 1978–1979. With the number of rivers consisting of flow area of 60 square miles totaling 22,000, six major basins dominate the country—**Yangzi**, **Yellow**, **Huai**, Hai, **Songhua**, and **Pearl**. The Yarkant-Tarim River in **Xinjiang**, which terminates in Taitema Lake, is the longest inland river at 2,030 kilometers (1,240 miles). Second only to **air pollution** in severity, the cleanup and restoration of polluted and degraded rivers is a top priority of environmental **policy-making**, with costs, including around major urban areas, estimated at RMB 1 trillion ($148 billion).

The greatest loss has been the disappearance of as many as 28,000 rivers since the 1950s, from a total of about 50,000, reductions brought about by excessive construction of dams and reservoirs blocking water flow that led to dry-up; excessive and unregulated extraction of groundwater (*dixiashui*); paving over rivers and canals in **cities** like **Guangzhou**; **soil erosion**; poor management; and, according to the Chinese government, **climate change**. Surveys by such international environmental groups as **Greenpeace** indicate that 70 percent of the country's extant rivers suffer from pollution, numbering 436 out of the largest 532 (81 percent), and 13 out of 15 sectors of the largest seven rivers, including the Yellow, Yangzi, and Pearl. Monitoring sites of **water quality**, numbering more than 12,000 nationwide, also indicate that 78 percent of the river water in 2010 was unfit for human consumption, while catches of freshwater river fish have dropped precipitously, especially in the case of the Yangzi, from 427,000 tons in the 1950s to 100,000 tons in the 1990s.

Agricultural runoff of **fertilizers** and **pesticides** remains the largest contributor to river water contamination, which since the early 1990s has included dramatic increases in toxic algae plumes (*zaolei yumao*), reducing oxygen levels and creating "dead zones" (*siqu*), particularly in the Yangzi and Pearl River deltas. With more than 20,000 petrochemical plants (*shihua chang*) in the PRC, many sited along rivers (14,000 on the Yangzi and Yellow alone), industrial discharge also remains a serious problem, along with untreated **sewage** from urban sprawl, as 80 percent of cities still lack adequate treatment facilities. Most seriously polluted, according to a 2018 report, are rivers located in the north, where slow water flow across relatively flat plains increases the concentration of industrial effluents, especially from older heavy industrial facilities, especially **chemical** plants. Included are the Hai River (**Beijing** to **Tianjin**) and the Liao River (Inner Mongolia to Liaoning), with the former containing high cadmium (*ge*) levels and the latter rated as one of the most seriously polluted in the country, primarily by chemical wastes and raw sewage from such local cities as Dalian, Fushun, and **Shenyang**. The Songhua River has shown improvement from a rating of "moderately polluted" in 2006, to "slightly polluted." To the south, the Yangzi and Pearl are rated as having "good water quality" but with pollution extending into some of China's most pristine areas, for instance, Guilin, Guangxi province, where **garbage** and rubbish can be seen flowing down the otherwise scenic Cha and Li rivers.

Following directions from China's top leaders, for instance, a declaration by President **Xi Jinping** that "clean water is a priceless asset," plans for improving water quality in rivers and **lakes** have been incorporated into major amendments to the country's Environmental Protection Law (2015), as well as the 13th Five-Year Economic Plan (2016–2020). Calls have also been made to reduce overall water consumption by 23 percent. Noting the "black and stinky" water infecting many of the country's vital waterways, commitments have also been made to construct of 400,000 kilometers (248,000 miles) of wastewater pipelines, while 160,000 highly polluting fish farms (**aquaculture**) are slated for shutdown. River clean-up has also followed a process of "resilient rivers" subjecting polluted waters to oxidation and diverting waters through plants, which filter out heavy metals and carbon dioxide while lining river banks with sandy beaches and vegetation.

Reacting to persistent bureaucratic conflicts and unclear leadership responsibilities related to river management, a system of "river chiefs" (*heliu qiuzhang*) was created in July 2018, with overall decision-making authority granted to individual leaders, numbering 300,000, to address serious pollution in their designated areas of responsibility. Major rivers in China (with the number of large tributaries in parenthesis) are as follows: Argun (3), Hai (6), Han (3), Heilong (15), Huai (13), Jialing (5), **Lancang** (3), **Min** (4), **Nu** (6), Pearl (7), Qiangtang (3), Qinhuai (2), Qingyi (7), Songhua (7), Tarim (8), Wei (5), Yangzi (33), **Yarlung Tsangpo** (8), Yellow (16), and Yuan (2). Dotting many riverscapes are such scenic waterfalls (*pubu*) as Huangguoshu, the largest in the country at 67 meters high (219 feet) on the Baishui River (Guizhou); Detian on the Guichun River in Guangxi near the border with Vietnam; Hukou on the Yellow River on the border of Shaanxi and Shanxi provinces; Luoping on the Jiulong River (Yunnan), and Nuorilang in the Jiuzhai Valley Nature Reserve (Sichuan).

RURAL WASTE (*NONGCUN FEIWU*). With inadequate or nonexistent waste management systems throughout large swaths of the countryside, rural waste has been a perennial problem in the People's Republic of China (PRC), especially since the inauguration of economic and agricultural reforms in 1978–1979, as rural solid waste (RSW) has increased exponentially, doubling in quantity from 6 million tons in 2008, to an estimated 12 million tons in 2020. Surpassing total urban waste in 2010, 65 percent of China's 600,000 villages engage in the collection of rural household trash and other forms of solid waste, while lacking expensive treatment facilities and trained personnel, with effective **recycling** and **transportation** capacity largely absent, especially in poorer areas. Endless multitudes of **plastic** packaging and wrappings, broken toys, used clothing and shoes, and food waste are either discarded in scattered dump sites situated along rural roads and river banks, and near **lakes** and ponds, or subjected to illegal burning, contributing to heightened **air pollution** and **water pollution**.

Despite such government declarations as Guidance on Rural Solid Waste (2015), lack of landfills and **composting** in the countryside has led to 79 percent of RSW being illegally incinerated, with only 5 percent disposed of safely, some of which is highly toxic medical waste, less than half of which ends up in sealed containers with proper biohazard (*shengwu weihai*) warning labels. Spurred on by widespread availability in the countryside of consumer goods in "modern" packaging, rural residents produce an average of 1.07 kilograms of waste per day, an amount equal to rates in more developed countries like Japan, with significant implications for the **health** of rural dwellers. With wealthier villages engaging in more comprehensive waste management and the adoption of the Law on the Prevention and Control of Environmental Pollution by Solid Waste (2007), current plans call for extending waste management to all the country's villages and expanding the construction of **waste-to-energy** plants to serve rural areas.

Most notable is the construction of biogas digesters in 30 million rural households for converting animal and human waste (an estimated 3 billion tons annually) into methane for use in cooking, lighting, and generating **fertilizers**. The use of biogas digesters reduces the burning of firewood and other heavily polluting fuels. Constituting 1.2 percent of the country's total energy needs with large-scale farms and wineries providing electricity to nearby **cities** and towns, biogas production in China yields a total savings of 8 million tons of **coal** and 13 million tons of wood annually, reducing **deforestation**, **soil erosion**, localized **desertification**, and indoor **air pollution** while increasing rural incomes and overall standard of living. Led by organizations such as the China Biogas Society, a goal of 50 million rural households with biogas digesters has been set for 2020.

RUSSIA. While environmental policies developed in the **Soviet Union** during the reign of Josef Stalin (1929–1953) had a profound and long-lasting impact on the People's Republic of China (PRC), especially during the period of central economic planning (1953–1978), since the establishment of the

Russian Federation in 1991, the two countries have diverged on important global environmental issues, most notably **climate change**, but cooperate in other areas, for example, the exploitation of natural resources. Following the endorsement of the 1997 Kyoto Protocol in 2002, China has generally supported international efforts to stave off climate change, while Russia, led by longtime leader President Vladimir Putin, has been hesitant to even acknowledge humanity's role in generating fundamental changes to the global climate.

Backed by a disbelieving Russian public on climate issues, the Russian government has refused to institute any efforts to rein in the country's emission of greenhouse gases (GHG), put at 5 percent of the global total, versus China's leading 23 percent. On other mutually influential environmental matters, the two countries have frequently joined forces, most notably in rejecting, along with Norway, the planned landmark Antarctica Ocean Sanctuary, which, applied to 1.8 million mi^2, would have banned all fishing and provided crucial protection for orcas (*nijijing*), blue whales (*lan jing*), seals (*haibao*), and penguins (*qi'e*), as all three countries prioritized the interests of their fishing industries.

A major importer of natural gas (*tianranqi*) to replace highly polluting **coal**-fired plants, China has signed a 30-year deal with Russia for natural gas imports, primarily through the Power of Siberia pipeline, under construction on the northeast border of the PRC. Prime **forest** land in Russia, for instance, the eastern Siberian taiga, which for years during the Soviet Union were protected from **logging** and other destructive practices, are now being felled at an accelerated rate by illegal Chinese commercial **timber** interests operating with abandon in the region, often with the connivance of corrupt Russian officials.

S

SEAGRASS (*HAICAO*). Defined as flowering plants with long blades growing in shallow, sheltered marine and estuary environments, and usually anchored in sand or mud, seagrass is a major food source for hundreds of fish and other species, while also stabilizing seabed sediment and reducing **coastal erosion** from storm and wave surges. Similar to **coral reefs** and **mangroves**, seagrass serves as a nursery ground for juvenile fish, while also storing carbon at twice the level of rainforests, along with nitrogen. With 22 species found in the People's Republic of China (PRC), seagrass meadows are spread along the northern, middle, and southern coasts, with the greatest concentrations in the **South China Sea**, specifically the provinces of Guangdong (Liusha Bay and Donghai Island), Guangxi (Hepu and Pearl bays), and Hainan (Li'an and Sanya bays).

Major threats to seagrass along China's coasts include **aquaculture** and algae cultivation ponds, large-scale nets, poisons, dynamite and electron capture by fishermen, shellfish digging, **water pollution** from industrial effluents and **sewage**, and port and shipping channel construction, along with intensive bottom trawling by fishing vessels. Detailed surveys of China's seagrass sites began in 2006, primarily at the Hepu Demonstration Site, Guangxi province, followed by an Action Plan for the Protection of Seagrass Resources in 2007, and the establishment of "ecological red lines" in 2017, to protect marine species, one of which is seagrasses, from development. International participation includes the China branch of Seagrass Watch, an international assessment and monitoring program, and the Nature Conservancy, with attention to raising public awareness on the many benefits of seagrasses. The PRC also hosted the International Seagrass Biology Workshop at Sanya Bay, Hainan, in 2014.

SEA-LEVEL RISE. A challenge to the entire world, including the People's Republic of China (PRC), with its more than 14,500 kilometers (9,000 miles) of coastline and 6,500 offshore islands, sea-level rise poses a major threat, primarily to low-lying plains and deltas, which are confronting inundation by the end of the 21st century. Combined with land subsidence (*dimian chenji-*

ang) resulting from the excessive extraction of groundwater (*dixiashui*) by such large **cities** such as **Shanghai**, persistent sea-level rise occurred from 1980 to 2017 at an average of 3.3 millimeters (mm) per year, hitting a record high in 2016, with the last six years reaching peak points over the last 37 years. Despite the existence of defensive sea walls (*haidi*) and dikes (*diwei*) protecting major urban areas under threat of sea rise, damaging effects to the **coastal** environment include increased **flooding** exacerbated by typhoons, along with increased **soil** salinization (*yanhua*), from more frequent saltwater intrusions. Lands lost to intruding sea waters total 18,300 km^2 (7,000 mi^2), mainly around the major cities of **Guangzhou**, Shanghai, and **Tianjin**.

Major causes of global sea-level rise stem from rising thermal levels in the oceans, with sea waters on the Chinese coast rising by 0.7 to 0.9 degrees Celsius in 2017, along with melting **glaciers** and ice caps at the poles, generally attributed to **climate change**. Long-term remediation involves restoring coastal **mangrove** swamps (*hong shulin zhaoze*), **wetlands**, and seagrasses (*haicao*) and reeds (*luwei*) destroyed by unregulated and single-minded economic growth and land **reclamation**, as current predictions project a rise of as many as 130 mm (5 inches) by 2037, compared to 2009. Major institutions involved with issues of sea-level rise include the recently dismantled State Oceanic Administration (SOA), with functions taken over by the **Ministry of Natural Resources (MNR)**, established in 2018, and such environmental nongovernmental organizations (ENGOs) as the Shenzhen Mangrove Wetlands and Conservation Foundation.

SEWAGE (*WUSHUI*). A major problem in both urban and rural areas in the People's Republic of China (PRC), with almost 300 **cities** still lacking adequate treatment plants, raw sewage from households, businesses, and **industry** increased tenfold between 2003 and 2013, making it a major target of remediation, with the country doubling sewage treatment capacity since 2000. Surveys indicate that 329 million people in China are still without access to improved sanitation (*weisheng*), with rural areas lagging behind, even as 450 million people gained access to improved **water resources** between 1990 and 2008. With 4 million tons of toxic slurry produced annually and less than 20 percent subject to adequate treatment, disposal often involves illegal dumping into **rivers and waterways**, as well as landfills, with sludge mountains cropping up in major cities, one of which is **Beijing**, along with use of untreated water for **irrigation**.

The construction of sewage treatment facilities in the PRC began in the 1950s, mainly to deal with rain waterlogging (*lao*), followed in the 1980s by major sewage treatment plants nationwide as urbanization accelerated, with benefits extended more to eastern cities and less to western and northeastern urban areas. Altogether, 700 treatment plants have been built in 661 cities, boosting overall **water quality** but with major urban centers like **Shanghai** undergoing an explosion in high-rise buildings with less attention paid to installing underground sewage systems. Significant government investment in sewage treatment facilities was incorporated into the 11th Five-Year Economic Plan (2006–2011), with a goal of eliminating toxic pollutants from 96 percent of urban sewage sludge by 2020.

With 10 million people moving into Chinese urban areas annually, sewage quantities continue to escalate, especially in such densely populated areas as the **Pearl River** Delta, where with millions of tons of untreated sewage dumped into local waterways daily, two-thirds of the rivers in Guangdong province fail to meet minimum water quality standards. Sanitation conditions have improved in 87 percent of urban areas in the country since 2005, as plans call for the construction of 126,000 km (28,293 miles) of sewage pipes by 2020, raising the treatment of urban sewage by 50 million m^3 per day. Similar improvements have also occurred in 64 percent of rural areas but with many villages still relying on traditional communal latrines without flushing capacity. Lacking the capital for such relatively expensive investments in treatment plants, some local governments in both rural and urban areas are turning to revenue-generating **waste-to-energy** facilities modeled on foreign cities to generate biofuels, along with digestate **soil** enrichment nutrients and organic **fertilizers**.

SHANGHAI. Located in the **Yangzi River** Delta on the east coast of the People's Republic of China (PRC), with a population in the municipality (*shi*) of 24 million people, one of the largest in the world, *Shanghai* ("on the sea") is a major financial and industrial center in the Chinese **economy** with rapid growth, including **energy** consumption, especially since the adoption of economic reforms in 1978–1979. Major environmental problems in the municipality include **water pollution**, excess **garbage** and rubbish, and the threat of impending **sea-level rise**.

With rapid population growth from 5 million (1949) to 14.2 million (2010) to the current level of 24 million, consisting of 14.2 million permanent residents and 8.9 million floating migrant workers, the annual generation of garbage and rubbish has increased from 960,000 tons (1950) to 8.9 million tons (2010), with disposal shifting from burying in overtaxed landfills to incineration. Annual night soil (*fenbian*) produced in the city reached a peak of 4.18 million tons in 1978, before declining significantly with the installation of septic tanks (*huafen chi*). While waste gas emissions have also increased to an annual 4.4 billion m^3 from 1991 to 2010, overall **air pollution** is relatively low, with few days registering high fine particulate matter (PM 2.5) concentrations, mainly because of the city's location on low-lying wide-open plains with largely unimpeded air circulation.

One of the additional major environmental effects of the city's rapid growth as the consumer capital of the country, with a large **ecological footprint**, is the dwindling and highly polluted **water resources**, as half of the city's approximate 43,000 **rivers**, streams, canals, creeks, and **lakes** suffer from severe pollution, with more than 1,800 undergoing immediate cleanup. Included are the Huangpu River and Suzhou Creek, which, flowing through the city center, are afflicted by substantial pollution from annual wastewater discharge of 56 million tons by **industry** and 67 million tons from residential sources. The major source of the city's drinking water and contributing to local **irrigation**, the Huangpu registers high biological oxygen demand (BOD/*shengwu xu yangliang*), with elevated levels of nitrogen (*dan*), phosphorous (*lin*), **oil**, and phenol (*benfen*). Hit by a major pollution event in March 2013, when 16,000 dead pigs came floating down the river, the Huangpu is also undergoing increased warming, evidently from overheated wastewater. A tributary of the Huangpu, Suzhou Creek, also known as the Wusong River, contains significant ammonia (*an*) and nitrogen pollution from both industrial and human waste, creating serious visual pollution and the release of foul odors.

With major portions of city waterways classified in a water census (2010–2012) as heavily polluted (Grade V and V+), remediation efforts such as the Suzhou Creek Rehabilitation Project have been pursued. The actions have included dredging rivers, removing floating pollutants, demolishing illegal structures, adding **garbage**-treatment facilities, and building separate **sewage** and rainwater systems with underground pipes to discharge city waste into the nearby **East China Sea (ECS)**. Overall **water quality** since 2013 has continued to decline, however, in part from the leaching of aging and chock-full landfills into local groundwaters, which are being cut back as a source of city water after years of overexploitation and resultant land subsidence (*dimian chenjiang*), which has subjected city waterways to increasingly frequent saltwater intrusions from **sea-level rise**.

The reduction of green space, for example, cultivated land, to make room for the expansion of buildings, many of them high-rise structures, especially in the Pudong financial center, built in 1993, and paved roadways have increased the city's Urban Heat Index (UHI) and caused drops in relative humidity. Countermeasures include 157 city parks and many tree-lined streets, especially in the old foreign quarter, and green roofs (*lüse wuding*) with flower and vegetable gardens that lower building-top temperatures slated to cover two million m^2 by 2020. Protective status has also been provided to Chongming Island, the largest alluvial (*chongji*) island in the world, located at the mouth of the Yangzi River, with **lakes**, **forests**, **wetlands**, and a national park. While the local **environmental protection bureau (EPB)** has ordered the closure of factories near intakes for the city's water supply from the Huangpu River, central **environmental inspections** and officials have criticized the Shanghai government for an inadequate pace of environmental cleanup and meager penalties and fees imposed on environmental offenders in a city overwhelmingly dominated by economic interests and high consumption habits.

SHENYANG. The provincial capital of Liaoning province in the northeast of the People's Republic of China (PRC) and once known as Mukden, Shenyang developed as a major center of heavy **industry** from the 1930s onward, notably during the period of central economic planning (1953–1978). Production in the central city of machine tools (*jixie gongjiu*), **motor vehicles**, and aerospace (*hangtian*) generated some of the most severe **air pollution** in the country, with fine particulate matter (PM 2.5) concentrations of 1,400 micrograms per m^3. With a population of 8 million people, the city relied heavily on the burning of **coal**, consuming 16 million metric tons annually for wintertime heating, which contributed as much as 30 percent of the air pollutants, along with motor vehicle emissions, especially from heavy diesel trucks and yellow dust blowing in from the nearby Gobi Desert.

Major amelioration efforts in the city began with shutting down many old heavy industrial and revenue-losing facilities with aging and highly polluting equipment, while building new plants outside the city center, employing less-polluting production processes and installing desulfurization equipment. Substantial improvements in **air quality** were also achieved by replacing coal with natural gas as the primary source of residential heating. Wastewater treatment from industrial and residential sources was also improved, reaching 77 percent of urban totals—slightly higher than the national average of 70 percent—and leading to a number-one ranking on the Urban Sustainability Index (USI) among 112 **cities** in the PRC. With a "greening ratio" (*lühua lü*) of 40 percent for its many urban parks and open areas, for example, the renowned South Canal Linear Park, featuring a rich array of flowers, trees, and vegetation, Shenyang was named a National Forest City in 2005.

SHENZHEN. Rated by the United Nations as one of the most sustainable cities in the world in 2002, Shenzhen with a population of 13 million people (2018) has made major advances in environmental amelioration especially of persistent and serious **air pollution** but with nagging problems of **water pollution** of area **rivers and waterways** and municipal solid waste (MSW). Designated as the first Special Economic Zone (SEZ) in the People's Republic of China (PRC) in 1980 by then paramount **Chinese Communist Party** (**CCP**) leader **Deng Xiaoping**, the city was long known as a center of highly polluting industries and a growing fleet of 3 million **motor vehicles** spewing clouds of toxic smoke that caused dangerous levels of air contamination. The city's dense concentrations of high rise buildings and population created a high urban heat index (UHI). Located in a sub-tropical monsoon climate, Shenzhen and the surrounding region has suffered from rising temperatures, increased precipitation with more frequent severe weather events, and fewer days of sunshine. More problematic is the multiple problems afflicting the city's **water resources** with 173 of the 310 rivers flowing through and nearby the city rated as unfit for human use (Grade V or below), including the Guanlan, Longgan, Pingshan, and Maozhou (the city's largest), all with high levels of ammonia, phosphorous, and nitrogen. Construction of 33 municipal sewage plants in recent years has failed to keep up with the enormous increase in wastewater from continued population growth and expanding consumerism with 930,000 tons discharged daily into local rivers.

Greatest success has come from improvements in the city's **air quality** with overall levels of air pollution dropping by 50 percent with reductions of PM 2.5 concentrations of 30 percent from 2012–2016. Under the zero-emissions **transportation** design, 30,000 highly polluting vehicles have been removed from city streets while an entire fleet of **electric vehicles** (**EVs**) has been introduced including 16,000 EV buses and an all-electric taxi fleet (2020) backed by state-subsidized charging stations along with dramatic expansion of "smart bicycles" employing GPS. Slated to become the first Chinese city to meet European Union-level **emission standards**, Shenzhen is also home to the BYD electric vehicle manufacturing company, a major producer of all-electric and hybrid automobiles which is devoted to an all-electric **energy** transportation sector for the entire country. Additional measures include a municipal **cap-and-trade** scheme, expansion of the city's underground metro system, and major restrictions on issuance of municipal licenses to fossil-fuel burning vehicles.

Crammed into a mere 2,000 km^2 with 150 urban skyscrapers, Shenzhen has also adopted policies for mitigating the environmentally deleterious effects of buildings and other concrete structures while also expanding green space and engaging in a major urban tree planting program. Plans call for 80 percent of all buildings to meet "green" certification by 2020 along with adoption of "**sponge city**" measures including improved urban drainage and water treatment programs. Most proactive is the Shenzhen Asian Cairn Farmscraper project sponsored by a French company involving construction of six mixed-use tower buildings equipped with open-air epidermis of photovoltaic and photo-thermal **solar** cells, wind turbines, and systems of wastewater **recycling** through viticulture and lagoons of phyto-puration with room for food production and office space reducing reliance on intra-city transportation. Southern China University of Science and Technology located in the city is a major source of expertise on environmental sciences and policies.

SHI LIHONG. A **journalist**, filmmaker, and founder of Wild China Film, with her husband, Xi Zhinong, a former Central China Television (CCTV) cameraman, Shi Lihong became interested in environmental issues during film studies in the United States at the University of California. A reporter for the English-language *China Daily*, Shi became interested, along with her husband, in the fate of the **golden snub-nosed monkey** in the face of planned **logging** in **Yunnan province**, the subject of their first documentary film. Turning their attention to the issue of proposed **hydropower** projects on the **Nu River**, one the last truly wild **rivers** in the People's Republic of China (PRC), successful efforts in opposing these plans by grassroots environmental **activists** and groups were featured in their film entitled *View of an Angry River* (2018). Along with appearances in *Waking the Green Tiger*, Shi also appears in *Search for Sacred Mountain*, documenting the efforts of Buddhist monks in **Tibet** to preserve the pristine environment from destructive logging and development projects, especially such fragile areas as the **Qinghai–Tibetan Plateau**, near the confluence of the headwaters of the **Yangzi**, **Yellow**, and **Lancang**/Mekong rivers, an area known as the "Water Tower of Asia" (*Yazhou shuita*).

SHIJIAZHUANG. The provincial capital of **Hebei province**, with a population of 10 million people, 4 million in the central city, Shijiazhuang has been rated one of the most polluted **cities** in the world, mainly from the concentration of such **energy**-intensive industries as **chemicals**, nonferrous metals (*youse jinshu*), **steel**, and coking plants (*jiaohua chang*). Located in the south central region of the province, Shijiazhuang suffered serious **air pollution**, primarily from **coal**-burning plants, along with **motor vehicle** emissions and dust from interminable construction sites, in a city with a population that has expanded fourfold since the establishment of the People's Republic of China (PRC) in 1949. Registering 322 days of heavily polluted air in 2013, with the Air Quality Index (AQI) reaching the maximum level of 500, with high levels of fine particulate matter (PM 2.5) concentrations, problems include **acid rain**, **ozone** buildup, inversion layers with air temperatures rising with altitude, heavy greenhouse gas (GHG) emissions, photochemical smog, and heat-island effect, conditions that Chinese scientists compare to a "nuclear winter" (*he dongtian*).

Also contaminated were the city's limited **water resources**, drawn almost entirely from groundwater polluted throughout the years by the releases from textile, chemical, and **fertilizer** plants developed from the 1950s onward. With major efforts at remediation beginning in 2013, spurred by **lawsuits** against the city by irate citizenry, antipollution measures have included the following: closure of industrial facilities and thermal plants with inadequate equipment; replacement of old city buses with new, cleaner-burning vehicles; and the creation of more urban green space. While **air quality** has shown improvement, a new action plan was issued in 2018, aimed at **transportation**, construction, and open-pit **mining**, after red alerts were declared in January of that same year in response to elevated smog levels.

SICHUAN PROVINCE. The second-largest province in the People's Republic of China (PRC), covering 487,000 km^2 (188,000 mi^2), with a population of 81 million people (2013), *Sichuan* (literally "four rivers") province consists of three distinct regions, the **Qinghai–Tibetan Plateau** in the west, the Red Basin (*Hong pendi*) in the southwest, and hills and **mountains** in the east, with climates ranging from alpine-arctic to wet subtropical. Major rivers in Sichuan include the namesake "four" of Jialing, Jinsha (upper reaches of the **Yangzi**), **Min**, and Tuo, along with the Dadu, Fu, Jin, and Yalong, with several mountain ranges, most notably Chola, Bayan Har, Daba, Daxiangliang, Daxue, Hengduan, Jinping, Qionglai, Shaluli, and Xiaoxiang, with Mt. Gongga, in the Daxue range, as the highest peak, at 7,556 meters (24,783 feet).

Located at 30 degrees latitude, the province is rich in **biodiversity** with an enormous variety of **plant species** and **wildlife**. Dubbed the "garden of China" by noted British horticulturalist E. H. Wilson, Sichuan contains 10,000 species of higher plants, one-third of the national total, with trees ranging from cypress (*bai*), palm (*zonglü*), pine (*song*), bamboo (*zhuzi*), tung (*tong*), and citrus (*ganju*) at low altitudes, to evergreens (*changlü*) and oak (*xiangmu*) at intermediate altitudes, and to coniferous (*zhenye*) and scrub (*caxi*) at higher altitudes. Rarities include the dawn redwood (*liming hongmu*) and Chinese dove (*gezi shu*), and, on Mt. Emei, such plants as nothophoebe omeiensis and rosa omeiensis.

With 40 percent of the 562 mammals in China, the province is home most notably to the giant **panda bear** (85 percent of the national total, including a rare albino), along with other species, many threatened, including the red panda, **golden snub-nosed monkey**, Chinese goral (*ban ling*), Himalayan marmot (*hanta*), giant flying squirrel (*juxing feishu*), bear cat (*xiong mao*), and antelope (*lingyang*), many ensconced in the more than 40 **nature reserves and national parks** in Sichuan, particularly the panda-rich **Wolong National Nature Reserve**, along with seven World Heritage Sites, four World Biosphere sites, and two geopark sites. Among the province's 779 **bird** species, 42 threatened and two endemic to the province, are several varieties of buntings (*caiqi*), cranes (*he*), cuckoos (*bugu*), doves (*gezi*), ducks (*ya*), geese (*e*), gulls (*ou*), herons (lu), egrets (*bailu*), magpies (*que*), minivets (*shanjiao*), owls (*maotouying*), partridges (*zhugu*), pheasants (*yeji*), swans (*tian'e*), terns (*yanou*), thrushes (*huamei*), tree creepers (*shu paxing*), and woodpeckers (*zhoumuniao*), including the highly prized golden pheasant (*jinji*) and black-necked crane (*hejing he*).

Scenic and heavily touristed sites include: Jiuzhai Valley Nature Reserve, with many crystal **lakes** and waterfalls (*pubu*), including Nuorilang, the widest in China; Daocheng-Yading Area, the country's most well-preserved alpine **ecosystem**; Huanglong Valley, with its multicolored pools from calcite deposits and snow-capped mountain peaks; Shu'nan Bamboo Forest, the largest bamboo grove in China, known as the "sea of bamboo"; Mt. Emei, with wild monkeys; Jinkou Grand Canyon, on the Dadu River on the northwest plateau; Siguniang ("Four Sisters") Mountain, with high peaks surrounding picturesque valleys, serving as giant panda sanctuaries; and Xuebaoding, a center of plant and birdlife, along with snow-capped Xuebaoding Mountain. The site of many **hydropower** stations, for instance, Baihetan and Xiluoduo on the Jinsha, with a total installed capacity of 75 gigawatts (GW), twice the level of the local power grid, Sichuan is scaling back planned dam projects for 2016–2020, to avoid wasting power.

One of the most serious environmental problems involves the extensive terracing of sloping lands, which has led to significant **soil erosion**, along with frequent **landslides**, particularly in the western region, which sits astride the Qinghai–Tibetan Plateau. The degradation of **grasslands**, prominent in the northwest region, apparently from overgrazing, remains a major environmental issue, along with sedimentation buildup and **water pollution** from agricultural and industrial effluents, especially into the Qionghai Sea, located in the mountains of southern Sichuan. An active seismic zone (*dizhen dai*), Sichuan has been hit by major **earthquakes**, most notably in 2008, with 90,000 deaths, many of them children caught in poorly constructed school buildings, along with another quake in 2017, although with few fatalities.

Subject to unannounced spot-check **environmental inspections**, especially in 2015, the province has inaugurated many greening projects, with an emphasis on reducing **energy** consumption through the widespread use of biogas, while also maintaining the ecological vitality of the fragile Jinsha River. A prime target of **deforestation** during the 1950s and 1960s, especially during the **Great Leap Forward (1958–1960)**, when total **forest** cover was reduced to 13 percent, major reforestation campaigns were begun in the 1990s, with more than 7 million hectares planted by 2014, when provincial forest cover reached 35 percent. **Social protests** have broken out in the province, especially in opposition to the construction of such potentially highly polluting industrial facilities as a planned copper alloy factory in the city of Shifang in 2012, which drew thousands of opponents onto the streets.

SLOPING-LANDS CONVERSION PROGRAM (SLCP/*PODI ZHUANHUAN JIHUA*). The largest land retirement and reforestation program in the developing world, pursued by the People's Republic of China (PRC) since 1999, SLCP has a goal of converting 14.67 million hectares (36 million acres) of marginal cropland on terraced slopes greater than 25 degrees into

forest and **grasslands**. Also known as the "Grain for Green" (*Guwu wei Lüse*) program, with a commitment of RMB 337 billion ($40 billion), SLCP was a direct response to major **ecological disasters** involving a 267-day dry-up of the **Yellow River** in 1997, followed by massive **flooding** on the **Yangzi River** in 1998, both attributed to significant **deforestation** and **soil erosion**. With 2 billion tons, 65 percent of total soil, lost to rivers annually coming from terraced lands (*titian*) on hillsides and mountainsides, the program is aimed at amelioration through direct government payments to farmers. Voluntary enrollments in the program number 32 million households from 400 counties in 25 provinces, with a goal of increasing the country's forest land by 10 to 20 percent, along with a concomitant decrease of 10 percent in cultivated farmland on marginal and fragile lands prone to large-scale soil loss.

Evaluation of program effectiveness has given mixed results, with 613 km^2 (236 mi^2) of cropland converted into forests and 528 km^2 (203 mi^2) into **grasslands**, along with significant reductions (22 to 24 percent) in silt runoff and substantial increases in carbon sequestration (*tanhui*), amounting to between 222 and 468 million tons in a 10-year period. Problems include a low survival rate for planted trees, with converted cropland often consisting of flatlands or low-sloping lands, with a majority of genuinely steep-sloping lands (*guahua di*) greater than 25-degrees located largely in western regions of the PRC. Rural households have also seen little or no increase in income, as promised by the program, which was launched in conjunction with the National Forest Protection Program.

SOCIAL PROTESTS (*SHEHUI WENTI*). Beginning in the city of Xiamen, Fujian province, in 2007, and the second-largest cause of mass action in the People's Republic of China (PRC) after illegal land seizures, environmental social protests have taken place in major **cities** like **Shanghai** and **Tianjin**, along with medium and small towns, including Maoming (Guangdong), Longkou (Shandong), and Shifang (**Sichuan**). Spontaneous in nature and generally limited to local communities afflicted by environmental issues involving a so-called NIMBY ("not in my back yard") reaction, social protests have involved hundreds or even thousands of people, often sparked by information spread on social **media**, particularly Weibo (China's version of Twitter) and WeChat, with little or no coverage provided by the government-controlled press and television.

While consisting of relatively well-off middle class increasingly concerned with threats to **health** by **air pollution** and **water pollution**, and especially the construction of **chemical** plants, protestors also include members of the working class and farmers reacting to similar concerns. Expressing a deep-seated and entrenched distrust of government officials, even including members of local **environmental protection bureaus (EPBs)**, protestors have frequently targeted planned production facilities of paraxylene (PX), a chemical used in the manufacture of polyester (*juzhixianwei*) and **plastics**, despite government assurances of plant safety. Other frequent targets include planned incinerators (*fenshao lu*) for waste and **garbage** disposal and **coal**-fired power plants, which, despite commitments to shift **energy** production to cleaner natural gas and other **renewable energy**, continue to be built typically in poorer areas of the country.

While generally peaceful although somewhat rowdy, with protests animated by such slogans as "protect our environment" (*baohu huanjing*), "let me breathe" (*rang wo huxi*), "oppose PX!" (*fandui* PX), and "PX get lost" (PX *zoukai*), violence has, on occasion, broken out, with protestors occupying government offices and even setting police posts ablaze. Heavy-handed police action aimed at "clearing the streets" has included scattering crowds with truncheons and arresting alleged protest leaders and **activists**. With a strong commitment to a policy of maintaining "social stability at all costs," protests are usually broken up in short order based on the fear that such mass actions might spread beyond localities but with basic demands usually addressed, although only in the short term, and with offending production facilities frequently moved to other more receptive and remote sites.

While China experiences between 90,000 to 150,000 protests annually concerning a variety of issues and social complaints, many participants in environmental mass actions are doing so for the first time. Provinces experiencing significant environmental social protests include Fujian, Guangdong, Liaoning, Sichuan, **Yunnan**, and Zhejiang, and the cities of Chengdu, Dalian, Kunming, Longkou, Maoming, Ningbo, **Shanghai**, Shifang, **Tianjin**, Wenchuan, and Xiamen.

SOIL EROSION (*TURANG QINSHI/SHUITU LIUSHI*). Afflicting one-third of lands, while posing a major threat to agricultural production and the long-term goal of **sustainable development**, soil erosion has been a perennial problem in China from centuries of continuous cultivation, and extending into the People's Republic of China (PRC). National land surveys conducted by the Chinese government indicate afflicted areas of 40 million hectares (96 million acres), primarily from **wind** and **water**, with total soil loss of 4.5 billion tons annually from both natural and human causes, the latter having the greatest effect. The most important human causes of soil erosion include **overpopulation** and overcultivation of marginal, sloping lands (*qingxie tudi*), a problem dating back to the imperial era and exacerbated by poor management of industrial land use, excessive large-scale development, and poorly and haphazardly planned construction projects, including roads and railroads. Deleterious environmental effects include widespread **deforestation**, with concomitant siltation of **rivers**, **lakes**, and reservoirs.

While every region of the PRC has experienced soil erosion, those most affected are the northwest and southwest, with the former experiencing **desertification** and the latter, at current rates of soil loss, confronting major threats to the livelihood of 100 million people. Areas with little conventional vegetation, including **grasslands**, shrubs (*guanmu*), and **forests**, are particularly susceptible, as demonstrated on the **Loess Plateau**, on the North China Plain, where barren lands denuded of vegetation have contributed to the proliferation of gullies (*shuigou*), tunnel inlets (*suidao rukou*), and the mass movement of lands, including mudslides and **landslides**, to the detriment of sustainable agricultural production.

Most affected are the middle reaches of the **Yellow River**, especially in northwestern Shanxi and northern Shaanxi provinces, which account for 25 percent of the annual sediment load from an area less than 5 percent of the total river basin. Between the 1950s and 1980s, sedimentation (*chenjiang*) increased by one-third, with the Loess Plateau losing one centimeter of topsoil per annum. Other highly afflicted areas include the upper reaches of the **Yangzi River** and the hilly provinces of Guizhou, Hunan, Jiangxi, and **Sichuan**, the last experiencing significant soil erosion on 44 percent of largely sloping lands. While the northeast and south are overall less afflicted, soil erosion has affected areas in Fujian, Guangdong, and Heilongjiang due to multiple causes, including such wasteful agricultural practices as traditional rotational or swidden **agriculture** (*lunxie*), slash-and-burn (*daogeng huozhong*) cultivation in **Xishuangbanna**, and the excessive exploitation of herbs and plants for **traditional Chinese medicine (TCM)**.

Major amelioration efforts include reforestation of denuded areas, primarily on fragile hills and **mountains**, along with stabilization of **water** runoff, especially during heavy rain storms through the widespread construction of drainage ditches and check dams for reducing water flow velocity. Bans on cultivating steep sloping lands (*guahua di*) were enacted by the Grassland Management Act (1982), followed by even tougher restrictions imposed by the 1999 **Sloping-Lands Conversion Program (SLCP)**, which according to official data reduced the rate of soil erosion between 2000 and 2010.

With soil erosion highly correlated with rural **poverty**, current government plans call for a reduction in soil loss of 1.5 billion tons by 2030, as some of the most severely affected regions require an end to human habitation and such activities as farming, especially on steep slopes, to ensure recovery. Institutions involved in addressing soil erosion include the Institute of Soil and Water Conservation and such international organizations as the International Development and Research Center of Canada.

SOIL QUALITY (*TURANG ZHILIANG*). The soil base of Chinese **agriculture** has been of low fertility throughout history, with an average organic matter of less than 1.5 percent and less than 0.7 percent in large areas.

Lacking nutrients crucial to productive agriculture, including manganese (*meng*), nitrogen (*dan*), phosphorous (*lin*), potassium (*jia*), and zinc (*xin*), as much as two-thirds of Chinese soil is alpine, allitic, regosol, and luvisol (rocky and loamy), while only one-third can be characterized as the more fertile black, red, loessial, paddy, meadow, and hydromorphic soils.

Since the establishment of the People's Republic of China (PRC) in 1949, soil quality has undergone further deterioration from the accumulation of industrial heavy metals and dramatic increases in **chemical fertilizers** and **pesticides**, with an estimated 10 million hectares (24.7 million acres) of cropland suffering serious pollution in 1992. Despite major efforts at amelioration and restoration by national and local government agencies, Chinese lands are increasingly denuded of significant vegetation, leading to widespread **soil erosion** and contamination of groundwater (*dixiashui*). Problems include the largest arsenic (*shen*) contamination in the world, at 70 percent, mostly from extensive **mining**, with 10,000 mines nationwide for mining copper (*tong*), gold (*jin*), and other minerals, along with industrially produced lead (*qian*). With extensive monitoring and multiyear soil surveys conducted, the latest from 2008 to 2014, soil pollution is considered more serious than **air pollution** and **water pollution**, with 40 percent of total lands suffering some level of pollution, 16 percent exceeding state pollution limits, and 5 percent deemed "heavily polluted."

Major Chinese government efforts at amelioration include a 2016 action plan for curbing soil pollution by 2020, with complete stabilization and improvement in soil quality by 2030, which was strengthened by the Law on the Prevention and Control of Soil Pollution (2018), with legal and financial accountability primarily attached to pollution-offending firms. Nationwide, 100 key soil remediation projects are in place, with the total cost of complete cleanup estimated at RMB 5 trillion ($760 billion), largely to be borne by local governments, as many offending companies have moved or gone out of business, doing so before detection of the problem. China is also adopting the principles of regenerative farming aimed at promoting healthier soils with greater water absorption and carbon sequestration capacity through reliance on natural processes over chemical inputs. Major institutions and organizations involved in detecting and ameliorating soil quality problems include the China Soil Association, Environmental Sciences Group, Institute of Agricultural Research and Regional Planning, and Institute of Soil Science, while public awareness of the problem has grown, especially after soil quality data was no longer withheld as a "state secret" (*guojia mi*).

SOLAR GEOENGINEERING. *See* GEOENGINEERING (*DIQIU GONGCHENG*).

SOLAR POWER (*TAIYANG NENG*). The largest market in the world for photovoltaic (PV/*guangfu*) and solar thermal **energy** since 2013, the People's Republic of China (PRC) also has the largest installation of PV capacity, with 154 gigawatts (GW) in 2018, along with 70 percent of global solar thermal capacity. Beginning in 2007, with substantial Chinese government subsidies for the production of silicon solar PV cells (originally invented in the United States in 1954) and the construction of large-scale solar farms in the sun-drenched western regions of China, notably the Tengger and Golmud deserts, solar power is a central component of the PRC's pursuit of substituting **renewable energy** for its heavily fossil-fuel dependency. Constituting less than 2 percent of the total electrical power generation in the country, solar power in China suffers from a low-capacity factor as a result of substantial loss of power generation from long transmission lines carrying power from remote western areas to urban areas, largely in the east.

With considerable overcapacity and substantial waste of solar-generated electricity, China is phasing out its heavy subsidization of the **industry**, while gradually shifting production from large-scale farms to rooftop installations in **cities** like **Dezhou**, Shandong province. Among the major solar power innovations in China is the construction of a more energy-efficient floating solar power farm located in Anhui province on a **lake** situated over an abandoned **coal** mine and pilot construction of a solar panel road in Jinan, Shandong Province, for generating electricity and recharging **electric vehicles (EVs)**. Unresolved problems include devising an environmentally sound process for disposal of used solar panels containing substantial toxins occurring sometime around 2040.

Major Chinese **companies** involved in solar power include CHINT Group; Hanwha SolarOne; JA Solar; Jinniu Energy; and Suntech, the world's largest producer of PV panels. While essential to energy conservation and other notable environmental goals, solar power has an environmental downside, as the production process of PV panels and the refinement of silicon produce considerable waste and pollution. Installation of solar heaters on homes and farms has grown substantially with the PRC the largest user in the world, resulting in substantial reductions in wood and coal burning, the latter by an estimated 14.5 million tons annually.

SOLID WASTE. *See* GARBAGE (*LESE*) AND GARBAGE DISPOSAL.

SONGHUA RIVER (*SONGHUA HE*). Flowing for 1,434 kilometers (891 miles) from the Changbai Mountains on the border with the Democratic People's Republic of Korea (DPRK/North Korea) through Jilin and Heilongjiang provinces in the People's Republic of China (PRC), the Songhua River, formerly known as the Sunggari, includes as a primary tributary the Second

Songhua. The site of a major **ecological disaster** involving a large and highly destructive toxic spill in November 2005, Songhua is one of the most polluted **rivers** in China, largely from the concentration of outdated heavy industrial facilities, including **steel**, machine tools (*jixie gongju*), locomotives (*jiche*), and aircraft (*feiji*) built during the period of central economic planning (1953–1978). With virtually no attention to pollution controls on industrial effluents, the Songhua was subject to serious contamination by organic mercury (*gong*) and other toxic pollutants from **chemical** factories located along the Second Songhua tributary, with the self-purification recovery process slated to take decades following government controls on continued contamination, especially by mercury.

The November 2005 spillage involved approximately 100 tons of benzene (*ben*), nitrobenzene (*xiaojiben*), and aniline (*ben'an*) from an explosion at a riverside petrochemical plant near Jilin city (Jilin province) owned by China National Petroleum Company (CNPC), the country's largest **oil** enterprise. Meandering down the slow-flowing waterway, an 80-kilometer-long slick 33 times the legal limit moved toward the city of Harbin, capital of Heilongjiang province and home to 4 million people, threatening the local **water** supply. With factory officials initially denying any spillage into the river and then attempting to dilute the slick by draining in reservoir water, the State Environmental Protection Administration (SEPA) in **Beijing** waited five days before issuing an emergency declaration and a full two weeks before informing United Nations officials of possible threats to neighboring **Russia**. Information on the spill was also withheld from the public, as local, regional, and central government authorities were evidently paralyzed by confusion and bureaucratic immobilism, with some top officials engaged in an attempted cover-up, prohibiting **journalists** from publishing stories about the spill.

Along with cutting off water supplies to the city of Harbin, 10,000 people were evacuated as swarms of dead fish floated in the waterway. Revealing serious deficiencies in emergency response and lack of coordination within the country's environmental protection bureaucracy, the crisis led to major administrative and personnel reforms, including the following: a reorganization in the relations between SEPA and regional **environmental protection bureaus (EPBs)**; the establishment of Regional Supervision Centers; revisions to the Water Pollution Control Act, especially involving emergency response; and the resignation of the then-head of SEPA, **Xie Zhenhua**.

SOUTH CHINA SEA (SCS/*NAN ZHONGUO HAI*). Extending from the Taiwan Straits in the north to the Straits of Malacca and Karimata Straits bordering the Java Sea to the south, the South China Sea (SCS) covers a tropical and semitropical area of 3.5 million km^2 (1.4 million mi^2), with lucrative **fisheries** and 30 percent of the world's **coral reefs** facing major environmental degradation. Referred to in the People's Republic of China

(PRC) as simply the "South Sea" (*Nanhai*), SCS is fed by several major **rivers** originating in China including, the Jiulong, **Min**, **Pearl**, and **Lancang** (flowing as the Mekong from Southeast Asia). Containing more than 250 small islands and atolls, SCS is home to 600 species of coral reefs, 3,000 fish species, and 1,500 sponge species, many of them **endangered**, with the Spratly (*Nansha*) and other island groups serving as major nesting and breeding grounds for a rich variety of **birdlife**, for example, seabirds, along with marine turtles (*hai gui*) and substantial **seagrass** reserves. The second-busiest sea lane in the world, including a major route for **oil** deliveries from the Middle East to Asian markets, SCS suffers from significant pollution from ships, with warming waters from **climate change** threatening coral reefs, along with island-building near and on top of reefs, primarily by the PRC, a violation of the United Nations Convention on the Law of the Sea (UNCLOS), to which China is a signatory.

Other environmental threats include excessive inflows of organic nitrogen and phosphates salt, as well as outbreaks of red tides, along with overfishing, especially by the PRC, with its huge fishing fleet. Equally damaging is the dredging of sand and gravel for island-building and illegal trapping of giant clams (*juha*) by poachers employing highly destructive excavation of the species' abodes in coral reefs. With substantial reserves of natural resources, including oil, natural gas, and methane clathrates, or fire ice (*huobing*), along with geostrategic considerations, SCS is a focal point of international conflict involving competing claims over sovereignty by several nations, including Indonesia, Malaysia, the Philippines, and the PRC, preventing enhanced action on many environmental issues. Several joint declarations and agreements, many nonbinding, have been reached on controlling pollution and regulating fisheries in the SCS, for instance, protection for the increasingly threatened giant clam.

SOUTH-TO-NORTH WATER DIVERSION PROJECT (SNWDP/*NANSHUI BEIDAO*). The largest water conservancy scheme in the world, transferring as much as 45 billion m^3 of **water** annually from the water-rich south to the parched north of the People's Republic of China (PRC), SNWDP consists of three separate diversion canals—eastern, central, and western—measuring more than 1,000 kilometers in length, with significant environmental impacts in both the water-providing and water-receiving regions. Diverting water primarily from the lower, middle, and upper reaches of the **Yangzi River**, along with waters from the Danjiangkou Reservoir on the Han River in Hubei province, the project is designed to bring multiple benefits to **agriculture** through expanded **irrigation** and to municipal water supplies in such major **cities** as **Beijing**, **Tianjin**, **Shijiazhuang** (**Hebei province**), and Zhengzhou (**Henan province**), along with restoring heavily depleted groundwaters throughout the northern region.

As with most large-scale projects, SNWDP will fundamentally transform **hydrology** and **ecology**, most significantly to the upper, middle, and lower reaches of the Yangzi River, especially in the river's massive delta (*sanjiao-zhou*). While the drawdown of the Yangzi will only amount to 4 to 5 percent of total river volume annually, major environmental impact will occur at the point source in the upper reaches, where estimates are for a 35 percent decline, most seriously during the summer dry season, with attendant deleterious effects on important tributaries including the Tongtian, Dadu, and Yalong **rivers**, located in Qinghai and **Sichuan** provinces. Impacts on the densely populated Yangzi River Delta include increased saltwater intrusion and **soil** salinization (*yanhua*) and alkalization (*jianhua*), along with sediment accumulation in the estuary stemming from reduced river flow, along with declines in aquatic life.

With the Yangzi already suffering from high levels of pollution, any water transferred to the north will contain significant containments, so much so that one-third of the project's entire cost is earmarked for 379 separate pollution control and water treatment projects, while the Grand Canal was considered too polluted with foul-smelling green gunk to be employed along the eastern route. Studies have also indicated that consumption for **irrigation** would overwhelm the other potential uses, robbing the north of the environmental restoration promised by the project. The potential impact on land use, regional climatic variations, and agricultural productivity along the planned routes is the subject of a joint Chinese–Italian project known as the Sustainable Water Integrated Management of the South-to-North Water Diversion Project East Route, launched in 2004.

SOVIET UNION (*SULIAN*). Providing substantial economic and technical assistance to the People's Republic of China (PRA) during a period of alliance and international solidarity (1950–1960), the Soviet Union had a profound influence on environmental policies in the PRC, especially as pursued during and immediately following the reign of Josef Stalin (1927–1953). With a $300 million loan and thousands of scientific and technical advisors provided by the Soviet Union, the inauguration of central economic planning in China (1953–1978) led to a series of policies based on Soviet experience with long-term environmental impact, including rapid development of heavy **industry**, especially iron and **steel**, **mining**, and the construction of large-scale **hydropower** projects for the generation of electrical power, along with collectivization of **agriculture**. Just as the Stalinist Plan for the Transformation of Nature (1948) had devastating effects on the Soviet Union's natural environment and resources, so too did Chinese policies prove highly disruptive of **forests**, **rivers**, **lakes**, **plant species**, **fisheries**, and **wildlife**, with little or no regard for environmental costs, including **air pollution**, **water pollution**, and serious **soil erosion**.

For both Josef Stalin and **Chinese Communist Party (CCP)** chairman **Mao Zedong**, nature was seen as an "enemy" to be conquered and shaped by human action, especially through such "gigantomania" projects as giant dams (Dnieprostroi/Gezhouba), huge canals (White Sea/Red Flag), and large steel-making facilities (Magnitogorsk/Panzhihua). Contemptuous of **nomadic pastoralism** and other forms of independent economic activity, both the Soviets and the Chinese mistreated herders, with concomitant degradation of valuable **grasslands** and waterways. Maintaining ideological orthodoxy and controls was also essential to these policies, as both Mao and Stalin opposed and persecuted outspoken scientists and environmentalists, with any attempt at independent environmental organization or action suppressed. While both leaders endorsed efforts at "greening" (*lühua*), for example, the creation of **nature reserves**, with the Soviet leader also calling for massive tree planting, as large green areas were set aside in Soviet cities like Moscow, these environmentally friendly goals were generally overwhelmed in both countries by destructive economic policies, along with the creation of highly toxic **chemicals** and radioactive waste (*fangshexing feiwu*) from industry and the development of **nuclear** weapons.

As tensions emerged between the PRC and the Soviet Union following Stalin's passing in 1953, and the emergence of Nikita Khrushchev as top Soviet leader (1955–1964), the outbreak of the Sino–Soviet split (1956–1966) and subsequent withdrawal of Soviet advisors from China had little effect on environmental policies pursued in the two countries, which remained substantially similar. Generally indifferent to the environment, both Khrushchev and Mao shared a hostile attitude toward nature reserves, while promoting radical agricultural policies (Khrushchev's Virgin Lands/ Mao's People's Communes) that proved enormously destructive to farmlands and other natural resources, including forests and soils. With both leaders prohibiting any public involvement in **policy-making** and inclined to boastful claims of dramatic economic gains using highly distorted statistics, the two countries continued to share serious environmental damage, including excessive **logging**, wasteful and inefficient **energy** projects, and degradation of grasslands and pastures, as short-term goals trumped long-term planning.

Similarities in environmental policy and continuing problems of pollution and contamination persisted throughout the 1964–1985 period, when both the Soviet Union and the PRC experienced substantial deterioration of environmental conditions. Just as Leonid Brezhnev (1964–1982) followed the removal of Nikita Khrushchev by giving top priority to programs of economic development that had substantially deleterious effects on the environment, especially in agriculture, so too did the PRC following the death of Mao Zedong (1976) pursue economic growth at all costs, notably degradation of the environment under the leadership of paramount leader **Deng Xiaoping**

(1979–1992). In both societies widespread environmental deterioration occurred from a multiplicity of poorly conceived policies, particularly the Brezhnev Food Program and the Deng Agricultural Responsibility System, which in China led to the enormous overuse of chemical **fertilizers** and **pesticides**, as well as continuing soil erosion, along with contamination by the rapid expansion of chemical and mining facilities, often located in or near densely populated urban areas.

While popular outcries and complaints about environmental contamination were often suppressed in both highly authoritarian Communist regimes, **media** and public attention to major environmental problems grew, as both countries were influenced by the increasingly global importance of environmental mitigation. Initial steps toward environmental amelioration were taken with such new policies and **laws** as the Soviet Main Guidelines (1981–1985) for environmental protection, especially of **soils** and **water resources**, and the PRC's Trial Environmental Protection Law (1979).

This process accelerated dramatically with the emergence of new leaders, most notably Mikhail Gorbachev (1985–1991) in the Soviet Union and Jiang Zemin (1989–2002) and particularly **Hu Jintao/Wen Jiabao** (2003–2013) in the PRC, when greater political openness in both societies led to the emergence of **civil society** with the rapid growth of **environmental awareness** and **activism** among the general public. Greater media attention to major problems like the contamination of Lake Baikal, the Caspian Sea, and the Volga River in the Soviet Union, and highly destructive logging along the **Yangzi River**, combined with serious **ecological disasters**, most notably the Chernobyl nuclear disaster in Ukraine (1986) and the devastating **floods** on the Yangzi (1998), brought environmental issues to the forefront. Nonstate environmental action groups emerged in both countries, including "green" social and political organizations in the Soviet Union and "econ-nationalist" groups in the increasingly independent republics, which would contribute to the breakup of the country in 1991, and a proliferation of environmental nongovernmental organizations (ENGOs) in China. Major national institutions of environmental protection were also created with greater bureaucratic authority, namely the Soviet State Environmental Protection Agency and China's **Ministry of Environmental Protection (MEP)**, with both organizations confronting persistent environmental problems from the legacy of breakneck economic and industrial development.

See also RUSSIA.

"SPONGE CITIES" (*HAIMIAN CHENGSHI*). An approach to urban **water** management and conservation aimed at reducing persistent **flooding** by capturing and retaining rainwater, especially during intense meteorological events, so-called sponge cities involve such nontraditional measures as the construction of urban **wetlands**, green rooftops, and rain gardens, the last consisting of shrubs and flowers in a small depression to hold and absorb rainwater. Devised in 2014, after serious flooding in such major **cities** as **Beijing**, Nanjing, **Shanghai**, Shenzhen, **Tianjin**, Wuhan, and Xiamen, as traditional drainage systems for channeling water out of the cities into nearby **rivers and waterways** were overwhelmed, the Sponge City Concept (SPC) is generally based on the Low-Impact Development (LID) and Sustainable Urban Drainage System models developed in Western countries, such as The Netherlands, confronting similar problems of persistent flooding. With urban areas in the People's Republic of China (PRC) growing fourfold since 1978, natural landscapes have been systematically denuded with impermeable pavement and concrete structures, primarily large buildings blocking the natural flow of water, while also creating widespread problems of waterlogging (*lao*) and surface **water pollution**. Cities situated on major flood plains such as Kunshan, Jiangsu province, have also protected against major flooding by enclosing low-lying tracts of land with dikes ("polders"), which allows for recirculation of canal water through precinct-scale wetlands nestled within open space.

Regular flooding of urban areas in the PRC has doubled since 2008, affecting 150 cities, increasing the incidence of **landslides** and other environmentally destructive and life-threatening events but with water shortages growing more serious, particularly in the north, as infiltration into urban groundwater has dropped precipitously to a mere 20 to 30 percent of total volume. Committing to a pilot program involving 30 major urban areas, one of which is Shanghai, major goals of the SPC approach include capturing 70 percent of rainwater, especially during the summer monsoon season, for reuse, with 20 percent of major urban areas developing SPC infrastructure, including permeable roads and pavements constructed of water-absorbing porous asphalt and concrete, raised walkways, bioswales, and underground storage tanks for temporarily storing, **recycling**, and purifying stormwaters. While urban authorities in places like Linggang, Pudong District, Shanghai, have taken an aggressive approach to installing water retention and reuse infrastructure, many localities suffering from accumulated debt and lack of central government financing are reluctant to invest in SPC infrastructure, which generates little, if any, revenue into city coffers. Among the major figures involved in designing the Chinese version of SPC and infrastructure is Kuang Xiaoming of the Shanghai Tongji Urban Space and Ecology Institute.

STEEL INDUSTRY (*GANGTIE GONGYE*). With the production of more than 1 billion tons in 2017, largest in the world and greater than the United States, **Russia**, and Japan combined, the steel **industry** in the People's Republic of China (PRC) is a major source of industrial pollution and a primary target of remediation efforts by the Chinese government. The third-largest producer in China of sulfur dioxide (SO_2), nitrogen oxide (NO_x), and fine particulate matter (PM 2.5) concentrates, steel production is also responsible for 10 percent of all wastewater afflicting the country's **rivers**, **lakes**, and groundwater sources, while consuming 10 percent of the nation's **energy**.

Many of the deleterious environmental effects of steel-making derive from the coking and iron-making processes, which release **air pollution** emissions, industrial wastewater, hazardous wastes, and solid wastes. From coking comes sulfur and coke dust (*jiaotan fen*), while the substantial amount of water needed to cool coke is polluted with slag wastes containing limestone and iron ore impurities. Major factors determining the relatively high pollution of China's steel sector compared to more developed countries include industry fragmentation, with large numbers of small mills engaged in energy-intensive production processes, usually involving the burning of **coal**, while lacking the sufficient capital for installing expensive pollution control equipment and waste water drainage systems; reliance on highly polluting and coal-burning basic oxygen furnaces (BOF) over the less-polluting electric arc furnaces (EAF); and relatively little production of less-polluting scrap steel.

Rapid expansion of the industry, with fourfold growth in production from 2000 to 2007, exacerbated pollution problems, with major steel-producing cities like Tangshan (**Hebei province**) suffering major environmental degradation, including airborne waste gasses, slag, and daytime darkness from smoke and ash emissions spewing forth from the city's 57 separate steel enterprises. Major government efforts at remediation as part of the larger campaign to reduce **air pollution** and **water pollution** include the closing of small, antiquated, and dirty mills, especially in 2017, described as the "largest closure of steel factories in history"; consolidating the industry into larger, more capital-rich enterprises with a financial capacity to install expensive pollution control equipment and waste water drainage systems; moving steel plants away from such highly polluting areas as **Beijing** and **Tianjin**; shutting down plants temporarily during the smog-prone winter months; shifting from BOF to EAF production, while also using modern sinter equipment for processing fine-grained raw materials into coarser-grained iron ore; and closing illegal and dirty induction furnaces on a permanent basis—all of which has been made more palatable with the current large-scale overcapacity in the industry, as a reduction of 255 million tons has been effected. Major remediation efforts are also being pursued to deal with the 100 million tons of industrial waste produced by steel production annually. Although one-third is recycled into cement, concrete, and other construction materials, the rest is dumped illegally or stacked up in thousands of slag heaps scattered throughout the country, contributing to air and water pollution and degradation of farmland. China is also a major producer of aluminum (*lü*), one of the three most heavily polluting industries, involving an energy-intensive smelting (*yelian*) process.

SUSTAINABLE DEVELOPMENT (*KE CHIXU FAZHAN*). Defined as the meeting of human development goals, while at the same time sustaining the capacity of natural systems to provide natural resources and **ecosystem** services upon which **economy** and society depend, sustainable development has been a priority goal of the People's Republic of China (PRC), incorporated as a major strategy of the country's modernization beginning with the Ninth Five-Year Economic Plan (1996–2000). Reluctant to accept sustainable development promoted by the international community at the United Nations Conference on Environment and Development (1992) as an undue restriction on the pursuit of economic development by poorer countries, sustainable development criteria are now applied to every major realm of Chinese development policy, including **agriculture**, **industry**, **water resources**, and, most importantly, environmental protection.

Embodied in China's adoption of **Agenda 21** protocols and the 2030 Agenda for Sustainable Development, both promoted by the United Nations, the pursuit of sustainable development in the PRC is driven by the severity of environmental pollution with attendant **health** problems, resulting primarily from the single-minded pursuit of rapid economic development beginning in 1978–1979. Included in the Chinese vision of sustainable development is a reduction in the gap between the wealthy eastern and backward western regions, along with such technological innovations as the creation of "smart cities" (*zhihui chengshi*) and **renewable energy** to meet the needs of the present without compromising the well-being of future generations.

T

TAI LAKE (*TAIHU*). Located in the **Yangzi River** Delta with a surface area of 2,250 km^2 (869 mi^2) and an average depth of only two meters (6.6 feet), Tai Lake is the third-largest freshwater lake in the People's Republic of China (PRC), which after decades of heavy pollution became a target of a major cleanup campaign. Revered by Chinese literati and poets throughout history for its pristine "emerald waters" (*bishui*) that reflected surrounding **mountains**, Tai Lake has been at the center of Chinese prosperity, evoked as the "land of fish and rice" (*yumi zhixiang*), crucial to the economic well-being of nearby cities like **Shanghai**, Suzhou, Wuxi, and Hangzhou. With 20 percent of the lake converted to farmland by major **reclamation** campaigns during the **Great Leap Forward (1958–1960)** and **Cultural Revolution (1966–1976)**, along with the proliferation of **aquaculture**, rapid economic growth and industrialization beginning with the economic reforms of 1978–1979 turned the lake into a poster child of environmental pollution and degradation. Major causes for the lake's enormous deterioration included agricultural runoff of **fertilizer** and **pesticide** residue, dumping of **garbage** and animal waste, and an outlet for untreated **sewage** from surrounding rural areas, along with the discharge of wastewater from shoreline industries, for instance, 2,800 **chemical** factories.

With lake circulation disrupted by the construction of **hydropower** projects and weirs (*yan*) on feeder **rivers**, by the 1990s Tai Lake was suffering from widespread eutrophication (*shuiti fu yingyanghua*) bringing about serious outbreaks of annual cyanobacterium algae blooms ("pond scum"), which, in 2007, covered the entire lake, causing major fish die-offs, a drop in local rice yields, and a cutoff of crucial water supplies to urban areas like Wuxi. Government efforts to remediate the problem include such short-term measures as river dredging and scooping up tons of surface algae, along with large-scale water diversion from the Yangzi River and more long-term efforts at removing the high concentrations of nitrogen (*dan*) and phosphorous (*lin*) lodged in lake bottom sedimentation, which feed the algae, as well as the construction of sewage treatment plants. Also deployed is the world's largest aquaponics floating **wetlands** project, four acres in size, which is designed to remove algae-producing nutrients from the lake's waters while also absorbing urban, industrial, and agricultural waste runoffs. While overall **water quality** in the lake has shown some improvement in the 2010s, scientists expect that a thorough cleanup of the lake will take decades, especially with the increasing deleterious effects of **climate change**. Major lake **fisheries** include the famous "three whites" (*san bai*) composed of white bait, fish, and shrimp along with the highly prized "hairy crab" (*dazhaxie*), a local delicacy.

See also WU LIHONG (1968–).

TANG XIYANG (1930–). A major pioneer of the environmental movement in the People's Republic of China (PRC), Tang Xiyang promoted the "Green Camp" initiative of introducing Chinese citizens, particularly the young, to the richness of the natural world. Tang is also a major opponent of reckless environmental practices, especially excessive **logging**, in such previously pristine regions as **Yunnan province**. Educated at Beijing Normal University and a practicing **journalist** for years, he was persecuted in the Anti-Rightist Campaign (1957–1958) and the **Cultural Revolution (1966–1976)**, when, "sent down" (*xia xiang*) to the countryside, he sought personal salvation in an embrace of the natural world.

The publisher of *Green Nature* magazine (1980) and author of *A Green World Tour* (1993), the latter regarded as the bible of China's nascent environmental movement, Tang has long argued for greater involvement of the Chinese public in dealing with environmental problems, rejecting the age-old reliance in China on a single "wise" leader. Arguing for a fundamental change in popular attitudes and values that feed the prevailing **"waste culture,"** Tang has also invoked Chinese traditional philosophy, seeing a need for more Daoism and believing he served as a "green sutra" (*lüse jingwen*) to China, not unlike ancient monks who delivered the teachings of Buddhism into Chinese culture. Bestowed the Ramon Magsaysay Award for Peace and International Understanding in 2007, he continues to write and lecture extensively on environmental issues in China, advocating for greater human rights and democracy in confronting the devastation of the once-pristine natural world.

TANNERIES (*PIGE CHANG*). At more than RMB 1 trillion ($196 billion) annual revenue, the leather-tanning **industry** in the People's Republic of China (PRC) has been a major source of environmental pollution, especially of **rivers and waterways**, in centers of production, particularly in **Hebei province**. With the proliferation of small-scale, family run tanneries since the introduction of economic reforms (1978–1979), deleterious environmental impacts have occurred, requiring a strong government response that began in 2015. With 200 to 3,000 liters (52 to 78 gallons) of wastewater produced per one finished leather piece, an enormous discharge of untreated wastewater was generated during the boom years for the industry (1990–2012), especially in major centers of production, for example, Wuji and Xinji townships (Hebei), the latter regarded as the "leather capital" (*pige du*) of China.

In addition to chromium (III) used as a major tanning agent, 30 percent of which is not absorbed by raw hides, major pollutants discharged unfiltered into waterways for decades included hexavalent chromium (*liujia ge*), a known carcinogen, along with acids (*suan*), alkali (*qiang jian*), sulfide (*liuhuawu*), lime (*suancheng*), and dyes (*ranliao*), with the Hutuo River running through Xinji and nearby **Shijiazhuang** overwhelmed by sludge that constituted 90 percent of river volume. Beginning with the promulgation of the Discharge Standards of Water Pollution for Leather and the Fur Markets Industry (2014), tanneries have been ordered to install treatment plants, with noncomplying operations forced to close down (70 in Xinji alone), while mergers and consolidations of the industry have been pursued to create more capital-rich enterprises. Remnants of the industry requiring continued cleanup include old dump sites, while some noncompliant **companies** bypass regulations by shipping waste to poorer provinces or building hard-to-detect underground pipelines, with some moving production to Africa.

TAXATION (*SHUISHOU*). Passed by the National People's Congress (NPC) of the People's Republic of China (PRC) in 2016, and enacted in January 2018, the Environmental Protection Tax Law imposes taxes on businesses and public institutions involved in discharging **air pollution**, **water pollution**, solid waste, and **noise pollution**. Replacing pollution discharge fees enacted in 1979, with RMB 17.3 billion collected in 2015, with 10 percent transferred to the central government, environmental protection taxes will be collected by local governments, specifically the **environmental protection bureaus (EPBs)** and the more powerful taxation authority, on 260,000 estimated entities, especially in **industry**. With a discretion to set a rate ranging from RMB 1.2 to RMB 12 per unit for air pollution and RMB 1.4 to 14 for water pollution, to reflect different regional conditions, funds will be retained at the local level. Labeled a "green tax system," with pollution control described by the Chinese government as a major "battlefield," previous exemptions for entities that produce less than 50 percent of the threshold pollution levels were abolished but with new exemptions for **agriculture**, **transportation**, and **sewage** and household waste. As a measure to encourage amelioration efforts by polluting entities, discharges reduced by 30 percent of the threshold will result in a 25 percent reduction in rates. **Cities**, most notably **Beijing**, have imposed the highest rates, at RMB 12 per unit of air pollution and RMB 14 per unit of water pollution, as a possible extension of the tax on carbon dioxide (CO_2) is being considered.

THIRD FRONT (*SANXIAN JIANSHE*). A massive transfer of industrial production and technical personnel to the interior regions of the People's Republic of China (PRC) largely for self-defense pursued from 1964 to 1980, the Third Front involved 483 factories and plants, many in highly polluting heavy and military-related industries, along with 92 research institutes. Dispersed throughout generally remote and isolated territories often near **mountain** ranges and even in caves, Third Front facilities were concentrated in western and southwestern provinces, primarily Guizhou, Qinghai, **Sichuan**, and **Yunnan**.

Bringing industrial production, including **steel**, **mining**, and power generation, to relatively poor and economically underdeveloped regions, Third Front production came at the expense of environmental degradation, including extensive **air pollution** and **water pollution**, in previously pristine regions of the country. Involving huge investments estimated at RMB 205 billion ($31 billion), one-fourth of the national total, into facilities, with half of the proposed plants left uncompleted and many poorly designed and highly inefficient, construction of Third Front facilities was brought to a halt by the introduction of market reforms (1978–1979). Emerging as key sectors in relatively backward, local economies, such Third Front projects as the Panzhihua Iron and Steel plant in Sichuan province proved difficult to close down despite high levels of pollution and environmental contamination.

THREE GORGES DAM (*SANXIA DABA*). The world's largest **hydropower** station, with an installed capacity of 22,500 megawatts (MW), the Three Gorges Dam, spanning the **Yangzi River** at Yichang, Hubei province, was completed as a major component in pursuit of **renewable energy** by the People's Republic of China (PRC) but with substantial negative environmental effects on the **river** and surrounding areas. With state-of-the-art turbines (*wolunji*) operating at full power, China avoids 100 million tons of greenhouse gas (GHG) emissions, including millions of tons of dust, 1 million tons of sulfur dioxide (SO_2), 370,000 tons of carbon monoxide (CO), and significant amounts of mercury (*gong*). Improved shipping along the reservoir between Yichang and the upstream inland city of **Chongqing** reduces the reliance on heavy polluting trucks for **transportation** of goods, with estimated additional reduction in carbon dioxide (CO^2) emissions of 10 million tons. Providing storage space in a 1-million-square-kilometer catchment area for as much as 31 million acre-feet of water, the massive reservoir, 660 kilometers (440 miles) in length and 1,100 meters (3,600 feet) in width, is also designed to control the incidence of major **flooding** downstream, as occurred, with devastating effects, in 1998.

An example of the engineering "gigantomania" inherited from the **Soviet Union**, the Three Gorges project, which was strongly endorsed by **Chinese Communist Party (CCP)** chairman **Mao Zedong** (1949–1976) and Premier

Li Peng (1987–1998), has had a multitude of deleterious effects on the environment, including life-threatening **landslides**, the destruction of river **fisheries** and **habitat** of **plant species** and **wildlife**, increased **water pollution**, alteration of the river's **ecology**, and local **climate change**. With reservoir water increasing pressure on already unstable river banks, along with seepage into nearby hills, weakening rocks and **soils**, landslides have become increasingly frequent, for instance, the devastating 700 million ft^3 rock slide into the Qinggang River, a tributary of the Yangzi, in July 2003, along with fears that the huge reservoir might trigger **earthquakes** in the geologically fragile region. Among the 300 fish varieties in the Yangzi River, the dam acts as a major barrier to spawning zones of salmon and other anadromous species, notably the river's diminishing number of sturgeon (*zhonghua xun*), numbering less than 1,000, while the prized **Yangzi River dolphin** has been declared extinct.

With less water flow into the lower reaches, saltwater intrusion has penetrated further up into the Yangzi River Delta, bringing in more jellyfish and creating an unchecked increase in snails (*luo*) carrying deadly schistosomiasis (*xuexichongbing*). Plant species, numbering 6,000 in the region, and wildlife have also suffered, with substantial habitat inundated by the reservoir, along with threats from the alteration of local weather patterns, mainly reduced rainfall. Less water flow downstream not only affects plant life and wildlife, but has also reduced inflows into **Dongting** and **Poyang** lakes, with both experiencing major dry-up.

With dramatic slowdown in water flow above the dam, carpets of **garbage** and algae plumes (*zaolei yumao*) have formed, as the reservoir has become a repository of waste by local cities and industries, adding to the pollution from flooded villages and old waste dumps. The natural processes of aeration (*tongfeng*)—the circulation of air mixed with or dissolved in water—and diffusion—the intermingling of substances—have also been disrupted, changing river **ecology**. Government amelioration efforts include the plant-rich Shennongjia National Nature Reserve, located near the dam, along with artificial-spawning programs for **endangered** fish species but with substantial losses to the region's **biodiversity**. Criticism of the dam project by leading hydrologists and scientists was published in *Yangzi! Yangzi!* edited by Ms. **Dai Qing**, subsequently banned in the PRC.

THREE RIVERS HEADWATERS AREA (TRHA/*SANJIANG YUAN*). Located in southern Qinghai province in the hinterlands of the **Qinghai–Tibetan Plateau** at an altitude of more than 4,000 meters (13,120 feet), the Three Rivers Headwaters Area (TRHA) contains the source waters of three major rivers—the **Yangzi**, **Yellow**, and **Lancang**—in the People's Republic of China (PRC). Considered a highly fragile "hot spot" of the planet's entire **ecosystem**, where a slight change in temperature can have a major global impact, at 150,000 km^2, the area is also the site of more than 16,000 **lakes**, along with the largest, highest, and most diversified **wetland** in the world (1,100 km^2). Subject to the deleterious forces of **climate change**, the area has experienced several problems, including a rise in average temperatures since 1958, fragmentation and loss of **habitat**, and overall ecosystem degradation—including **desertification**, threatening local **plant species**, **wildlife**, and the well-being of millions of people in Central and Eastern China, along with populations throughout Southeast Asia.

Among the nationally protected wildlife in the region are the **Tibetan antelope (*chiru*)**, wild yak (*ye maoniu*), Tibetan wild ass (*yelü*), snow leopard (*xue bao*), white-tipped deer (*baitou lu*), and golden eagle (*jinying*), along with 40 **endangered** vascular plants out of a total of 2,238 in the region, with 705 species endemic to the Qinghai–Tibetan Plateau. With degrading wetlands and peatlands reducing the capacity for storing melting snow and ice, degeneration of the area's permafrost (*yongjiu dongtu*) has also brought about a dramatic increase in environmentally destructive rodents (*niechi dongwu*) with the concomitant decline of pastoral areas for local herders.

Government amelioration efforts include the establishment of the Three Rivers National Nature Reserve (2000), along with a General Plan for Ecological Protection and Restoration aimed at establishing "ecological security" (*shengtai anquan*), defined as conditions of physical surroundings providing for the needs of inhabitants without diminishing natural resources for the region. Included is the large-scale outbound migration of the local populations, mostly Tibetans, with concomitant bans on grazing and wetlands and grasslands protection.

TIANJIN. The largest **coastal** city in North China and a major port city on the **Bohai Sea**, *Tianjin* ("heavenly ford") is a center of industrial production undergoing rapid urbanization, while confronting serious environmental problems, including **air pollution**, **water pollution**, solid waste disposal, growing scarcity of **water resources**, and attendant **health** problems. Built on low-lying marshland (*zhaoze di*) and situated on the badly polluted Hai River flowing through the city, with the threat of **flooding** controlled by the construction of diversion channels, Tianjin is the fourth-largest city in the PRC after **Shanghai**, **Beijing**, and **Guangzhou**, with a population of 15 million people. With many factories located within close distance of residential areas, which have expanded into industrial areas with a population growing by hundreds of thousands annually, Tianjin was the site of one of the worst **ecological disasters** in the history of the People's Republic of China (PRC) in August 2015, when a hazardous **chemical** storage facility exploded. Killing 173 nearby residents and firefighters, and leveling surrounding buildings, the explosion also released large amounts of highly toxic sodium cyanide (*qinghua na*) into the city's air and water supply.

Other environmental problems include poor **water quality**, with only 15 percent of the city's surface water meeting national standards, with the Hai River rated "badly polluted" and many city canals covered with toxic green runoff from local petrochemical plants. Air pollution is also serious, with local smog described as thick enough to taste and exacerbated by dust storms from the Gobi Desert, as **air quality standards** have declined in recent years with increased fine particulate matter (PM 2.5) concentrations. Contamination of **soils** include significant levels of hexachlorocyclohexane, used in insecticides (*sha chongji*), while scattered across the municipality are large numbers of open-**sewage** pits.

While amelioration measures were adopted in 2014, including restrictions on **motor vehicles**, halts to many construction and manufacturing projects, and school closures, the city was nevertheless harshly criticized for poor enforcement of environmental **laws and regulations** by central government **environmental inspections** carried out in 2017. With more than 4,000 local enterprises ordered to clean up or confront permanent shutdown, Tianjin is shifting its industrial base to "green" production like **wind power** turbines, while engaging in a major program of soil remediation and reduction of pollution from ships in the port of Tianjin. Along with **Beijing** and **Hebei province**, Tianjin is a top priority of national environmental remediation, featuring the creation of Tianjin Eco-City for 350,000 people, slated for completion by 2020.

TIBET (*XIZANG*). Situated in the southwest region of the People's Republic of China (PRC), the Tibet Autonomous Region (TAR) is commonly known as the "roof of the world," with an average elevation of more than 4,000 meters (12,000 feet). The site of five major **mountain** ranges, including the Kunlun, Karakoram-Tanggula, Himalayas, Hengduan, and Gangtise-Nyenchen-Tanglha, with the **Qinghai–Tibetan Plateau** constituting a major portion of the TAR, Tibet is rich in **forest** cover, amounting to 25 million hectares (62 million acres), largest in the country, and an estimated 24,000 **glaciers**, estimated at 14 percent of the world's glacial mass.

Serving as major feeders into the region's major **rivers** and **lakes**, Tibet is home to the **Yarlung Tsangpo River** (the upper reaches of India's Brahmaputra) and such sacred and pristine **lakes** as Manasarovar, near the equally sacred Mt. Kailash in the southwest. Numerous salty and alkaline lakes dot the countryside with windswept alpine **grasslands** and deep valleys, where the region's few **cities** and towns, many built around major monasteries and temples, are located, notably the capital, Lhasa. Translated in Chinese as *Xizang*, meaning "western treasure house," Tibet is home to 5,700 varieties of **plant species** and a rich variety of **birdlife**. This includes the ruddy shell duck (*hongke ya*), also considered sacred; Tibetan-eared pheasant (*Xizang yeji*); snowbird (*xueniao*); black-neck crane (*heijinghe*), revered in Buddhist traditions; and brown-headed gull (*zongtou'ou*). There is also such **wildlife** as the prized **Tibetan antelope (*chiru*)**, along with wild yak (*ye maoniu*), wild ass (*ye lü*), and other species, some threatened, along with the wild and ferocious Tibetan mastiff dogs (*cang'ao gou*).

Designated a prime **mining** center for the PRC in the 12th Five-Year Economic Plan (2011–2015), Tibet has substantial deposits of 126 minerals, some of which are copper (*tong*), gold (*jin*), graphite (*shimo*), lithium (*li*), magnesite (*lingmeikuang*), silver (*yin*), sulfur (*liu*), titanium (*tai*), vanadium (*fan*), and uranium (*you*). With the construction of the Golmud (Qinghai)–Lhasa rail link enhancing export, mineral production in the region has ballooned, which, along with excessive **logging**, has brought about enormous degradation of the Tibetan landscape. Equally destructive has been the impact on the region's **water resources** from the dumping of 100 million tons of wastewater (2007) and 18.8 million tons of solid waste (2009) containing such toxins as cyanide (*qinghuawu*), mercury (*gong*), and heavy metals (*zhong jinshu*). Casually discarded from mines, many illegal, the region has suffered from increasingly frequent and devastating **landslides**, mud/rock slides, and glacial avalanches. With rising temperatures (0.4 degrees per decade) and increased precipitation, major **climate change** is occurring throughout Tibet, most notably accelerated permafrost and glacier melt, the latter amounting to a loss of 18 percent of the total ice cover since 1950. Environmental changes in TAR have global influence, especially in altering monsoon and weather patterns, ranging from heat waves in Europe to warmer winters in Canada.

Following the devastating 1998 **floods** on the **Yangzi River**, that were traced to deteriorating environmental conditions in Tibet, the Chinese government has pursued major amelioration efforts, most significantly the establishment of numerous **nature reserves and national parks**, which constitute 20 percent of TAR territory, for instance, the massive Qiang Tang National Nature Reserve, the largest in the country, and the Lalu Wetlands National Nature Reserve, located in Lhasa and a favorite **habitat** of TAR birdlife. Widespread bans on **logging** have been instituted, with the Himalayan mountain region in TAR declared a **biodiversity** hot spot (2005), along with protection for endemic vultures (*tuying*), tigers (*hu*), and wild water buffalo (*yesheng shuiniu*). Yet, emphasis on development continues, with plans for major **hydropower** projects to provide electricity, including 11 hydropower plants planned for the Yarlung Tsangpo River and eight on the Jinsha, the upper reaches of the Yangzi.

Ongoing in-migration of Han Chinese, primarily to urban centers, is generating growing municipal wastes, which have exceeded the local capacity for treatment and disposal. While limits on livestock grazing have been imposed as a method for restoring deteriorating grasslands, traditional economic livelihoods based on **nomadic pastoralism** have been seriously undermined.

Native Tibetan environmentalists, one of whom is the exiled and outspoken Dalai Lama, especially on the issue of climate change, are tagged by Chinese authorities as a threat to local "stability" (*wending*).

TIBETAN ANTELOPE (*CHIRU*). A medium-sized bovine native of the **Qinghai–Tibetan Plateau** in the People's Republic of China (PRC), the Tibetan antelope was driven to near-extinction, from an estimated 1 million in 1990, to less than 150,000 in 2018, with numbers now increasing due to their protected status domestically and internationally. Beginning in the 1980s, the *chiru* was hunted by poachers, many from outside **Tibet**, for the animal's warm and soft fur, referred to as "soft gold" (*ruan jin*) and used in the production of luxurious shawls known as *shahtoosh*, especially popular in India.

Local patrols, for example, the **Wild Yak Brigade**, in the 1990s, were formed to ward off hunters who violated fundamental Lamaist–Buddhist prohibitions against the killing of any animal life. The Chinese government created the Qiang Tang Nature Reserve in northern Tibet, which, at 334,000 km^2, is the second-largest and highest reserve in the world, where the majority of *chiru* now roam with class I protected status and are listed as a near-threatened species by the United Nations. International trade in the fur, including for the luxurious shahtoosh, has been declared illegal. Assisted by such international environmental groups as the World Conservation Society and the World Wildlife Fund (WWF), threats from poachers and illegal traders in the animal's highly prized fur have been substantially constrained, although other hazards to the species exist from extensive rangeland fencing for livestock and gold **mining**, along with disruptions in **habitat** brought about by the construction of the Golmud–Lhasa railroad. In herds numbering in the hundreds, Tibetan antelope inhabit alpine and cold steppe in the semi-arid Qinghai–Tibetan Plateau at altitudes between 3,250 to 5,500 meters (10,000 to 18,000 feet), with limited numbers also found in southern **Xinjiang** and western Qinghai provinces. The largely successful effort by local villagers at halting destructive poaching is dramatized in the film *Mountain Patrol: Kekexili*, produced by the National Geographic Society.

TIMBER (*MUCAI GONGYE*). A major **industry** that has undergone rapid growth, especially since the introduction of economic reforms (1978–1979), timber in the People's Republic of China (PRC) is currently under strict state supervision, with **logging** banned in natural growth **forests**, leading to heavy reliance on imports of raw wood, largely from other developing countries in Southeast Asia and Africa. While timber and wood production were carried out exclusively by state-owned enterprises (SOEs) prior to 1980, market liberalization has led to the creation of numerous private firms, largely small-

and medium-sized, as the PRC has emerged as one of the largest timber markets in the world, with demand from construction, wood-paneling, furniture, plywood, and **paper** pulp producers.

Relying heavily on fast-growing plantation forests, many collectively owned, or household forest farms and consisting of coniferous (*zhenye*) and cone-bearing (*zhuixing*) trees, including larch (*luoyesong*), Korean pine (*hong song*), and Chinese fir (*shanmu*), and deciduous trees (*luoye*), including oak (*xiangshu*), birch (*huamu*), and especially poplar hybrids (*yangshu zaijiao zhong*) in the northeast, southwest, and northwest, China consumes more than 400 million m^3 of timber annually for both the domestic and export markets. With natural forests in the PRC spared the excessive clear-cut logging that occurred prior to the enactment of the National Forest Protection Plan (1998), the major environmental impact has shifted to forests in the Amazon and Congo river basins, and to Siberia, in countries like Brazil, Gabon, Guinea, Indonesia, Papua New Guinea, and **Russia**. While Chinese construction firms assist in the opening up of remote forests to commercial logging, 50 percent of internationally traded wood is headed for the PRC, much of it (16 to 24 million m^3 annually) from illegal operations. In a market dominated by small- and medium-sized timber firms, many illegal in the PRC, pollution and waste increased, leading to a government crackdown on offending enterprises in July 2016.

TRADITIONAL CHINESE MEDICINE (TCM/*ZHONGYI*). Built on a foundation of 2,500 years of medical practice and generally supported by the government of the People's Republic of China (PRC), traditional Chinese medicine (TCM) uses a rich variety of animal parts and **plant species** with deleterious effects on the environment within the PRC and abroad. Among the most prized animal parts are tiger bone (*hugu*), rhinoceros and antelope horn (*xiniu, lingyang jiao*), deer antler (*lu lujiao*), jaguar fangs (*meizhou hu*), bear penis (*xiong yinjing*), Totoaba fish bladder (*yu pangguang*), and pangolin scales (*chuanshajia*), with devastating effects on these animal and marine populations both in China and abroad, particularly in Southeast Asia, Latin America, and Africa. Medicinal plant species used in various forms of TCM number more than 11,000, with 1,800 to 2,100 facing extinction as a result of excessive exploitation by untrained and scientifically illiterate collectors, combined with a general lack of government regulation and scientific planning.

Growing demand for relatively expensive TCM products from an increasingly well-off Chinese middle class, coupled with an effort to increase TCM exports, has led to an expansion of total area devoted to herbal cultivation from 400,000 hectares (988,000 acres) in the 1950s to 9.3 million hectares (23 million acres) in 2017. Most plants are still obtained through "wild collection," which when overdone can have devastating environmental effects, as was the case with wild licorice (*ye gancao*) in Inner Mongolia, where excessive digging led to the near-extinction of the plant, along with **desertification**. With much of artificial cultivation posing significant problems, Chinese growers have shifted to "natural fostering" that integrates the production of herbal medicinal plants with **biodiversity** conservation, avoiding such destructive impacts as **deforestation** and the overuse of **fertilizers** and **pesticides**.

Among some of the most widely used herbal products in TCM (and in Indian traditional medicine) are Himalayan yew, snakeroot, red sander, elephant's foot, and desert cistanche, along with the anti-cancer taxus chinensis and the wild lamiophlomis. Accused of being "fraught with pseudoscience," particularly the belief in enhancing male sexual potency through the consumption of bear or tiger body parts because of the animals' inherent ferocity, China has banned the sale and import of **endangered wildlife**, including tigers and pangolins, both made virtually extinct in the wild in the PRC, but with body parts of internationally threatened species like jaguars from Latin America imported to fill the void. Among the marine species threatened by TCM is the vaquita porpoise, endemic to the Sea of Cortez off the coast of Mexico, whose numbers have dropped to fewer than 20 in the wild, a result of being swept up by gill nets deployed to capture the endangered Totoaba fish, nicknamed "aquatic cocaine" for its prized bladder fetching high prices in Asian markets and believed to hold aphrodisiac and curative properties. It is often used in fish maw.

TRANSPORTATION (*YUNSHU*). Consisting of every conventional mode, the most common being **motor vehicles**, intercity and intracity rail and subway systems, and aviation and shipping on inland waterways, transportation in the People's Republic of China (PRC) has undergone substantial growth, especially since the introduction of economic reforms (1978–1979), with substantial environmental impacts. The greatest increases in transport volume in recent years have come from the growth of privately owned automobiles from both foreign and domestic manufacturers, along with buses, trucks, and other heavy vehicles, which have created increasing **air pollution**, especially in China's **cities**, where automobile ownership and heavy trucks involved in the delivery of commercial goods and construction projects are concentrated.

Lacking a significant regulatory framework on motor vehicle emissions in the 1980s and early 1990s along with generally poor maintenance by Chinese drivers, ambient **air quality** in major cities suffered accordingly, with dramatic increases in carbon dioxide (CO_2), carbon monoxide (CO), nitrogen dioxide (NO_x), and **ozone**, especially near such highly congested traffic areas as the second ring road in **Beijing**. With road and highway construction unable to keep up with the growth in motor vehicles, including cars, trucks, and motorcycles, from 34 million in 1998, to 300 million in 2017, including 200 million automobiles, with six cities (Beijing, Chengdu, **Chongqing**, **Shanghai**, Shenzhen, and Suzhou) having more than 3 million registered automobiles each, 10 of the world's 25 most vehicle-congested cities are now located in the PRC. While the burning of **coal** and industrial air pollutants affecting ambient air quality in most Chinese cities have stabilized, increases in urban air pollution derive primarily from the growth in motor vehicles, which are responsible for 86 percent of CO_2 emissions, with a growth of 160 percent between 1994 and 2007.

Major government efforts at reducing the environmental impact of motor vehicle transportation began in the 1990s with improvements in **fuel and emission standards**, including phasing out leaded gasoline (1992), promoting ethanol (2001), and introducing a gasoline and diesel consumption tax (2009). Emission standards have also been tightened on light-duty vehicles (1999) and new automobiles (2000–2016), while older vehicles were subject to mandated retrofitting (1998). Moreover, inspection and maintenance programs have been upgraded by such measures as enhanced training of personnel.

The introduction of pollution-free **electric vehicles (EVs)**, including automobiles, buses, and scooters/e-bikes, for urban transportation has also been pursued, with pilot programs involving the latter introduced in cities like Shenzhen (2009), along with the construction or expansion of intercity and intracity rail services. The application of Intelligent Transportation Systems (ITS), in which innovative information and communication technologies are applied to road transport, is also being pursued, while economic incentives like lower transit fares and the highly restrictive issuance of license plates in major cities are designed to increase reliance on less-polluting public transport. The ecological and environmental impact of other modes of transportation, notably railways, is increasingly important, as in the case of the Golmud–Lhasa railway, where dramatic and expensive measures were taken to minimize the effect of the powerful trains on local **wildlife** and permafrost (*yongjiu dongtu*).

U

UNITED NATIONS ENVIRONMENT PROGRAMME (UNEP). An agency of the United Nations with headquarters in Nairobi Kenya, UNEP is charged with coordinating international policies on the global environment promoted by the United Nations while assisting such developing nations as the People's Republic of China (PRC) in formulating and implementing sound environmental policies, for example, the challenge of **climate change**. Founded in 1972, UNEP was an outgrowth of the United Nations Conference on the Human Environment in Stockholm, Sweden, attended by a PRC delegation authorized by Chinese premier **Zhou Enlai** and headed by Chinese environmentalist **Qu Geping**, who would subsequently work for UNEP.

Joint efforts by UNEP and the PRC include assisting developing countries, especially in Africa, in confronting environmental problems through such organs as the International Eco-System Management Partnership, with offices in **Beijing**, drawing on Chinese experiences in dealing with **biodiversity**, pollution, and **chemical**-related problems, with monies provided by the China Trust Fund. In the wake of the withdrawal by the United States from the Paris Climate Accords in 2017, UNEP urged China to take a leading international role in confronting climate change and the PRC was chosen to co-chair a UN Conference on Climate Change slated for September 2019. Major joint studies between UNEP and Chinese institutions include *China Green Economy Initiative* and, with the PRC's **Ministry of Environmental Protection (MEP)**, *Green Long March*, with a focus on photovoltaics (PV/*guangfu*) used in **solar** energy, along with **wind power** and **bioenergy**. UNEP has also contributed to the Task Force under the China Council for the International Conference on Environment and Development, while also assisting in the publication of *Modelling China's Green Economy, 2010–2050* and the *Green Energy Textbook*, prepared for policy-makers. Chinese assistance to developing countries on environmental issues was also established in 2017, by Li Ganjie, China's minister of the environment, and UNEP head Erik Solheim. UNESCO has established 53 World Heritage Sites under UN auspices in China (second most in the world), of which 13 are natural, including prominent **mountains** Mt. Tai (Shandong), Mt. Huang (Anhui), Mt.

Wutai (Shanxi), and Mt Wuyi (Fujian/Jiangxi) and other natural wonders, such as Jiuzhai Valley (**Sichuan**), Three Parallel Rivers (**Yunnan**), and Dujiangyan Irrigation System (Sichuan).

URÜMQI. The Capital city of the **Xinjiang**-Uighur Autonomous Region, with a population of 4 million in 2015, Urümqi faces severe environmental problems, including significant **air pollution** and growing **water** scarcity, as the number of city residents is expected to reach 5 million by 2020. With an **economy** heavily reliant on heavily polluting **energy** production, notably **oil**, along with petrochemicals (*shihua*) and cement (*shuini*), the city was rated as having the worst **air quality** in the country in 2016, just ahead of Baoding and Xingtai in the industrial heartland of **Hebei province**. In Hebei, air quality has improved in recent years, but Urümqi has suffered a net decline.

Pollutants from industrial emissions and the burning of **coal**, especially during winter months, along with sulfate-rich dust from agricultural fields to the south, combine to produce below-standard levels of air quality in Urümqi for half of the days from January to March. Substantial increases in the city's population, especially by migrants from the east, along with rapid economic development, have undermined overall environmental quality, exacerbated by growing water scarcity. The drying up of major portions of the nearby Urümqi River, brought about by lower feeder waters from the shrinking "Glacier Number One" in the Tianshan ("Heavenly") **mountains** to the north and a contraction of nearby Say'opa Lake, have contributed to the problem of inadequate water supply.

Once touted as a "green city" of trees and flowers, Urümqi is also affected by the inexorable process of **desertification** afflicting the entire region, although major countermeasures have been pursued since the 1960s by the Xinjiang Institute of Ecology and Geology, located in the city. Pollution mitigations of the city's electric power plants and the iron and **steel** industry have relied primarily on technological and engineering transformation, and structural emission measures, with antiquated cement and other heavy polluting factories moved to the region during the **Third Front** of industrial decentralization in the 1960s and 1970s now slated for shutdown. The site of many **wind power** farms, for example, Dabancheng, located outside the city, as well as **solar power** plants, Urümqi is central to the pursuit of **renewable energy** in China but with significant energy losses due to the flawed national grid transmitting power to other regions of the country, especially during the winter months. The construction of so-called "passive houses," with substantially less energy consumption, employing solar power during summer and natural gas in winter, has also been inaugurated as part of the city's effort to pursue more energy-efficient buildings.

V

VERTICAL FARMING (*CHUIZE NONGYE*). Farming techniques employing controlled environment agriculture (CAG) technology, with fruits and vegetables grown indoors in vertically straddled layers or columns as much as 20 feet high in nutrient-rich water or hydroponics (*shuipei*) and under light-emitting diodes (LED), vertical farming is being vigorously pursued in the People's Republic of China (PRC) by government research institutes and private **companies**. With virtually no need for **soil**, **fertilizers**, **pesticides**, or even natural sunlight, vertical farming facilities can be sited in or near major **cities**, providing China's growing urban population, especially the well-off middle class, with clean and fresh food, especially leafy vegetables, avoiding the food safety problems that have afflicted the country in recent years.

Using planned and fast-growing production, with optimized space and less climate effects, vertical farms are planned for set-aside districts in cities like **Shanghai** (Sunqiao district), with a demonstration plant also planned for **Beijing**. Supported by foreign investors, one of whom is Amazon's Jeff Bezos, whose company Plenty is helping to finance the establishment of 300 vertical farms in the PRC, vertical farming employs artificial intelligence (*rengong zhineng*) to achieve dramatic increases in productivity as much as 10 times greater than conventional farms, with substantially less labor. Problems include relatively high capital and **energy** costs, with the practice limited to only a few food plants, mainly fast-growing leafy greens.

W

"WAR ON NATURE" (*DUI ZIRAN DE ZHANZHENG*). Invoked as the militaristic clarion call for carrying out the great transformation of nature advocated by **Chinese Communist Party (CCP)** chairman **Mao Zedong** during the late 1950s and 1960s, the "war on nature" led to major environmental degradation in the People's Republic of China (PRC), severely impacting **agriculture**, **forests**, **rivers**, **lakes**, and **wetlands**. Diametrically opposed to the traditional Chinese concept of "heaven and humanity in harmony" (*tian ren heyi*) and borrowed heavily from the rhetoric promulgated by Josef Stalin (1927–1953) in the great transformation of nature proclaimed in the **Soviet Union** (1948), the Chinese "war" posed a separation between humans and nature, while calling for bending the physical world to fit human needs and seeing human will as a more powerful force for altering the natural order than objective scientific laws that were frequently violated. Proclaiming a utopian urgency to remake the environment, wartime rhetoric and slogans were underlying themes in the radical policies advocated by Mao Zedong, especially during the **Great Leap Forward (1958–1960)** and **Cultural Revolution (1966–1976)**, as well as the construction of the **Third Front**, when massive **deforestation** and widespread destruction of wetlands and **birdlife** were carried out with a blind ideological fervor, often by young people, "sent down" (*xia xiang*) to remote regions of the country, ostensibly to boost production.

Acting on such slogans as "get grain from mountaintops and lakes" (*cong shanding he hubo huoqu liangshi*), extensive land **reclamation** campaigns were carried out, along with cultivation of marginal lands, especially in mountainous areas and newly reclaimed river and lake beds generally unsuitable for grain production. Major casualties in this "war" included rainforests in **Xishuangbanna** in **Yunnan province**, the Haigeng wetlands north of **Dian Lake** near Kunming (Yunnan), and countless forests decimated in the notorious campaign to supply wood as fuel for the ill-conceived **steel**-producing **backyard furnaces** promoted in the Great Leap. In the context of the broader mass campaigns of "class struggle" (*jieji douzheng*) and political repression pursued throughout the Maoist era (1949–1976), the "war on nature" left a legacy of environmental damage, including widespread **air pollution** and **water pollution**, which are now a major focus of amelioration and restoration, this time in a new "war on pollution" (*dui wuran de zhanzheng*), proclaimed by Premier **Li Keqiang** in March 2014.

"WASTE CULTURE". Evident in both urban and rural areas in the People's Republic of China (PRC), waste culture, more popularly known as the "throwaway society" (*rengdiao shehui*), has become increasingly prevalent, with concomitant environmental damage to the country's **rivers** and **lakes**, along with the appearance of trash dumps in such major **cities** as **Beijing** and **Tianjin**, as well as in many municipalities and even small, remote villages. Composed of food discarded by households, with university students admitting to filling their cafeteria plates with more than they can eat because it "looks good," popular throwaway items include the ubiquitous **plastic** and Styrofoam food containers, with 60 million thrown out daily, known as "white pollution" (*baise wuran*), along with disposable **chopsticks** casually tossed onto streets, along roadways and rail lines, and into rivers and waterways, even in the most remote regions of the country, for example, the **Qinghai–Tibetan Plateau**.

Among the factors leading to such unsavory personal habits are the rapid growth of a consumer **economy** dominated by cheap products, often in elaborate plastic packaging, along with convenient food delivery services, with great appeal to fast-paced young workers. A country once known, particularly during the 1950s and 1960s, for thrift, symbolized by the wearing of secondhand clothes and reusable cloth diapers, has been turned into a society of throwaway paper towels and Pampers, with an estimated 26 million tons of clothes discarded annually, as **recycling** by charity organizations devoted to redistributing used items remains underdeveloped. National tidiness, previously enforced by ubiquitous neighborhood committees and transport wardens, has given way to a free-for-all market- and convenience-driven culture where even Tibetan monks, known for their respect for nature, have been observed casually tossing food packages and other detritus out of bus windows.

Measures for counteracting such evidently deeply engrained habits include prohibitions on the distribution of food containers and disposable chopsticks by street vendors and takeout food outlets by local authorities but without national guidance or coordination by the central government. Private programs have emerged, for instance, the Clean Plate Initiative, advocating for zero-food waste, and "Bring Your Own Chopsticks" (BYOC), aimed at reducing the use of disposable chopsticks, at 80 billion annually, with great costs to the county's **forests**, along with the installation of recycling machines for bottles and cans, with school programs focused on cultivating responsible habits of personal waste disposal by the younger generation.

WASTELANDS (*HUANGDI*). In a country where 58 percent of the land mass is composed of **mountains** and 28 percent of deserts, the People's Republic of China (PRC) is also afflicted by vast swaths of wasteland, especially from abandoned industrial and **mining** sites, along with unusable agricultural lands with significant **soil erosion**, solid waste, and **water pollution**. Villages located close to large **cities** have become primary wasteland sites after years of serving as dumping grounds for accumulated urban **garbage** and trash, especially **plastics** and **e-waste**. Official categories of wasteland range are "highly polluted," "not highly polluted," "not suitable for arable land," and "new planned cultivated land."

Returning polluted wasteland with excessively salty soils, cracked and subsided lands, and widespread deterioration to previous viable conditions is both technically difficult and relatively expensive. Among the major restoration efforts involving consolidation and **reclamation** is the conversion of barren lands to **forests**, as in the case of the Saihanba National Forest Park in **Hebei province**, a man-made woodland converted from wasteland and desert. Other important measures include the creation of so-called **ecocities**, of which more than 100 are planned for the PRC, for example, the ecocity project jointly sponsored by the PRC and the government of Singapore to be built on wasteland consisting of saltpan, degraded beaches, and wastewater ponds 40 kilometers (25 miles) outside of **Tianjin** Municipality.

WASTE-TO-ENERGY (*FEIWU ZHUANHUA WEI NENGYUAN*). Considered a long-term solution to the growing problem of municipal solid waste (MSW) in the People's Republic of China (PRC), waste-to-energy (WTE) plants involve incineration at temperatures as high as 850 degrees Fahrenheit to eliminate toxins and hard-to-burn food residue contained in urban **garbage** and rubbish, while also producing electricity. Generating 175 million tons of MSW annually, the largest in the world, and growing 8 to 10 percent annually, waste in the PRC is projected to grow to 500 million tons by 2025.

With most landfills, numbering 650 nationwide, at near-capacity and only 11 **composting** plants in the entire country, China relies on 231 operating WTE plants to burn approximately 33 percent of MSW, with plans to reach 54 percent by 2020, by constructing 103 additional plants, one of which is slated to be the largest in the world outside the southern city of Shenzhen, where daily MSW has grown from 50 tons to 15,000 tons. Providing 18.7 billion kilowatts of electrical power in 2014, constituting 1.2 percent of **renewable energy** in the PRC, WTE plants confront increasing public opposition and **social protests** based on the fear of released pollutants, including carbon dioxide (CO_2) and methane (*jiawan*), with plants built in second- and third-tiered **cities** often skirting environmental regulations requiring the installation of expensive filtration technology to reduce construction costs.

Alternatives include systematic campaigns to reduce MSW, along with reusing and **recycling** trash, practices contrary to the **"waste culture"** so prevalent in Chinese society. Producing 40 million tons of sludge annually, China has also introduced pilot programs involving sludge-to-energy (STE) plants for converting **sewage** through anaerobic digestion into biogas, with five projects planned for **Beijing** to deal with the 30 sludge piles surrounding the city.

WATER-CONTROL PROJECTS (*SHUILI GONGCHENG*). A central concern of state authority and policy in China since the imperial era and extending into the People's Republic of China (PRC), "controlling the waters" (*zhi shui*) in a country with 27,000 **rivers** in 2017 (a reduction from 50,000 in 1950) has involved major water-control and transfer projects, including dams, reservoirs (*shuiku*), and other large-scale infrastructure. Major functions of such projects include **flood** control and prevention; the provision of **water resources**, especially to urban areas and for **irrigation**; shipping; and **hydropower** generation. During the era of rule by **Chinese Communist Party (CCP)** chairman **Mao Zedong** (1949–1976), water control and conservancy was a top government priority, especially during the **Great Leap Forward (1958–1960)**, when thousands of dams and reservoirs were constructed, many small in scale and hastily built, often of dubious quality and effectiveness.

Pursuing more scientifically based policies on water control, especially since the 1990s, the PRC is confronted with problems of growing water scarcity, both surface and groundwater resources, especially in the arid and semiarid north and northwestern regions, including such major **cities** as **Beijing** and **Tianjin**, along with increasing problems of **water pollution** from rapid economic growth, particularly since the inauguration of economic reforms (1978–1979). Shifting to a greater focus on water usage, water-usage efficiency, and water contamination, inaugurated in 2011, existing water-control projects have been rejuvenated, with degraded systems restored to deal with persistent problems of flooding, specifically along the **Huai** and **Yangzi** rivers, brought about by massive upstream **deforestation** and **climate change**.

Traditionally dominated by powerful government bureaucracies headed by the Ministry of Water Resources (MWR), water control and conservation **policy-making** now involves greater public participation, especially by farmers, through the newly established Farmers Drainage and Irrigation Association (FDIA), set up to better coordinate competing interests on decisions involving distribution of often-scarce waters, particularly for irrigation. Major water-control projects involving embankment and gravity dams include Zupingpu (**Min River/Sichuan**), Nierji (Nen River/Heilongjiang), Baise (You River/Guangxi), Linhuaigang (**Huai River**/Anhui), Zaoshi (Xiayu River/Hunan), and Mania (Nyangqu River/**Tibet**), designed to improve irrigation, generate power, and provide for better flood control and water supply. Controversies involving water-control projects persist, especially involving pristine and largely untapped water resources like the **Nu River** and in regions of political tension, for instance, **Xinjiang** and Tibet, along with the massive **South-to-North Water Diversion Project (SNWDP)**.

WATER POLLUTION (*SHUI WURAN*). Second only to **air pollution** in severity and public concern, water pollution afflicts major **water resources** throughout the People's Republic of China (PRC), notably surface waters, especially **rivers**, **lakes**, and crucial groundwater (*dixia shui*) supplies suffering contamination. Major causes include unfiltered industrial effluents and poorly treated **sewage** discharged into local waterways, estimated at 3.78 billion tons nationwide in 2015; **chemical** and **oil** spills; and along with agricultural runoff of **fertilizer** and **pesticide** residue. Dangerous amounts of toxins include arsenic (*shen*), cadmium (*ge*), chromates (*gesuanyan*), cyanide (*qing huawu*), lead (*qian*), mercury (*gong*), oil, phenol (*benfen*), and nitrosodimethylamine (*yaxiaoan*), with substantial proportions of water supplies to major **cities** (15 out of 27 in 1990) virtually undrinkable. This includes **Shanghai** and **Beijing**, where 57 percent and 40 percent, respectively, of available surface waters are rated grade V and V+ (unfit for any use, including **industry** and **agriculture**), a situation forcing many cities to draw on rapidly depleting and often equally contaminated groundwater supplies.

With environmental **activists** like **Ma Jun** warning of an impending water crisis in the early 2000s and the **Ministry of Environmental Protection (MEP)** issuing a comprehensive report on **water quality** in 2008, state action has included a commitment of more than RMB 700 billion ($106 billion) for dealing with water pollution (2015), followed by the Water Pollution Prevention and Control Action Plan, also known as the Water Ten Plan (2015), with a stated goal of reducing overall water pollution levels from 30 to 50 percent. Declaring "three red lines," or target years, for 2015, 2020, and 2030, major goals include achieving a grade III water quality standard for 70 percent of the nationwide watersheds and 93 percent of drinking water sources. Provincial governments throughout the country have also been ordered by MEP to meet improved water quality targets set for every five years, as half of the provinces failed to meet their goals for 2011–2015, with some provinces (Shanxi, **Sichuan**, and Inner Mongolia) suffering net declines. Confronting both increasing water demand and water scarcity, the country relies on more than 12,200 monitoring sites located throughout the nation, which currently rates 35 percent of the nation's water supplies as good quality, 32 percent as suitable for drinking, 20 percent solely for industrial and agricultural use, and 13 percent virtually useless.

With a commitment by government authorities to release comprehensive water pollution data to the public by 2020, greater public participation in reporting major cases of pollution is also being pursued through hotlines and websites of environmental enforcement agencies, although continued cross-bureaucratic and territorial conflicts often prevent vigorous enforcement, especially by local authorities. While real reductions in heavy metal contamination have been achieved in recent years, pollution from organic sources continues to increase, including phosphates (*linsuanyan*), chemical oxygen demand (COD/*huaxue xu yangliang*), ammoniacal nitrogen (*an dan*), and dissolved oxygen (*rongjie yang*).

See also SOUTH-TO-NORTH WATER DIVERSION PROJECT (SNWDP/*NANSHUI BEIDAO*).

WATER QUALITY STANDARDS (*SHUIZHI BIAOZHUN*). The scientific evaluation of water quality in the People's Republic of China (PRC) relies on two separate water quality indexes (WQIs) for surface and groundwaters. Surface waters (*dibiao shui*) are evaluated on the basis of several indexes, including dissolved oxygen (DO), chlorophyll-a (CHL), chemical oxygen demand Mn (COD Mn), ammoniacal nitrogen (NH^4t-n), total nitrogen (TN), and total phosphate (TP), with rankings from grade I to grade V+. Groundwater (*dixia shui*) is rated on the basis of 26 parameters, including, among others, carbonates, total dissolved solids (TDS), chemical oxygen demand (COD), nitrates, total acidity, arsenic, ammonia, total alkalinity, ammonia nitrogen, and aluminum, with rankings from 1 to 5 (excellent, good, medium, poor, and extremely poor). Studies of major **rivers** (**Yellow**, **Yangzi**, and Liao), **lakes** (**Tai** and **Poyang**), reservoirs (**Three Gorges Dam**), and aquifers employing the two WQI have been carried out in the PRC, along with studies of surface waters employing the Single-Factor Pollution Index (SFPI).

WATER RESOURCES (*SHUI ZIYUAN*). A country with 20 percent of the world's population but only 7 percent of global freshwater resources, the People's Republic of China (PRC) is confronted with enormous environmental challenges in maintaining adequate supplies of water for **agriculture**, **industry**, and human consumption in both urban and rural areas. Sixth largest in the world in total volume, water resources per capita for a population of 1.38 billion (2017) is 2,100 m^3 annually, one-fourth of the global average, with annual water shortages exceeding 50 billion m^3 per year.

Surface waters (*dibian shui*) in the country's more than 27,000 **rivers** and 2,800 large **lakes** constitute 82 percent of total supply, with a highly uneven geographical distribution between the water-rich south (80 percent of the total) and the arid and semiarid north and northwest, where low precipitation

(151 millimeters annually) has created greater reliance on rapidly depleting groundwater (*dixia shui*) supplies. Storage capacity of surface waters increased 14 percent from 1949 to 1989, with greatest demand (84 percent) from rapidly expanding **irrigation** and such industries as **mining** and **steel** production. Widespread inefficiencies and waste in agriculture and industry alike have created difficulties, for example, in rice production in such water deficit areas as the North China Plain.

While 63 percent of the country's territory lies north of the **Yangzi River**, the region's share of national water resources is only 19 percent, with 11 provinces and municipalities suffering from exceedingly dry conditions, including such economically vital places as **Beijing**, **Hebei**, **Henan**, Jiangsu, Shandong, **Shanghai**, and **Tianjin**. Since the vast majority of rainfall occurs in the summer months (60 to 80 percent of the annual totals), spring **droughts** are common, with shortfalls made up by withdrawals from the country's 98,000 reservoirs (*shuiku*), as allocated by the Ministry of Water Resources (MWR). Major river basins in China, including the **Yellow**, Hai, **Songhua**, Liao, and **Huai**, receive less rainfall than parched regions in Africa, while 80 percent of the country's water resources are provided by the Yangzi River, which itself has also undergone shrinkage of surface waters by 13,000 km^2 (5,018 mi^2) from the 1950s to 1980s, with the length of all inland waterways in the country declining by 30 percent.

Substantial water resources also exist in the country's rich array of **glaciers** (48,000 in total and 36,000 on the **Qinghai–Tibetan Plateau**) but with increased melting rates having shrunk total glacier volume by 30 percent in the past 100 years, notably the vital "Glacier Number One" north of **Urümqi**, the capital city of **Xinjiang** province. Water diversion projects from the Yellow River and other sources to **cities** like Beijing, Tianjin, Qingdao, and Dalian were also carried out, with the planned **South-to-North Water Diversion Project (SNWDP)** involving even larger water transfers. Measures to increase the total volume of water resources include desalinization (*tuoyan*), with pilot programs begun in the water-short city of Xiangshan, Zhejiang province.

Groundwater sources provide 11 percent of national water resources—especially in north China, where 50 percent of total consumption is supplied by underground sources. This water is used for **coal** and agricultural production (70 and 13 percent, respectively) and to supply semi-arid cities like Beijing, which relies on 40,000 private wells for 60 percent of its water. Primary groundwater sources in the country include three large aquifers: North China Plain Aquifer System (encompassing Hebei, Henan, and Shandong and Beijing and Tianjin), Song-Liao Basin in the northeast, and Tarim Basin in Xinjiang, all of which are undergoing serious depletion with a total groundwater deficit in north China of 180 billion m^3. Aquifer-recharging projects employing rainwater, reclaimed water, and desalinized water that began in such select areas as Shanghai in the 1960s have intensified in the 2010s to combat increasing problems of ground subsidence and compaction brought about by excessive groundwater withdrawals. Inaugurated in 2014, 172 water-conservation and irrigation projects have been concentrated in the agricultural sector in the water-short central and western regions. Heavy water-consuming industries, such as fracking for shale gas, are developing new technologies that use less water, but have been warned that the country will be dangerously low in water resources by 2030. Major institutions involved in water resources work include Beijing Water Science and Technology Institute and the Peking Institute of Water Sciences.

WEN JIABAO (1942–). Premier of the People's Republic of China (PRC) during the presidency of **Hu Jintao** (2003–2013), Wen Jiabao was a strong advocate of balancing economic growth with environmental protection. Trained as a geologist and serving in the geology bureaucracy in far western Gansu province (1968–1978), Wen was appointed a vice premier in the Ministry of Geology and Mineral Resources (1998), where he oversaw policies on the environment, along with **agriculture** and finance. Advocating for greater government transparency on national policies concerning such matters as the environment, he made a personal visit to **Sichuan province** in the immediate aftermath of the devastating **earthquake** in May 2008, while he represented the PRC at the United Nations Conference on Climate Change in Copenhagen, Denmark, in 2009, where a last-minute, nonbinding agreement on pollution emission controls (*wuran paifang kongzhi*) was reached with the United States. Noting the failure of the PRC to achieve environmental protection targets laid out in the 11th Five-Year Economic Plan (2006–2011), Wen called for holding local leaders responsible for future environmental problems. In his last public address to the National People's Congress in March 2013, he warned that the country was being faced with unchecked environmental degradation and should engage in a "determined effort to solve pressing environmental problems."

WETLANDS (*ZHAOZE DI*). Defined as the interface between land and water with groundwater lying on or near the ground surface normally at a depth of no more than two meters (6.4 feet), wetlands in the People's Republic of China (PRC) constitute 5.58 percent of national territory, covering 65.9 million hectares (163 million acres), the fourth largest in the world and 10 percent of the global total. With the first wetland **nature reserve** established in the 1950s, the PRC has 1,557 wetland protection areas, 57 "wetlands of international importance" (Ramsar Convention sites), 602 wetland nature reserves, and 898 national wetland parks.

Home to 4,220 **plant species** and 2,312 animal species, wetlands serve as the "kidneys of the landscape," with such benefits as the following: contributing to ecological conservation by cleaning and purifying **water resources** and storing highly polluting "blue carbon" (*lantan*), carbon captured by the oceans; mitigating large **flood** events; recharging groundwater aquifers; providing **habitat** for **fisheries** and **birdlife**, including migratory waterfowl; supplying resources of **timber** and peat lands (*nitan di*) and overall **biodiversity** habitat; and protecting **coastal** areas from the destructive effects of typhoons and hurricanes. Reflecting the complex geological and varied climates of the PRC, from tropical and subtropical in the south to arid and semiarid in the northwest, natural wetlands in the PRC come in several types, including marine (*haiyang*), estuarine (*hekou*), riverine (*hebian*), lacustrine (**lakes**), and palustrine or swampy bogs (*zhaoze*), covering 10 million hectares, primarily in the northeast.

Major losses of wetlands, including 50 percent in coastal areas, began in the 1950s, with land **reclamation** for conversion to farmland and **aquaculture** sites, and such development projects as airports and industrial zones, along with the growing impact of **climate change**. Joining the Ramsar Convention, the international agreement on wetlands conservation, in 1992, China followed up with the establishment of the first national wetland reserve, the Sanjiang Marsh Nature Reserve, Fuyuan, Heilongjiang province. Destruction continued despite these efforts, with total loses of 2,883 km^2 (1,112 mi^2) between 1990 and 2010, largely to unrelenting urban expansion and intruding farmland, mostly in the country's eastern region, and exacerbated by the weak enforcement of prohibitions against wetland destruction. Loss rates accelerated in the 2000s, with 23 percent shrinkage between 2003 and 2013, primarily of reservoirs (*shuiku*), ponds (*chitang*), and marshlands (*zhaoze di*), according to the China National Land Cover Database and two national wetland surveys conducted in 2003 and 2014. By 2013, total wetland area in the PRC had decreased by 3.4 million hectares (8.3 million acres), at an annual loss rate of 1 percent, with concomitant pressures on fisheries and birdlife, especially in coastal areas of Zhejiang and Jiangsu provinces, as well as **Shanghai** and **Tianjin** municipalities, where only 24 percent of coastal wetlands are legally protected, with 16 million hectares (40 million acres) protected nationwide.

Chinese government measures aimed at halting wetland loss and restoring wetlands from farmland and other uses is an expensive and long-term process that began with a Conservation Action Plan (2000), followed by Regulations for the Conservation and Management of Wetlands (2013). Declaring "red lines" of protection for the existing 65.9 million hectares of wetlands in 2015, drastic curbs were also imposed on local authorities, forbidding development projects in coastal wetlands (January 2018), a policy laid out in the *Blueprint for Coastal Wetlands in China*. Financial commitments include

RMB 7 billion ($1.1 billion) for wetland protection and the restoration of farmland to wetlands of 51,000 hectares (125,000 acres) from 2011 to 2015, a rate insufficient to reach the goal of stabilizing wetlands areas at 53 million hectares (130 million acres) by 2020, as the country still lacks a comprehensive National Wetland Protection Law.

A member of the International Convention on Wetlands (1992), the PRC has joined with several international groups in addressing wetland protection and restoration, including the United Nations Development Programme (UNDP), the World Wildlife Fund (WWF), Paulson Institute (United States), and such multinationals as HSBC and Coca-Cola. Major domestic agencies involved in wetland protection include the Office of Wetland Conservation and Management, under the State Forestry Administration, along with such research organizations as the Institute of Wetland Research, **Beijing**. Among the most prominent wetland reserves are the following: Zoige (Ruoergai) Wetland (**Sichuan province**), with peat bogs (*nitan zhaoze*), lakes (*hu*), wet grass (*shi cao*), and carex swamp (*taicao zhaoze*), the largest in the country; Bayanbulak wetland (**Xinjiang**), with the Swan Lake Nature Reserve, the only swan reserve in China; Liao River Delta Wetland (Liaoning province), one of the largest reed coastal areas, with marshes and a post for migratory birds; Yellow River Delta Wetland (Shandong province), a rare estuary wetland with swag grasses (*zangwu cao*), ponds, and reservoirs; XiXi Wetland (Zhejiang province), with ponds, lakes, and swamps, located within the city of Hangzhou; Argun Wetlands (Inner Mongolia and Heilongjiang), one of the biggest in Asia and providing shelter and feeding grounds for migratory birds during the dry season; and Wulagai Wetland (Inner Mongolia), the largest in the province and currently threatened by the reduction of river feed waters by **hydropower** projects and diversions to the province's many **coal** mines. Named an International Wetland City, Haikou city, Hainan province, is site of Sanjiang Provincial Wetland Park, the largest **mangrove** wetland in China, and Yangshan Wetland Park, the only volcanic lava wetland in the country.

China has also deployed floating treatment wetlands (FTWs) in which small platforms on open bodies of water, such as lakes and ponds, allow aquatic emergent plants to grow in water that is typically too deep with roots spreading through the floating islands and down the water creating dense columns. Rated 200 times more efficient than conventional wetlands, FTWS are multifunctional, removing nutrients that feed algae blooms and tackling pollutants from urban and industrial wastes and agricultural runoff. Fish **habitat** and spawning grounds are also provided with crops including rice, samphire, and herbs grown on floating reed beds with plant roots absorbing excess nitrogen and phosphorous assist in denitrification to prevent offshore "dead zones."

WILD YAK BRIGADE (*YE MAONIU LǓ*). Founded in 1992, in Qinghai province, as a defense against marauding poachers in the Hoh Xil (Kekexili) region, the largest uninhabited area in the People's Republic of China (PRC), the Wild Yak Brigade consisted of approximately two dozen members and was instrumental in halting the massive killing of **Tibetan antelope (*chiru*)**, which roamed the area, measuring 45,000 km^2 (17,370 mi^2). Following the massive influx of prospectors into Kekexili searching for newly discovered gold deposits, the slaughter of Tibetan antelopes began first for food and then for their highly valuable fur, sold as shahtoosh, primarily in India. Along with capturing suspected poachers and their weapons, the brigade (labeled the "Guardian Angels of China") also confiscated huge amounts of antelope furs, denying their sale by middlemen and brokers to lucrative international markets.

Although supported by **Friends of Nature (FON)**, one of China's most notable environmental nongovernmental organizations (ENGOs) and its prominent leader, **Liang Congjie**, the brigade suffered from inadequate economic and logistical support by the Qinghai provincial government, even as they dealt with the harsh and often dangerous conditions in the high-altitude region when tracking down the illusive and dangerous poachers. Bestowed a major environmental award in 2000, the Wild Yak Brigade was disbanded one year later, with most members folded into the generally ineffective Kekexili District Protection Administration, while a few were charged with illicit selling of antelope pelts to meet their meager economic needs. The exploits of the brigade are featured in the film *Mountain Patrol: Kekexili*.

WILDERNESS AREAS (*HUANG YE*). Long considered a place of uncultivated land of little value populated by wild and dangerous animals, wilderness in the People's Republic of China (PRC) still carries a negative connotation, often referred to as "wasteland" (*huangdi*) untamed by human activity. While **national parks** in China, numbering 244, are an important component of the country's strategy of creating an "ecological civilization" (*shengtai wenming*), promoting **ecotourism** and recreation generally trumps scientific and conservation purposes, with perennial problems of tree vandalism, loss of **biodiversity**, and **air pollution** and **water pollution** stemming from the extensive construction of access roads, hotels, restaurants, stores, and other commercial ventures.

Depicted as frightful and treacherous, as evoked in the traditional notion of China's northeastern region as the "Great Northeast Wilderness" (*Bei Dahuang*), designated wilderness areas are considered isolated and desolate areas lacking in beauty and culture where people were sent for punishment, as occurred most recently during the **Cultural Revolution (1966–1976)**. With some environmental nongovernmental organizations (ENGOs) and filmmakers like Wild China Films promoting a more positive sense of wilderness as valued for its own sake, the PRC has been the selected as the site of the 11th World Wilderness Congress, slated for late 2019, even as the country still lacks any sites designated as "wilderness" under category Ib of the International Union for Conservation of Nature (IUCN).

WILDLIFE (*YESHENG DONGWU*). One of 12 megadiverse countries in the world, the People's Republic of China (PRC) has a rich variety of wildlife on land and in inland and **coastal** waters, the most **biodiverse** country outside the tropics. Of 7,516 vertebrates (2,340 terrestrial), 4,936 are fish, 1,269 **birds**, 562 mammals (*buru dongwu*), 403 reptiles (*paixing dongwu*), and 246 amphibians (*liangqi dongwu*), with many, for example, the giant **panda bear**, red panda (*xiao xiongmao*), **Yangzi River dolphin** (declared extinct), Yangzi River alligator (*eyu*), and **golden snub-nosed monkey**, endemic to China. Among all species, more than 840 are **endangered** or seriously threatened, including one-sixth of all mammals and two-thirds of amphibians, mainly due to human activity involving the loss of **habitat**, **desertification**, pollution, **hunting and poaching**, and commercial exploitation, including procurement from breeding farms for **traditional Chinese medicine** (**TCM**) and culinary delicacies, especially in South China.

Generally neglected throughout the period of rule by **Chinese Communist Party (CCP)** chairman **Mao Zedong** (1949–1976) and during the initial years of economic reform (1978–1993), wildlife species in China were first provided significant protection with the Wildlife Protection Law (1998), which provided for two classes (I and II) of endangered or threatened species, although problems of enforcement remain, especially by local authorities. The most revered species include the giant panda and red panda, **Tibetan antelope (*chiru*)**, golden snub-nosed monkey (*jinsehou*), white/pink dolphin (*fenhongse haitun*), golden pheasant (*jinji*), snow leopard (*xue bao*), South China and Siberian tiger (*hu*), and Bactrian camel (*shengfeng tuo*). Conditions and changes in the country's wildlife are documented by national surveys conducted every 10 years, while a major investigation into wildlife resources was conducted in 2004.

Among mammals, primates in China include gibbons (*changbiyuan*), macaques (*mihou*), leaf monkeys/langurs (*yehou*), snub-nosed monkeys, and loris (*huhouzhilei*), with all gibbons subject to class I or class II protection, while macaques are the most common monkey in the country. Ranging in habitat from Hainan Island to Shanxi in Northwest China, the number of gibbons in China has fallen substantially, from approximately 254,000 (1998) to 70,000 (2004), largely due to the inadequate enforcement of protection **laws and regulations**, as monkey brain remains a delicacy in southern Chinese cooking, now increasingly popular throughout the country, especially among the nouveau-riche, who often pride themselves on the consumption of endangered wildlife.

The most famous and revered of the gibbons is the golden snub-nosed monkey, also known as the **Sichuan** golden-haired monkey, ranging in the **forests** of Southwestern and South China at altitudes of 1,500 to 3,400 meters (4,900 to 11,100 feet) and numbering an estimated 15,000, most remaining in national **nature reserves** as the species is threatened by loss of habitat. More threatened is the gray snub-nosed monkey, of which only 700 remain in the southwestern province of Guizhou; the Hainan black-crested gibbon (*heiguan changbiyuan*), of which only about 20 are left in the Hainan Bawangling National Nature Reserve; and the white-headed langur monkey (*baitou yehou*) in Guangxi province, at present found only in Vietnam. Nearing extinction is the northern white-cheeked gibbon (*baijia changbiyuan*), which has been overhunted for its reputation as a good luck charm, while the **Yunnan** lar gibbon (*Yunnan changbiyuan*) has been declared extinct. The nocturnal slow loris (*yejian manxing louhou*) can also be found in Yunnan and is considered threatened by extensive wildlife trading and habitat loss due to development projects. Historically, gibbon sounds were said to have inspired songs written by ancient Chinese poets.

Big cats in the PRC include the snow leopard (*xue bao*), Turkestan lynx (*sheli*), and clouded leopard (*yun bao*), with tigers (*hu*) previously hunted for body parts used in TCM and commercial fur. There are an estimated 50 tigers remaining in China, and as such they are considered critically endangered, residing solely in nature reserves. Included are the South China and Siberian or Amur tiger, the latter the world's largest, with numbers dropping by 92 percent between 1975 and 2009, while the Caspian tiger in **Xinjiang** has been declared extinct. Snow leopards reside primarily in **Tibet** and Xinjiang with a few also found in Gansu tracked by cameras in the Yanchiwan Nature Reserve in habitat at 3,000 to 4,500 meters (9,800 to 14,800 feet). China is home to the largest number of snow leopards in the world. Numbering between 2,000 and 2,500, poaching of the animal, prized for its attractive spotted fur, remains a serious problem and is addressed by such international groups operating in China as the Snow Leopard Trust.

Gray wolves (*huilang*) in China include the Tibetan and Mongolian subspecies, with habitat on the **Qinghai–Tibetan Plateau**, as well as in Gansu, Heilongjiang, Inner Mongolia, and Xinjiang provinces. While numbering more than 12,000, the species is under assault, even as the Chinese government requires licenses for hunting, including foreigners, as a measure to stabilize the population of threatened wildlife. Critical portraits of previous Chinese government efforts at eliminating wolves from the Mongolian **grasslands** is the subject of the book and film *Wolf Totem*.

Along with the widely proliferate red fox (*honghu*), which is found throughout China, except for in the northwest, the country is also home to the equally numerous civet cat (*limao*), which considered a culinary delicacy in dishes like "dragon-tiger-phoenix soup," may have been responsible for the spread of the SARS virus crisis, largely in South China (2003). Sables (*zidiao*), with their enormously attractive fur, are found in limited areas of Northeast China, where the animal is considered one of the "three treasuries of Manchuria," along with ginseng (*can*) and deer antler (*lulu jiao*). The pangolin is a scaled anteater. As a favorite culinary delicacy and with its scales used in TCM, it has been driven to near-extinction in China, turning the animal into one of the most internationally trafficked mammals in the world.

One of only two countries with four species of bears (*xiong*), China is home to the Asiatic black bear and brown bear (*hei*, *zong xiong*), both of which are found in mountainous and forested regions of the country and subject to class II protection. The rare sun bear (*taiyang xiong*), one of the smallest bears in the species, is found largely in tropical regions and subject to class I protection, along with the iconic giant panda bear, endemic to China. A key ingredient in TCM, bile from bear gallbladder (*dannang*) is a major **industry** in China, with approximately 12,000 brown and black bears kept on bear farms, although restrictions on the practice were instituted by the Chinese government in 1994.

While rhinoceroses (*xiniu*) roamed China in ancient times, the species became extinct in 1920, although rhino horn and tiger bone constitute key ingredients in TCM, with imports into China, banned because of declining global species in 1993 and reaffirmed in 2018. Elephants (*daxiang*) were also found in the country in ancient times, roaming as far north as **Beijing**, but they are now restricted in limited numbers to the rainforest in **Xishuangbanna** (Yunnan) in the Mengyang Nature Reserve, with the PRC imposing a ban on ivory imports (2018). Bactrian camels (*shengfeng tuo*), with their distinctive two humps, are the only truly wild camel, roaming the rocky Gobi Desert, with a capacity to survive brutal conditions of extreme heat or cold, with numbers declining 90 percent, largely due to hunting.

Deer (*lu*) in China consist of three major types, the musk deer, water deer, and Père David deer, the last so named in English for the animal's discovery in China by a French missionary and rendered in Chinese as *milu*. Numbering about 2,000, most are kept in captivity, as the deer was a prime target of hunting, with their antlers used in TCM, while also suffering from the effects of land **reclamation**, with approximately 700 left in the wild. Musk deer (*she lu*) come in five different species, alpine, black, Himalayan, Siberian, and dwarf, the last classified as endangered, and are poached for their musk gland, used in perfumes and TCM, with many now raised on breeding farms. Water deer (*shui lu*) are so named for living in close proximity to rivers and swamps, primarily in the lower reaches of the **Yangzi River**. Like musk deer, they lack antlers and are small in size.

Other varieties—many of them threatened—include the Thorold's white-lipped deer (*baichun lu*); black muntjac (*heiji lu*), also known as the barking deer (*feilu*); and Sika or spotted deer (*bandian lu*), extinct in most areas of the country except for the south, where a few hundred live in the wild. The small-tufted deer (*xiaocu rong lu*), along with the much larger sambar deer (*shuilu lu*), are native to South China, also the habitat of the wild hog (*yezhu*) and Asian badger (*Yazhou huan*). Limited numbers of moose (*tuolu*) are found in Northeastern China, near the border with **Russia**, their major habitat site. Native to Tibet are wild yaks (*ye maoniu*), weighing as much as one ton, along with the Bharol, or blue, sheep (*lanyang*), and the Mongolian gazelle (*Menggu lingyang*), also found in Inner Mongolia, driven to near-extinction by overhunting during the Great Famine (1959–1961). Wild goats include the Argali mountain goat and the Sichuan takin (*lingniu*), a goat-antelope, along with the mainland goat and antelope-like serow (*sumenling*) and gorals (*banling*).

Among China's 403 major types of reptiles, the Yangzi River alligator (*Yangzi eyu*), endemic to eastern China, is listed as "critically endangered," with a 90 percent drop in their numbers from 1955 to 2010, resulting primarily from habitat loss to rice paddies. Considered a "menace" by Chinese farmers, alligators are a source of meat, constituting a major ingredient in TCM and a culinary delicacy. Once endemic throughout the Yangzi River, few alligators remain in the wild, estimated at 130, and can be found mainly in the Chinese Alligator Nature Reserve (Anhui), with more than 10,000 held in captivity on breeding farms. Five of the seven threatened turtles (*gui*) in the world are in China, including the soft-shell, hawksbill (*yuan*), and Zhou's box turtle (*he gui*). Used for both food and medicine, consumption of the species is believed to promote longevity. Turtle shells have been employed as a traditional source of divination and also made into jewelry.

Major aquatic life included the Yangzi River dolphin (*baijitun*), nicknamed the "Goddess of the Yangzi," with the last sighting recorded in 2002, with the animal declared extinct in 2006. Also inhabiting the Yangzi River and **Dongting** and **Poyang** lakes, along with coastal waters, is the finless porpoise (*jiangtun*), or "river swine," which is also threatened by the overuse of waterways by shipping and **water pollution**, although conservation and protected status for the species has led to a recent stabilization of numbers estimated at more than 1,000 with three finless porpoise nature reserves established in traffic-free zones of the Yangzi. Also endangered is the giant salamander (*rong*), the largest salamander and amphibian in the world, at 1.8 meters (5.9 feet), which as a culinary delicacy is being driven to extinction. The degradation of coastal waters is also threatening the white/pink dolphin (*fenhongse haitun*), along with the dugong, a medium-sized herbivorous marine mammal nicknamed the "miraculous fish," located off the west coast of Hainan Island and considered endangered from the widespread loss of **seagrass** to pollution.

While major efforts to preserve and restore wildlife in **wilderness areas** have been pursued since the late 1990s, commercial breeding farms numbering 3,000 and wildlife parks in the PRC also play an important role in maintaining threatened species. Begun in the 1950s, commercialized wildlife is a RMB 100 billion ($14 billion) annual **industry** in China, for which there are currently no plans for termination, even as cases of animal abuse and mistreatment at the facilities have been revealed. Animals declared extinct in the PRC are as follows: Cabot's tragopan or horne pheasant (*zhiji*); hog deer (*zhulu*); long-nosed antelope (*changbi lingyang*); Oriental ibis (*Dongfang yibisi*); Siberian crane (*Xiboliya he*); Swinhoe pheasant (*yeji*); white-headed leaf monkey (*baitou yehou*); Xinjiang tiger (*hu*); Yangzi River dolphin (*baiji*); and Yunnan lake newt (*rongyuan*).

The impact of China on international wildlife has varied from enormously deleterious effects during the 1990s and 2000s when illicit global trade in wildlife ranked just below drugs and weapons to a more protective regime in recent years under the guise of the Convention on International Trade in Endangered Species of Wild Fauna and Flora (CITES). African elephants were especially hard hit with an estimated 30,000 killed annually primarily in Tanzania and Kenya for the valuable tusks that fed heavy demand in China for traditional ivory carvings craved as a status symbol by the country's growing middle class and banned in 2017. Similar deleterious effects afflicted the African rhinoceros population from a dramatic rise in demand for the animal's horn in China (and Vietnam) considered as a potential, but specious, cure for serious diseases including cancer. Sea turtles, sharks, and the endangered Totoaba fish (located off the coast of Mexico) also suffered as China's nouveau riche increased consumption of exotic culinary fare. Bans have been imposed on imports of ivory (2018) and rhino horn (1993/2018) and celebrities have campaigned against consumption of as shark fin (2011). Because the complex smuggling networks operating in Hong Kong and Fujian province are backed by organized crime and often involve Chinese government and diplomatic officials, the actual impact of these efforts on the supply of costly animal products is indeterminant.

"WIND AND WATER" (*FENGSHUI*). The ancient art of geomancy using almanacs, charms, and complex compasses to minimize risk from the surrounding environment and prevent any interruption to the free flow of energy (*qi*) considered essential to all life, fengshui consciousness involves a sensitivity to environmental matters, especially involving land, **water**, and **forests**. Closely tied to the ancient concepts of yin (receptive, shaded, north side of hill, south side of **river** bank) and yang (active, sun, south side of hill, north side of river bank), fengshui while recognizing the capacity of humans to transform their surroundings, stresses the importance of honoring the environment and harmonizing the relationship between humanity and nature by recognizing the intricate connectedness between the two as expressed in the traditional notion of "heaven and humanity in harmony" (*tian ren heyi*). In the countryside, this has involved the creation of so-called "fengshui forests" as sacred ground, which has led rural residents to preserve forests from development and other purely commercial interests. Emphasis on a fengshui consciousness has been at the center of efforts in China to cultivate greater **environmental awareness** among the public, especially involving the dangers from **air pollution** and **water pollution**, and the general degradation of the natural world.

WIND POWER (*FENG LI*). A major component, along with **solar power** of **renewable energy**, wind power in the People's Republic of China (PRC) produced 149 gigawatts (GW) of electrical power in 2016, with a goal of reaching 250 GW by 2020 and 400 GW by 2030. Wind power capacity grew from 1,260 megawatts (MW) in 2005 to 31,000 MW (30.1 GW) in 2010, when China became the world's largest wind power producer. Benefiting from large land mass and a long coastline of 14,500 kilometers (9,000 miles), the highest capacity is located in the windswept north and western regions of the country, mainly Gansu, Inner Mongolia, **Tibet**, and **Xinjiang** provinces. With potential wind resources estimated at about 2,380 GW of exploitable capacity on land and 200 GW at sea, China is also a major manufacturer of wind turbines and parts, as the wind power **industry** has been promoted as a key to economic growth, with a mandate requiring 70 percent domestic content in wind turbines.

The Gansu Wind Farm Project, under construction in the far west, is one of six national wind power megaprojects approved by the Chinese government, with an expected capacity of 20,000 MW (20 GW) by 2020, making it one of the world's largest collective wind farms. Chinese developers also unveiled the world's first permanent maglev wind turbine at the Wind Power Asia Exhibition, held in **Beijing** in 2006, with the construction of a base for the turbine generators begun in 2007. Offshore wind farms include the 102 MW Donghai Bridge Wind Farm (the first in China) located off **Shanghai** (2010), the 150 MW Longyuan Rudong Intertidal Wind Farm (2012), and the 300 MW Huaneng Rudong Wind Farm (the largest in China) with 70 turbines (2017), both off Jiangsu province, and the 250 MW Pingtan Wind Farm with 42 turbines (slated for completion in 2020), Pinghai Bay, Fujian province. Altogether, offshore wind power capacity in China is 1.5 GW (1,500 MW) and includes the Mingyang Super Compact Drive (SCD) wind turbine, the largest two-bladed turbine in the world, with peak electrical power generating capacity of 6.5 MW.

The development of offshore wind **energy** has not come as fast as expected, as the country had only installed 389.6 MW of offshore wind capacity by the end of 2012, still far from the target goal of 5 GW by the end of

2015. In 2011, a moratorium on building new wind farms was declared, as almost half of the installed wind turbines were sitting idle, unconnected to the national power grid, because of ineffective transmission lines over long distances from wind farms in the west to major consumers in the east. Competitive bidding for future construction of wind farms was introduced in 2018, as energy waste was cut in half, although restrictions remain on further construction in the far west. While horizontal axis wind turbines (HAWTs) are the dominant form of large-scale and commercially viable wind farms, vertical axis wind turbines (VAWTs) with fewer components capable of catching wind in any direction are very popular serving as small wind turbines (SWTs) usable on roof tops and in dense urban areas. A major manufacturer of VAWTs and heavily involved in research on the new technology, China has between 400,000 to 450,000 units ranging between 10 to 20 kW capacity as two-thirds of the country's land mass is suitable for SWTs. Problems of VAWTs include low power generation, operation within a relatively narrow range of wind speeds, poor stability (shortening turbine life span), and substantial maintenance requirements.

WOLONG NATIONAL NATURE RESERVE. Located in the Qionglai **mountains** in **Sichuan province** and a primary home to the giant **panda bear**, numbering between 100 and 150 animals, 10 percent of the national total, the Wolong National Nature Reserve was established in 1963, with the assistance of the World Wildlife Fund (WWF), and covers 200,000 hectares (494,000 acres). The core of the Sichuan Giant Panda sanctuaries, numbering seven **nature reserves** and nine scenic parks, Wolong is a World Nature Heritage Site, with a warm and humid environment that is favorable to arrow and square bamboo, the staple diet of the giant panda. The site of the Shenshuping Panda Bear Center, built after the devastating 2008 Sichuan **earthquake**, Wolong is a major breeding ground, employing somewhat unorthodox methods for the notoriously difficult reproduction capacity of the panda.

Also known as a "biogene bank" because of the wide variety of other **wildlife** and rich **plant species**, the reserve includes 56 rare and **endangered** animal species. Examples include the red panda, also known as the lesser panda (*xiao xiongmao*); **golden snub-nosed monkey**; snow leopard (*xue bao*); and white-tipped deer (*baitou lu*). These are in addition to precious Himalayan yew (*hongdoushan*) and beech plants (*shanmao ju zhiwu*). Environmental threats to the reserve include gravel and sand **mining** in local streams and the impact of increased and often intrusive tourism, with the pandas viewed as a money-making economic resource, practices that led the WWF to drop its support to Wolong in the 1990s.

WORLD BANK. A member of the World Bank since 1980, and one of the largest borrowers and technical recipients from the bank, the People's Republic of China (PRC) has benefited from loans for environmental amelioration, along with major studies of the country's environmental conditions, including **air pollution**, **water pollution**, and land degradation. Providing technical assistance, financing, and knowledge for improving **air quality** and **water quality**, the bank also promotes **sustainable development**, while assisting in the improvement of environmental protection governance.

Major environmental protection programs sponsored by the World Bank with cooperation of Chinese institutions include the following: cleaning up the seriously polluted **Pearl River** Delta by financing the expansion of wastewater treatment facilities for the **cities** of Foshan and Jiangmen; providing grants from the Global Environmental Facility for increasing water productivity and reducing pollution discharges from inland **rivers** like the Liao, Hai, and **Yellow** into the **Bohai Sea** off the Northeast China coast, where freshwater inflows have dropped dramatically; ameliorating land degradation in the upper watersheds of the **Yangzi** and Pearl rivers; improving water quality for small towns like Zhuji city, the world's largest manufacturer of socks, along the Qiantang River, Zhejiang province; and promoting environmental services to improve **health** in Qinghai province, financed by the International Bank for Reconstruction and Development (IBRD), the leading World Bank institution. Major publications dealing with the Chinese environment are as follows: *The World Bank and China's Environment: 1993–2003* (2005); *Shanghai Environmental Project: Sustainable Development for Shanghai* (2010); *China 2020: Clean Water, Blue Skies, China's Environment in the New Century* (1997); and *The Cost of Pollution in China: Economic Estimates of Physical Damage* (2007), the last in conjunction with the State Environmental Protection Administration (SEPA) of the PRC.

WU LIHONG (1968–). An environmental **activist** who, beginning in the 1990s, documented pollution in **Tai Lake** during the course of 17 years, Wu Lihong was arrested on charges of attempted bribery and incarcerated for three years (2007–2010). Formerly a factory salesman, Wu investigated collaboration between local government officials with business leaders of the hundreds of enterprises, many in the **chemical** sector, engaged in the substantial discharge of industrial effluents into Tai Lake, threatening water supplies to 2 million people. Initially lionized as an "environmental warrior" (*huanjing zhanshi*) by the National People's Congress (NPC) in 2005, for his antipollution revelations, Wu and his family were put under enormous pressure by local authorities, including the loss of his job, even as conditions in the lake deteriorated, with the appearance of highly polluting cyanobacterium algae blooms ("pond scum"). Upon his release from prison, he renewed his antipollution efforts by submitting formal complaints, without success, up the administrative hierarchy of the PRC, from the Jiangsu provincial **environmental protection bureau (EPB)** to the NPC gathering in March 2014, where Premier **Li Keqiang** announced, with great fanfare, a "war on pollution" (*dui wuran de zhanzheng*), which included cleaning up Tai Lake.

X

XI JINPING (1953–). Emerging as the top leader of the People's Republic of China (PRC) in 2012 and 2013, President Xi Jinping is a strong advocate of environmental protection and conservation, promoting the concepts of an "ecological civilization" (*shengtai wenming*) and a "beautiful China" (*meili Zhongguo*), while positioning the PRC as a major global leader on the issue of **climate change**. Trained as a chemical engineer with considerable experiences in the Chinese countryside as a youth "sent down" (*xia xiang*) to Shaanxi province during the **Cultural Revolution (1966–1976)**, Xi has made the environment a top priority of his administration, replacing the longtime notion that rapid economic growth and environmental protection are diametrically opposed with an emphasis on **sustainable development** and the pursuit of **renewable energy** as centerpieces of national economic strategy.

Calling on the Chinese government and people to respect nature and invoking the traditional notion of "heaven and humanity in harmony" (*tian ren heyi*), Xi, in a series of major speeches on the environment in 2017–2018, outlined four areas for national action: 1) promoting "green" development, especially clean technologies and **recycling**, reducing **energy** and water consumption; 2) solving major environmental problems, including **air pollution** and **water pollution** from industrial and **agricultural** runoff, with improved **river** basin and offshore area management, with greater government transparency and enhanced public participation; 3) protecting **ecosystems** with stricter boundaries imposed between the natural environment and farmland and urban development, along with measures to halt and roll back **desertification** and restore **wetlands** and natural **forests**; and 4) reforming environmental **laws and regulations**, while creating new and more powerful government agencies, for instance, the Ministry of Ecology and Environment (MEE/2018). Putting "ecological civilization" at the center of his plans for continuing Chinese reforms, he has also stressed the importance of fighting the proliferate **"waste culture"** of irrational and excessive consumption, linking the creation of a “beautiful China” to the country’s national pride, while targeting 2035 as the year the PRC will become a true “ecological civilization.”

XIE ZHENHUA (1949–). A leading figure on domestic and international environmental policy in the People’s Republic of China (PRC), Xie Zhenhua has held several important positions, notably director of the State Environmental Protection Administration (SEPA/1998–2005), from which he was forced to resign following the highly destructive **chemical** spill into the **Songhua River** in November 2005. Trained in engineering physics at Tsinghua University with a master’s in environmental law from Wuhan University, Xie helped establish the carbon-emissions trading system in the PRC, while also overseeing issues involving **nuclear power** safety and the establishment of uniform environmental standards for agricultural produce. The lead PRC negotiator at the United Nations Conference on Climate Change in Copenhagen, Denmark, Cancun, Mexico, and Durban, South Africa, in 2009, 2010, and 2011, respectively, he also helped negotiate the Paris Climate Accord (2015) and was an outspoken critic of the withdrawal from the agreement by the United States (2017). The recipient of several awards for his work on the environment, most notably the Lui Che Woo Prize (2017) in Hong Kong, Xie was named a cochair of the Global Climate Action Summit in July 2018, devoted to the worldwide prevention of **climate change** and the enactment of the Paris agreement.

XINJIANG. Located in the northwestern region of the People’s Republic of China (PRC) and officially known as the Xinjiang-Uighur Autonomous Region, *Xinjiang* (“new territory”), at 1.6 million km^2 (640,000 mi^2), is the largest administrative division in the PRC, constituting one-sixth of the nation’s territory. Home to the Taklamakan Desert, the largest and hottest place in the country, located in the Tarim Basin in the south, along with large-scale **glaciers**, notably the huge “Glacier Number One,” a major water source for the region’s capital city of **Urümqi**, Xinjiang is emblematic of the conflict between uncontrolled economic growth and environmental protection and degradation.

A major center of **water**-guzzling fossil fuel and cotton production, coined the “black-and-white **economy**,” Xinjiang suffers from persistent water shortages, with 96 percent of surface and underground sources consumed by **irrigation** for **agriculture**. Major **rivers** in the territory, for example, the Hotan, Ili, Tarim, and Yarkand, are endorheic, disappearing into deserts or terminating in salt lakes, with only the Irtysh River ultimately emptying into the sea, while many waterways have been diverted for irrigation, with little consideration of environmental consequences. Equally destructive especially to the region's **grasslands** have been the large herds of livestock, numbering 10 million, with environmental remediation pursued by the Gansu/Xinjiang Pastoral Development Project. Major **mountain** ranges include Kunlun in the south; Altai in the northeast; Tianshan ("Heavenly") in the center, separating the Junggar and Tarim basins; and Pamir and Karakoram in the southwest, with landscapes consisting of thick pine **forests**, alpine meadows, and dry grasslands. Several grand canyons also dot the landscape including the 5.5 kilometer (3.4 mile)-deep Tianshan Kuqa in the mid-west and the 20-kilometer (12 mile)-long and 400-meter (1,321 feet)-deep Anjihai (or Red Mountain) in the northwest. Other notable geological formations include Shipton's stone "arch," known in China as "Heaven's Gate" (*Tiananmen gong*), the highest natural arch in the world at 460 meters (1,500 feet) in the far west. An active seismic zone, Xinjiang is also home to many yardang land forms, streamlined protuberances carved from bedrock by wind and dust, with locations at Bailongtan and Sanlong. Among the 731 types of vertebrate **wildlife** found in the region, many threatened or nearing extinction, are argali sheep (*panyang*), brown bears (*zong xiong*), and mustang horses (*yema*), most held in captivity, along with saiga antelopes (*saijia lingyang*) and white storks (*bai guan*).

A primary target for economic and resource development beginning in the 1950s, and led by the quasi-military Xinjiang Construction and Production Corps, Xinjiang was also a primary site of many highly polluting **Third Front** facilities in the 1960s and 1970s. With 40 percent of the country's total **coal** reserves, Xinjiang is a center of **mining**, many of the mines illegal, along with the substantial production of **oil** and natural gas concentrated in the Turpan depression (*pendi*) in the eastern part of the Tianshan mountains and Tarim basin, with power plants and cement factories adding to the region's industrial base and contributing to the environmental degradation of **air** and **water**. Pursuing ambitious economic goals, often at the expense of environmental protection, many of the region's 26 **nature reserves**, including seven at the national level, are poorly managed, allowing illegal mining and ill-conceived **hydropower** projects, along with water-guzzling luxury sports facilities.

While ambient **air quality** in the region has improved with reductions in sulfur dioxide (SO_2) and smoke emissions, the latter evidently from the establishment of smoke-control zones, Xinjiang is, according to inspectors from the **Ministry of Environmental Protection (MEP)** in 2016, still engaged in the overexploitation of natural resources, with elevated levels of industrial and agricultural wastewater discharges and a continuing loss of groundwaters, including from the underground *karez* system established by the **minority** Uighurs (*Weiwu'er ren*). With its wide-open and sunbaked spaces, Xinjiang is a major site for **renewable energy** projects, including **wind power** plants in Urümqi and Turpan, and a large-scale photovoltaic

(PV/*guangfu*) **solar power** plant at Hami, while major coal-conversion-to-natural-gas projects are planned for the region to accelerate **energy** transfers to east coast **cities** via pipelines. The restoration of dried-up **lakes** and damaged **wetlands**, replenishment of aquifers (*hanshuiceng*), establishment of subterranean storage pools to collect glacial runoff, and major anti-desertification afforestation projects in the northwest Kekeya region along with the largest populus euphratica forest in the world encompassing more than 12 million hectares, are all being pursued as evidence that **climate change** is accelerating throughout the region.

XISHUANGBANNA. Located in the southernmost region of **Yunnan province**, with large rainforests in a tropical monsoon climate, Xishuangbanna is known as the "kingdom of plants and animals," with 20 percent of the mammals, 33 percent of **birdlife**, and 4,000 **plant species**, on only 0.2 percent of territory in the People's Republic of China (PRC). Prominent **wildlife** includes most of the last remaining elephants (*xiang*) in the country, numbering between 200 and 250, and, among bird species, the peacock (*kongque*), with both considered symbols of peace and happiness in Chinese culture. Among plant species are 800 medicinal varieties used in **traditional Chinese medicine (TCM)**, along with flowers the likes of the black berry lily, bougainvellia, camellia, and cape jasmine, as well as the extremely rare parasitic Rafflesiaceae, the largest flower on earth.

In an area of 19,233 km^2 (7,423 mi^2), consisting primarily of forested **mountains** and hillsides, along with valleys and **river** basins, featuring the **Lancang River** and its many tributaries, Xishuangbanna is home to the Dai, Hani, and other **minorities**, who espouse a strong reverence for nature. Especially venerated are the region's **forests**, of which there are five dominant forms: tropical evergreen, tropical semievergreen, montane evergreen broadleaved, 80 percent of natural forest cover, and local vegetative forests (limestone, pine, bamboo, river gallery, and deciduous), with 100 tree species in the region. Large-scale rubber (*xiangjiao*) and sugar cane (*ganzhe*) plantations, and small rubber tree farms, introduced in the 1950s, cover 20 percent of the territory, with serious detrimental effects on the environment, for instance, the depletion of **water resources** and **soil** nutrients, with the loss of 22.5 tons of topsoil annually for every one hectare (2.47 acres) of rubber trees. Local **agriculture** includes paddy rice and the famous Pu'er tea, grown on 1,700-year-old wild tea trees, along with a rich variety of tropical fruits the likes of sugar apples, eggfruits (canistel), salaks (snake fruit), mangosteens, sapodillas (chico fruit), langsats, mangoes, and passion flowers. The region is also the site of substantial **geothermal** resources.

Eight **nature reserves** have been established in the region, covering 14 percent of territory, with the first established in 1958, along with several protected areas, as total natural forest cover, excluding rubber trees, dropped from 69 percent (1976) to less than 50 percent (2013). Territory committed to rubber trees has expanded from 99,000 hectares (222,000 acres) in 1992 to 410,000 hectares (1 million acres) in 2010, as the PRC has become the largest rubber-consuming country in the world. With the deep-rooted rubber trees (*xiangjiao shu*) acting as virtual "water pumps," Xishuangbanna has become both hotter and drier, with 100 fewer foggy days annually. Along with the loss of 600 plant species, wildlife has also been negatively affected, with the disappearance of several species, including more than 40 percent of birdlife, largely from **hunting and poaching**, while migratory habits of the local elephant and tiger populations have been severely restricted by increasingly fragmented **habitat**, along with continuing danger from poachers and ongoing conflicts with local farmers concerning food sources. Among the government programs aimed at reconverting rubber trees back to natural forests is the **Sloping-Lands Conversion Program** ("Grain for Green"), but with few takers, as global prices for natural rubber remains relatively high. Research projects in the area include those being carried out by the Kunming Institute of Botany, the Yunnan Institute of Environmental Science, and the Asian Elephant Breeding and Rescue Center, where young, injured elephants have been restored to health and ultimately returned to the wild. Several **ecotourism** sites also exist, including Menglan Botanical Garden, Primeval Forest Park, Tropical Flower Garden, and Wild Elephant Valley, the last consisting of rainforests and habitat for trained and wild elephants to roam freely, often in competition with tourists.

Y

YANG XIN (1963–). Renowned environmentalist Yang Xin is founder of Green River, an environmental nongovernmental organization (ENGO) devoted to the protection of **rivers and waterways**, especially the **Yangzi River** and its headwaters in the Tanggula **mountains**, Qinghai province, where Yang has visited on multiple occasions. A professional photographer who first visited the site in 1986, for the start of a rafting trip down the entire river course, in which 10 members of the group lost their lives, he became acutely aware of the environmental damage in the remote region, including retreating **glaciers**, loss of **wetlands**, and **desertification**. A pioneer in defending **Tibetan antelope (*chiru*)** from **hunting and poaching**, Yang founded Green River following the brutal killing of Sonam Dorje of the **Wild Yak Brigade** for defending the animals, establishing the first privately run national protection station in the Hoh Xil Nature Reserve, Qinghai province.

Conducting surveys of the Tibetan antelope population, along with the distribution of waste and pollution in the area and throughout the **Qinghai–Tibetan Plateau**, Green River volunteers have constructed stop lights on both the Qinghai–Tibet highway and the Golmud–Lhasa railway as a measure of protection for the animals, with Yang submitting a report on the problem to the State Environmental Protection Administration (SEPA). With a second protection station planned for the upper reaches of the **Min River** in **Sichuan province**, Yang is also devoted to removing **garbage** and other wastes, notably proliferate **plastic** packaging, from glaciers and wetlands in the upper reaches of the Yangzi, earning him the moniker "garbage cleaner at the source of the Yangzi."

YANGZI RIVER (*CHANGJIANG*). Known in China as simply "long river" (*changjiang*), flowing 6,350 kilometers (3,943 miles), with a basin more than 1.8 million km^2 (698,000 mi^2) and headwaters in the **glaciers** of the Tanggula **mountains** in southern Qinghai province, the Yangzi River is the longest in Asia and third-longest **river** in the world. With 700 tributaries, for example, the Gan, Han, Huangpu, Jialing, **Min**, Tuo, Wu, Xiang, Yuan, and Zi, the Yangzi supplies 50 percent of **water resources** for the People's

Republic of China (PRC), primarily for **irrigation**, and contains major **fisheries** of 361 species, some of which are commercial carp (*liyu*), bream (*fang*), and Chinese perch (*qixi*), constituting 48 percent of the country's freshwater catch. **Endangered** species brought about by overharvesting and the loss of freshwater **habitat** include the Chinese sturgeon (*Zhonghua xun*), Yangzi River alligator (*eyu*), giant soft-shelled turtle (*juxing jiayu*), and **Yangzi River dolphin** (*baijitun*), the last declared functionally extinct.

Spanning 1,700 miles of the river's length from **Chongqing** to **Shanghai**, the Yangzi serves as a major cargo shipping route before spilling into the **East China Sea**. With tremendous water volume, reaching 194,000 ft^3 per second at seasonal water highs, Yangzi River **floods**, primarily in the spring monsoon season, have caused enormous physical damage and numerous human fatalities, most recently in the highly destructive 1998 flood. Known in the upper reaches as the *Jinsha* ("River of Golden Sands"), in its lower reaches the Yangzi enters a complex system of **lakes** (including **Dongting**, **Tai**, and **Poyang**), marshes, and multiple channels, where the plains in Hunan, Hubei, and Jiangxi provinces serve as a natural flood basin that has been reduced in size by land **reclamation**, which also limits the river's capacity to detoxify pollutants.

Severe environmental damage to the river includes **acid rain** and an annual inflow of 25 billion tons of **sewage** and industrial effluents from a watershed population of 400 million people, along with 5,000 petrochemical facilities and thousands of other factories. Agricultural runoff from **fertilizers**, including high levels of phosphorus (*lin*), and **pesticides** add to the river's environmental degradation, along with residue from rural waste dumps, animal manure and human wastes, and possibly radioactive materials. Waters once described as "so clear a pen could be seen sinking to the river bottom" are now, in many areas of the river, rated grade IV or above, unfit for human contact.

The site of major **hydropower** projects, including the Gezhouba, **Three Gorges**, Xiangjiaba, and Xiloudu dams, with more than 25 stations on major tributaries, along with 50,000 reservoirs, river **hydrology** has been substantially altered, especially by the giant 643-mile reservoir created by the Three Gorges Dam. Major slowdown of river flow has resulted, causing outbreaks of eutrophication (*shuiti fu yingyanghua*) on the river and associated water bodies in its flood plain, leading to algae plumes and cyanobacteria, posing serious **health** hazards and causing fish die-offs. A dramatic drop in the annual sediment load has also occurred, from 450 million tons to 150 million tons, with only the Red River (*Chishui*) tributary on the upper reaches remaining as a free-flowing river.

Disruption of functional **ecosystems** due to pollution and insufficient water flow has been reported by all 10 provinces and major municipalities through which the Yangzi flows, from Qinghai to Chongqing to Shanghai.

Destructive practices include sand **mining** of the river bottom, which, after leading to land subsidence resulting in bridge and riverbank collapses around Shanghai, has been banned. Smog is also prevalent throughout the basin from the large number of heavily polluting factories operating on or near the river's banks, along with a heavy concentration of shipping, especially in the Yangzi River Delta.

Major proposals for and efforts at cleaning up the Yangzi include the Yangzi Forum, established in 2005, composed of representatives from 20 countries and Chinese environmental nongovernmental organizations (ENGOs), and the Yangzi River Economic Belt (YREB), announced by President **Xi Jinping** in 2017, prohibiting the construction of new heavy industrial or **chemical** factories within one kilometer of the river, along with a mandate to move heavily polluting industrial facilities to new locations, often in remote and poorer areas. Also proposed is a reconnection of the Yangzi River mainstream with shrinking basin lakes, two-thirds lost since 1950, to facilitate the expansion of spawning grounds for migratory fish. Termination of the proposed Xiaonanhai hydropower project on the river's upper reaches retains one of the few free-flowing areas of the Yangzi near a crucial fish preservation site, while river volume will be substantially affected by proposed diversions by the **South-to-North Water Diversion Project (SNWDP)**.

YANGZI RIVER DOLPHIN (*BAIJITUN*). A freshwater dolphin endemic to China and found only in the middle and lower reaches of the **Yangzi River**, including the two river **lakes** of **Poyang** and **Dongting**, the species was officially declared extinct in 2006, with the last sighting coming in 2004. Also known as the Chinese River dolphin and the whitefin dolphin, and nicknamed the "goddess of the Yangzi" (*Yangzi nüshen*) by local fishermen and boatmen based on historical mythology, the *baiji* is evidently the first Cetacean species driven to extinction by humans. Numbering approximately 6,000 in 1950, the species underwent significant declines, with estimates of a few hundred in 1970, down to 400 in the 1980s and a mere 13 in 1997, the last figure based on an extensive river survey conducted that same year.

Major causes for the decline and ultimate extinction of the species in the Yangzi include the loss of **habitat** due to the dramatic growth in shipping and fishing on the river during the period of rapid economic growth and industrialization from the early 1950s onward. Hunted for flesh and skin during the **Great Leap Forward (1958–1960)**, when traditional reverence for the species was denounced in state propaganda, the species underwent significant declines from the growth of fishing following the introduction of economic reforms (1978–1979), when despite making hunting of the species illegal (1983), large numbers were inadvertently killed by rolling hooks and other fishing gear, for instance, electrofishing. Also contributing to the *baiji's* decline was the construction of large-scale **hydropower** projects on the

Yangzi, most notably the Gezhouba Dam (1989) and the massive **Three Gorges Dam** (2010), along with extensive **water pollution**, that led to the spread of infectious diseases to the vulnerable species.

Chinese government efforts to stave off the decline of the species by establishing such special in situ **nature reserves** as the Tian-e-Zhou Oxbow **wetlands**, as well as a lake and a Conservation Plan for Cetaceans of the Yangzi River (2001), were too little and too late. Capable of swimming as fast as 60 kilometers an hour (37 miles per hour), *baiji* suffered from poor vision, relying on sonar for navigation (echolocation), making it highly susceptible to disorientation and collision with increasingly powerful ships, which exponentially increased in numbers on the Yangzi, especially from 1978 onward. The last living *baiji*, named *qiqi* and housed at the Wuhan Institute of Hydrobiology, died in 2002. *Baiji* should not be confused with other dolphin species in China, including the white/pink dolphin (*fenhongse haitun*) and the finless porpoise (*jiangtun*), the latter also found in the Yangzi River.

YARLUNG TSANGPO RIVER. Flowing for 2,839 kilometers (1,760 miles) west to east in **Tibet**, with a total basin area of 913,000 km^2 (352,000 mi^2), Yarlung Tsangpo in Tibetan (Yalu Zangbo in Chinese) is the highest major **river** in the world, at altitudes between 4,500 meters (14,760 feet) and 3,000 meters (9,840 feet). With headwaters in the Angsi **glacier** near the sacred sites of Mt. Kailash and Lake Manasarovar, the river flows through the South Tibet Valley, bounded on the south by the Himalayas and in the north by the Kang Rinpoche and Nyen Tangiha **mountains**, before entering the Yarlung Tsangpo Grand Canyon, the world's deepest, at 4,800 meters (16,000 feet). Following the Great Bend, the river turns southward and enters India, where it becomes the Brahmaputra River and then the Jumuna in Bangladesh, before spilling into the Bay of Bengal.

Riverbank terrain in Tibet at the highest elevations of cold desert and arid steppe consists of vegetation limited to deciduous scrub, while at lower elevations there are conifer trees and rhododendron forests (*dujian shulin*). While traversing one of the most pristine areas in the People's Republic of China (PRC), the river is not without environmental problems, most notably heavy metal residue from **mining** operations of rich deposits of copper (*tong*), gold (*jin*), and silver (*yin*). **Hydropower** projects on the Yarlung Tsangpo include, most notably, the Zangmu dam, completed in 2015, located near the Indian border, with plans for several more dams, as Indian officials claim the PRC has plans for major water diversions (including a 1,000-kilometer-long [651 miles] tunnel from the Yarlung Tsangpo into parched **Xinjiang**), while citing Chinese iron (*tie*) and bacterial (*xijun*) pollutants in the river, allegations denied by China.

YELLOW RIVER (*HUANG HE*). The second-longest **river** in the People's Republic of China (PRC), after the **Yangzi River**, at 5,464 kilometers (3,395 miles), with a basin covering 752,546 km^2 (290,560 mi^2), the Yellow River was long considered the "cradle of Chinese civilization" (*Zhonguo wenming de yaolan*), offering crucial **water resources** to farmland on the North China Plain from ancient times to the present, where half of all wheat is grown in the country. With headwaters in the Bayangela **mountains** in Qinghai province in the **Three Rivers Headwaters Area** (*Sanjiang yuan*) fed primarily by the **glacier**-filled Ngoring and Gyarding **lakes**, and underground sources, the Yellow River is so named for the massive sediment load picked up from the **Loess Plateau**, through which the middle reaches of the river flows, creating an elevated riverbed that in the lower reaches is higher than the surrounding countryside. Cutting new channels, the Yellow River has shifted course on 26 occasions, adopting the present course in 1897, while causing massive **floods**, the last in 1931, with large numbers of human fatalities, sometimes in the millions, dubbing the river "China's sorrow" (*Zhongguo de beishang*).

Carrying as much as 1.6 billion tons of sediment annually, with 1.4 billion tons deposited in the basin, the "river above ground" (*dishang he*) reaches an elevation above the surrounding surface of 10 meters (33 feet) at the city of Kaifeng (**Henan province**), with the entire lower reaches from Zhengzhou (Henan) to the mouth at the **Bohai Sea** off Shandong province lined with levees, requiring constant maintenance and repair, now overseen by the Yellow River Conservancy Commission. Heavy withdrawals for **irrigation**, constituting 80 percent of the total river volume, have increased fivefold since 1950, expanding from 800,000 hectares to 7.3 million hectares (1.9 million acres to 18.3 million acres), along with drinking water for 155 million people, primarily in such expanding riverbank **cities** as Kaifeng and **Lanzhou**, with water scarcity becoming an increasing problem. Exacerbated by declining precipitation in the basin and extensive reservoir construction from the 16 major dams on the river, notably the controversial Sanmenxia, the river has suffered frequent dry-ups (*ganhe*), the first in 1972, and 32 overall, including nine of the 10 years in the 1990s, until timely releases from reservoirs maintained seasonal flows in the 2000s.

Pollution is also a serious problem, as 34 percent of the river's water is rated grade V and V+, making it unsuitable for **agriculture** or **industry**. As much as 4.2 billion tons of wastewater is dumped into the river annually, 70 percent from mostly riverbank **industry**. At times turning the waters into various colors, including pink, green, red, brown, and black, major pollutants in the Yellow River include arsenic (*shen*), copper (*tong*), **oil**, and oxygen-absorbing organic waste (*youji feiwu*). With 30 percent of the river's 150 fish species declared extinct, the annual fish catch from the once-rich waterway has been reduced by 40 percent. Conservation efforts involving extensive reforestation and reduction of **soil erosion**, along with large-scale sand excavation, have reduced sediment content since 2000, making for clearer river water, especially in the middle reaches. While seven of the eight provinces through which the river flows suffer from water scarcity, the Yellow River is slated for replenishment by the **South-to-North Water Diversion Project (SNWDP)**, along with benefits from enhanced water efficiency measures.

YELLOW SEA (*HUANG HAI*). Located between the People's Republic of China (PRC) and the Korean peninsula, constituting the northern part of the **East China Sea** with an area of 380,000 km^2 (150,000 mi^2), the Yellow Sea is a shallow body of water with an average depth of 44 meters (144 feet) and a maximum depth of 152 meters (499 feet). It is the site of the world's largest continuous mudflat seashore. Semienclosed and connected to the **South China Sea** and Japan through the Tsushima and Korean straits, the Yellow Sea is one of the most degraded marine areas on earth. The entire **ecosystem** of the sea has been declared **endangered**, as significant damage has been incurred by natural **habitat** along the **coastal** areas, along with serious **water pollution**, bringing harm to the 1,600 fish species and other marine life, some of which include cephalopods or squids (*wuzei*), crustaceans (*jiaqiao*), shellfish (*beilei*), clams (*ha*), and rich seaweeds (*fengfu haicao*).

With the innermost bay known as the **Bohai Sea**, into which the sediment-laden Hai and **Yellow** rivers flow, sand and silt amounting to 1.6 million tons annually are deposited into the Yellow Sea, which, along with periodic sandstorms, contribute to the sea's yellow tint. One of the largest areas of the continental shelf with waters rich in marine life, including dugongs, porpoises, marine turtles (*haigui*), and whales (*jingyu*), and a major flyway for migratory **birds**, the Yellow Sea has suffered from substantial overfishing by Japan and the PRC. The annual fish catch by China has grown from 425,000 tons in 1986 to 2.4 million tons in 2003, as several species, including the Pacific sandlance (*yujin yuke*), sandspotted seal (*ban haibao*), finless porpoise (*jiangtun*), large turtles (*da gui*), and some whale varieties, are considered endangered.

Pollution levels in the sea have also risen from **sewage** and industrial effluents, along with dramatic increases in seaweed production on waterborne rafts in Jiangsu province since 2005, which contribute to the periodic outbreak of algae plumes (*zaolei yumao*) and serious damage to coastal tidal flats and intertidal areas, with the latter considered endangered, having suffered 65 percent loss since the 1950s to **reclamation**. With international assistance from the United Nations Development Programme (UNDP), **World Bank**, and World Wildlife Fund (WWF), the Yellow Sea Large Marine Ecosystem Project, involving the PRC and the two Koreas, was created in the early 2000s to facilitate sustainable management practices, including marine protected areas and joint projects to reduce environmental stress by limiting fishing and cutting nutrient discharges into the waters. The Yellow Sea is also the site of a large offshore **wind power** farm undergoing construction off Yancheng city, Jiangsu province, with a capacity of producing 870 million kWhs annually.

YUNNAN PROVINCE. Located in the southwest of the People's Republic of China (PRC), *Yunnan* meaning "south of the clouds," so named for its location south of the heavily clouded and foggy **Sichuan province** basin, has a varied landscape, ranging from snow-capped **mountains** with **glaciers**, deep valleys, and gorges, to subtropical and tropical areas featuring dense rainforests, primarily in southernmost **Xishuangbanna**. At 394,000 km^2 (152,000 mi^2), including the Yunnan–Guizhou Plateau, and with a population of 45 million, many **minorities**, including Dai, Hani, Yi, Zhuang, and Miao, Yunnan has the largest diversity of **plant species** in the PRC, numbering 17,000 out of a national total of 30,000, with 2,500 species classified as endemic. With many of these species "discovered" and documented in the 19th century by foreign explorers from Europe and the United States, notably American botanist Joseph Rock, Yunnan was considered the "land that time forgot," with landscapes and **wildlife** largely unaffected by human activity.

Constituting a mere 4 percent of the country's land mass, the province contains 50 percent of mammals and **birdlife** species, some of which are the Asian gaur (*yeniu*), also known as the Indian bison, South China tiger (*Huanan hu*), Asian elephant (*Yazhou daxiang*), Zhou's box turtle (*hegui*), and various types of **golden snub-nosed monkeys**, with many of these species threatened or **endangered** and the Yunnan lar gibbon (*Yunnan changbiyuan*) declared extinct. Home to 620 fish species constituting 40 percent of the country's freshwater fish, many species are found in single **lakes** in the province, for example, **Dian** and Erhai, where they are endangered or are confronted with extinction.

Six major **rivers** flow through the province, including the **Yangzi**, **Pearl**, Red, **Nu**, Salween, and **Lancang**, with the last serving as a major site for **hydropower** projects that have been successfully blocked on the more scenic Nu ("angry") River, with its rushing waters and deep gorges. Along with several World Heritage Sites, for instance, the Three Parallel Rivers of Yunnan Protected Areas, Yunnan has several **nature reserves and national parks**, including Xishuangbanna Tropical Nature Reserve, Lijiang Spruce Plain, Napa Hai Nature Reserve, Mt. Cangshan and Erhai Lake Scenic Resort (Dali), and Dongchuan Redland.

Major environmental problems include the deleterious effects of extensive **mining**, as the province has 86 mineral varieties, some of which include the richest deposits in the country of zinc (*xin*), lead (*qian*), cadmium (*ge*), indium (*yin*), thallium (*ta*), and crocidolite (*qingshimian*). With little arable land, **agriculture** is limited to 5 percent of the province and is heavily concentrated in such cash crops as coffee and tea, along with wild mushrooms (*yesheng mogu*) and fungi (*junlei*), 87 percent of the nation's total, especially the highly prized Matsutake mushroom. Wild plants collected for **traditional Chinese medicine (TCM)** are also available in abundance, for example, the Himalayan yew (*hongdoushan*), used in the production of antimalaria artemisinin and anticancer Taxol, and the overexploited caterpillar fungus (*chongcao*), also known as the "Himalayan Viagra," which is rapidly becoming unsustainable. Covering 53 percent of the province, **forests** largely consist of commercially grown trees, particularly rubber, especially in Xishuangbanna, along with eucalyptus (*anshu*) and fir (*lengshan*) on plantations, with only 9 percent composed of original, old-growth forests.

Highly uneven rainfall patterns afflict the province, with many valleys suffering **deforestation** and resultant high aridity. Water scarcity and winter-to-spring **droughts**, as in 2009, have become major problems, as commercial trees, particularly rubber, soak up groundwater, while substantial **soil** loss and little vegetative undergrowth on tree plantations hinder the **level** of water retention previously provided by original dense forest land sacrificed to mass cuttings. Efforts at remediation include construction of a 375-kilometer (232 mile) tunnel from the water-rich Yunnan-Guizhou plateau to more arid regions in the province begun in 2017.

With inadequate rural and urban waste disposal systems, rubbish and human and animal **sewage** is frequently dumped untreated into rivers, as 70 percent of water samples in the province fail to meet **water quality standards**. Along with lax enforcement of environmental **laws and regulations**, especially by local officials, the province also suffers from inadequate **environmental statistics and assessment** and a lack of environmental planning for major infrastructure projects, an example being the proposed One Belt, One Road Initiative (OBOR). Illegal construction of buildings and other facilities on protected lands is also a problem, along with increasing pressure on nature reserves from growing tourism. Major amelioration plans include the Yunnan Environmental Program and the Yunnan Province Biodiversity Conservation Strategy and Action Plan (2012–2030).

Z

ZERO WASTE (*LING LANGFEI*). A concept originally developed by American Bea Johnson and officially adopted in the People's Republic of China (PRC) in 2019, zero waste is being pursued by individuals, villages, and entire **cities** with a goal of substantially reducing the generation of solid wastes, at 10 billion tons annually, with an accumulated total of 60 to 70 billion tons scattered throughout the country. With 10 cities selected by the State Council as national models of zero-waste policies, volunteer organizations the likes of Zero Waste Alliance have hosted citizen forums on the issue, promoting the concept's major principles to villages and individual families, joining in and coordinating their efforts under the rubric of the GoZeroWaste lab, begun in 2016.

The major principles of zero waste include the following: refusing single-use, nondegradable materials, especially **plastics**; reducing unnecessary consumption; reusing materials; repairing rather than replacing broken materials and items; **recycling**; and **composting** organic wastes. There are four major types of solid wastes, with national hot spots indicated in parentheses: general industrial wastes (Ordos, Inner Mongolia, Panzhihua, **Sichuan**); industrial hazardous wastes (Suzhou, Jiangsu, Yantai, Shandong); medical wastes (**Beijing**, **Shanghai**, **Guangzhou**); and urban living **garbage** (Beijing, Shanghai, Chengdu, Guangzhou, Hangzhou, Shenzhen). Reinforcing the zero-waste concept is the traditional Chinese lifestyle of frugality, as opposed to China's more recent infatuation with "fast fashion" and other elements in the increasingly pervasive **"waste culture."** Among the private **companies** promoting zero waste is Bulk House, where Chinese consumers can bring in containers for refills and trade in single-use items.

ZHANG KAIYUAN. One of the first environmental billionaires in the People's Republic of China (PRC), Zhang Kaiyuan chairs Beijing Environmental Production Tech, a major producer of environmental protection equipment, founded in 2001. Major products include flue gas desulfurization equipment employing both wet and dry processes for **coal**-burning power plants and industrial facilities, a growing market as the Chinese government imposes **air pollution** controls on thermal plants, which supply 70 percent of the country's electrical power.

ZHOU ENLAI (1898–1976). The longtime premier of the People's Republic of China (PRC) from 1949 to 1976, and foreign minister from 1949 to 1958, Zhou Enlai, while heavily involved in foreign relations and domestic political issues, was also an early advocate of environmental protection. Beginning with warnings on environmental degradation in the 1960s and alarmed by a massive dieoff of shellfish near Dalian, Liaoning province, Zhou appointed a PRC delegation to the United Nations Conference on the Human Environment in Stockholm, Sweden, in 1972, led by the country's then-leading environmentalist, **Qu Geping**. Zhou also authorized the first National Conference on the Environment (1973), where delegates from the country's provinces and municipalities reported widespread environmental damage of **rivers**, **lakes**, **soils**, and **coastal** areas, along with expanding **desertification**. With administrative chaos and dysfunction during the **Cultural Revolution (1966–1976)**, many of his directives dealing with environmental protection were effectively ignored and not carried out, leading to further environmental degradation, which was not addressed until after Zhou's passing in 1976.

Glossary

chu sihai	“wipe out the four pests”
daxiongmao	giant panda bear
Dayu zhishui	“Yu the Great controls the waters”
gaitian huandi	“reform the heavens and change the earth”
guahuadi	steep, sloping lands
heshan yise	“change the face of rivers and mountains”
huangmohua	desertification
huanjing baohu	environmental protection
huanjing quntixing shijian	environmental mass incidents (social protests)
huanjing waijiao	environmental diplomacy
jianhua	alkalinization
jiulong zhishui	“taming the waters with nine dragons”
haihuang zaotian	“open up the wasteland to farmland”
mei banfa	“there is no way”
muqin dadi	mother nature
ren ding sheng tian	“man must conquer nature”
ren duo liliang da	“with many people, strength is great”
rengdiao shehui	throw-away society
san dafa	three great cuttings
shahua	sandization
shamohua	desertification
shangren chu, huanghe qing	“when a great man emerges, the Yellow River will run clear”
sipang	“four arounds,” meaning planting of multipurpose tree species on farms
tian ren heyi	“heaven and humanity in harmony”
weihu zaotian	“encircle the lakes, create farmland”
xianwuran houzhili	“pollute first, control later”

yanhua	salinization
yao gaoshan ditou, heshui ranglu	"make the mountains bow and the rivers yield"
yifa zhiguo	ruling the country by law
Yugong yishan	foolish old man who removed the mountains

Appendix A
Chinese and Foreign Environmental Activists

***Boldface** indicates dictionary entry.*

CHINA

Bayar (Inner Mongolia/grasslands protection and restoration)
Chai Jing (Beijing/filmmaker)
Chen Guangbiao (Jiangsu Huangpu Renewable Resources)
Chen Jie (Beijing/photojournalist)
Chen Jiqun (Inner Mongolia/grasslands conservation)
Chen Liwen (Beijing/Zero Waste Beijing)
Dai Qing (Beijing/journalist)
Deng Fei (Kunming, Yunnan/journalist; established China Water Safety Foundation)
Duan Shumin (Shandong/exposing polluting factories)
E Yunlong (Beijing/Peking University Ecocivilization Research Center)
Feng Yongfeng (Beijing/founder, Green Beagle Environment Institute)
Guo Yongqi (Jinan, Shandong/founder, Green Qilu Transportation Development Group)
Hao Xin (Zhejiang/head, Green Zhejiang)
Hu Jia (Beijing/human rights and environment)
Hu Kanping (Beijing/editor, *China Green Times*)
Huo Daishan (Henan/Huai River pollution)
Jin Jiaman (founder, Global Environmental Institute)
Li Bo (Beijing/head, Friends of Nature)
Li Xueliang (Xinjiang/photojournalist)
Liang Congjie (founder, Friends of Nature)
Liao Xiaoyi (Shandong/founder, Global Village)
Liu Shu (Hunan/imprisoned for exposing government information on the environment)
Liu Yi (Xiamen, Fujian/China Mangrove Protection Project)
Liu Yonglong (Shanghai/Shanghai Rendu Ocean NPO Development Center/coastal cleanups)
Ma Da (Hebei/China Zero-Waste Alliance)

Ma Jun (Beijing/journalist)
Ma Zhong (Heilongjiang/wetlands conservation)
Qu Tianwen (Yunnan/fish restoration in the Yangzi River)
Shen Chunyi (Sichuan/student environmental leader)
Shi Yan (Beijing/Community Supported Agriculture)
Sun Xiaodi (Gansu/investigation of mines)
Tan Kai (Zhejiang/founder, Green Watch)
Wang Ding (Wuhan, Hubei/Yangzi River dolphin specialist)
Wang Jiuliang (Beijing/filmmaker)
Wang Song (Beijing/zoologist, compiled China's first red list of endangered species)
Wang Yongchen (Beijing/founder, Green Earth Volunteer)
Wei Dongying (Zhejiang/photojournalist)
Wen Bo (Beijing/journalist, Greenpeace East Asia)
Wu Dengming (Chongqing/founder, Green Volunteer League)
Wu Lihong (Jiangsu/Lake Tai pollution)
Wu Yiqun (Beijing/founder, Eternal Green)
Xi Zhinong (Yunnan/wildlife photographer)
Xie Yan (Beijing/head, Wildlife Conservation Society of China)
Xu Gang (Beijing/Research Academy of Environmental Sciences/soil science specialist)
Xu Xiujuan (known as the "red-crowned crane girl" for her pioneering work in breeding the rare birds in Yancheng Nature Reserve, Jiangsu province, where she unfortunately drowned)
Yan Baohua (Shenzhen/wetlands conservation)
Yang Xin (Qinghai/head, Green River)
Yu Dan (Beijing/Confucian environmental traditions)
Yu Kongjian (Beijing/author, *Art of Survival*, calling for ecological rethinking)
Yu Xiaogang (Yunnan/founder, Green Watershed)
Zeren Pingcuo (Tibet/nature photographer)
Zhang Chunshan (Yunnan/forest protection)
Zhao Qikun (Kunming, Yunnan/zoologist)
Zou Ji (Energy Foundation, member PRC climate change negotiating team)

INTERNATIONAL

Joseph Rock (United States/botanist, geographer)
George Schaller (United States/Wildlife Conservation Society)
Peter Scott (United Kingdom/World Wildlife Fund)
Ernest "Chinese" Wilson (United Kingdom/botanist, plant collector)

Appendix B
Chinese and Multinational Environmental Companies

CHINA

Air Visual (pollution prediction)
Beijing Orient Landscape and Ecology
Beijing Originwater Technology
Beijing SDL Technology Company
Beijing SPC Environmental Protection Technology Company
China Environmental Energy Construction and Environmental Protection Group
China Environmental Holdings
China Environmental, Ltd.
China Environmental Resources Group
China National Environmental Protection Group (municipal and industrial waste treatment)
Chongqing Kangda Environmental Protection
Create Technology and Science (environmental protection equipment)
Dynagreen Environmental Protection Group
Fresh Ideas (pollution prediction)
Fujian Longking (air pollution equipment)
Gangzhou Yueshou
Green Digital Finance Alliance
Guangxi Bosco Environmental Protection Technology
Huaxin Green Spring Environmental Protection Development Company
Jiangsu Huangpu Renewable Resources
Jiangsu LVHE Environmental Technology Company
Sailhero Environmental Protection High-Tech (environmental monitoring equipment)
Thunip Corporation (fiber rotating disk filters)
Top Resources Conservation Engineering (natural gas supply and heating equipment)
Turenscape (ecological security)
Tus-Sound Environmental Resources
Wuhan Huade Environmental Engineering and Technology Co.
Xindi Consulting

Zhongyang Yinyin

MULTINATIONAL

Anguil Environmental Systems Incorporated
Arcadis Global Design and Construction
Cambridge Environmental Resources Construction (CERC)
Environmental Resources Management
First Carbon Solutions
Green Stream Network
King and Capital Recycled Materials
Lakes Environmental Software
Navigant Consulting
Reach 24th Construction Group
RPS Group
Thinkstep Sustainability Software
Trinity Consulting
Veolia Group
Verantis Environmental Solutions Group
Yulong Environmental Protection Company

Appendix C
Environmental Laws and Regulations

LAWS (BY YEAR ENACTED/AMENDED)

Trial Environmental Protection Law (1979)
Marine Environmental Protection Law (1982/2000)
Grassland Management Act (1982)
Water Pollution Prevention and Control Law (1984/1996, 2008)
Forestry Law (1984/1998)
Grassland Law (1985/2003)
Rangeland Law (1985)
Fisheries Law (1986/2010)
Mineral Resources Law (1986/1997)
Land Administration Law (1986/1988, 1998, 2004)
Prevention and Control of Atmospheric Pollution Law (1987/1995, 2007)
Water Law (1988/2002)
Environmental Protection Law (1989/2015)
Wildlife Conservation Law (1989/2017)
Urban and Rural Planning Law (1989/2008)
Water and Soil Conservation Law (1991/2011)
Survey and Mapping Law (1992/2002)
Nature Reserve Law (1994)
Prevention and Control of Environmental Pollution by Solid Waste Law (1995/2005)
Electric Power Law (1996)
Law on the Coal Industry (1996/2011)
Noise Pollution Law (1996)
Flood Control Law (1997)
Protection against and Mitigation of Earthquake Disaster (1997)
Energy Conservation Law (1997/2008)
Meteorology Law (1999)
New Wildlife Protection Law (1999)
Desertification Prevention and Control (2001)
Administration of Use of Sea Areas (2001)
Agriculture Law (2002)
Cleaner Production Promotion Law (2002/2012)
Environmental Impact Assessment Law (2003)

Sand Control and Management Law (2002)
Radioactive Pollution Prevention and Control Law (2003)
Rural Land Contracting Law (2003)
Renewable Energy Law (2005/2010)
Prevention and Control of Environmental Pollution by Solid Waste Law (2007)
Property Rights Law (2007)
Promotion of Circular Economy Law (2008)
Protection of Offshore Islands Law (2009)
Environmental Protection Tax Law (2018)
Law on Prevention and Control of Soil Pollution (2018)
Protection and Control of Water Pollution Law (2018)

REGULATIONS AND DIRECTIVES (BY YEAR ENACTED/AMENDED)

Regulation of the Protection of Mineral Resources (1956)
Directive Concerning the Active Protection and Rational Use of Wildlife and Natural Resources (1962)
Directive for the Protection and Rational Utilization of Wild Animals (1964)
Regulations on Forest Conservation (1964)
Regulations for the Protection of Wild Animals (1973)
Provisional Regulations on the Prevention of Pollution of Coastal Waters (1974)
Interim Regulations of Pollutant Emission Permits (1980s/2016)
Regulations on the Management of Nature Reserves (1993/2013)
Measures for Disclosure of Environmental Information (2008)
Regulations on Pollution Discharge Fees (2013)
Guidance on Rural Solid Waste (2015)
Land Reclamation Control Measures (2016)
Pesticide Management Regulations (2017)
Technical Specifications for the Application and Issuance of Pollutant Permits—Iron and Steel Industry (2017)

Appendix D
Environmental Nongovernmental Organizations (ENGOS)

***Boldface** indicates dictionary entry.*

DOMESTIC

Alibaba Foundation
All-China Environmental Federation
Baotou City (Inner Mongolia) Environmental Federation
Beijing Animal Protection and Education Base
Beijing Environment and Development
Beijing Environmental Network
Beijing Environmental Protection Foundation
Beijing Forestry Society
Beijing Rainbow Peace Environmental Research Center
China Biodiversity Conservation and Green Development Foundation
China Biodiversity Observation Network
China Blue Sustainability Institute
China Carbon Forum
China Civil Climate Action Network (CEAN)
China Energy Conservation and Environmental Protection Group
China Environmental Protection Foundation
Center for Biodiversity and Indigenous Knowledge
Center for Environmental Education and Communication
Center for Legal Assistance to Pollution Victims
China Council for International Cooperation on Environment and Development (CCICED)
China Forum of Environmental Journalists
China Green Beat
China Low-Carbon Forum
China Mangrove Construction Network
China New Energy
China Safety Foundation
China Society of Forestry

China Water Risk Foundation
China Wild Plant Conservation Association
China Wildlife Conservation Association
China Wind Energy Association
China Youth Climate Action Network
China Zero-Waste Alliance
Chinese Society of Environmental Sciences
Chongqing Green Volunteers Association
Clean Air Alliance of China
Dalian University of Technology (Liaoning).
Ecological Society of China
Energy Foundation
Environmental Protection and Ecological Construction Coordinating Committee
Eternal Green
Friends of Nature
Future Generations China
Global Environmental Institute
Global Village of Beijing
Green Anhui
Green Beagle Environment Institute
Green Camel Bell
Green Camp
Green Cross
Green Development Foundation
Green Earth Volunteers
Green Environmental Advisory Center (Chongqing)
Green Environmental Volunteers
Green Eyes
Green Home
Green Hunan
Green Keepers
Green Longjiang
Green River
Green Stone
Green Student Forum
Green Volunteer League
Green Watch
Green Watershed
Green Web
Greener Beijing
Greenpeace East Asia
Guardians of the Huai River

Hengduan Mountains Research Society
Institute of Eco-environmental Science and Technology (Guangdong)
Institute of Environment and Development
Institute of Public and Environmental Affairs
International Center on Small Hydropower
International Ecological Economic Promotion Association
International Ecological Safety Collaboration Organization
Lead China
Liangqiang Voluntary Service
Partnership on Sustainable Low-Carbon Transportation
Qingdao Institute for Bio-Energy and Bioprocess Technology
Rural Development Institute
Saving the Spoon-Billed Sandpiper (Shanghai)
Sichuan University Environmental Volunteers Association
Sustainable Network Society
Taizhou Environmental Foundation
Tibetan Antelope Information Center
Wildlife Conservation Society
Wuhu Ecology Center Anhui
Xinjiang Conservation Foundation
Zero Waste Beijing

INTERNATIONAL

Animals Asia Foundation
Conservation International
Dutch Global Centre on Adaptation
Environmental Defense Fund
Environmental Investigation Agency
Forest Stewardship Council (FSC)
Global Footprint Network
International Crane Foundation
International Fund for Animal Welfare
International Rivers Network
International Water Association
Natural Resources Defense Council
Nature Conservancy
Pacific Environment
Snow Leopard Trust
Wetlands International
World Resources Institute

World Wide Fund for Nature (previously World Wildlife Fund)
Worldwatch Institute

Appendix E
State Environmental Organizations

Agro-Environmental Protection Institute (Chinese Academy of Agricultural Sciences [CAAS])
Beijing Climate Center
Biodiversity Committee (Chinese Academy of Sciences [CAS])
China Academy for Environmental Planning (CAEP)
China Climate Center Info-Net (Department of Climate Change, National Development and Reform Commission)
China Council for International Cooperation on Environment and Development (CCICED)
China Environmental Culture Protection Association (CECPA)
China Institute of Water Resources and Hydropower Research
China Meteorological Administration
China National Environmental Monitoring Center
China National Urban Air Quality Real-Time Publishing Platform
China Oceanic Info-Network, State Oceanic Administration (abolished 2018)
China Population and Development Research Center
Chinese Academy of Agricultural Sciences (CAAS)
Chinese Academy for Environmental Planning (CAEPP)
Chinese Academy for Environmental Sciences Planning (CAESP)
Chinese Research Academy of Environmental Sciences (CRAESP)
Institute of Botany (CAS)
Institute of Environment and Sustainable Development in Agriculture (CAAS)
Kunming Institute of Botany (CAS)
Ministry of Agriculture (MOA)
Ministry of Ecology and Environment (MEE)
Ministry of Natural Resources (MNR)
Ministry of Water Resources (MWR)
National Laboratory for Clean Energy
Shenyang Academy of Environmental Sciences
State Forestry Administration
Yellow River Conservancy Commission

Appendix F
National Nature Reserves and National Parks

NATURE RESERVES (PROVINCE/CONSERVATION FEATURES AND WILDLIFE)

Arjin Mountain (Xinjiang, Kumukuli Desert/Saker falcons, Tibetan fox and antelope, jasmine and lotus plants)
Baihe (Sichuan/golden-hair monkeys)
Baili Azalea (Guizhou/rhododendrons)
Baishuijiang (Gansu/giant panda bears, golden-hair monkeys)
Baiyin Aobao (Inner Mongolia/grasslands, golden eagles, black storks)
Caohai (Guizhou/lake and wetlands, bird sanctuary, black-neck cranes, bar-headed geese)
Cyathea (Guizhou/Danxia red rock landform, plateau canyons, stone pillars)
Daocheg Yading (Sichuan/three sacred mountains of Buddhism, pastures, lakes)
Dinghu Mountain (Guangdong/subtropical forests, waterfalls, "green pearl")
Dujiangyan (Sichuan/panda bears)
Fanjing Mountain (Guizhou/virgin forests, grey golden-haired monkeys, Buddhist holy site)
Foping (Shaanxi/giant panda bears, monkeys, Chinese goral)
Hoh Xil/Kekexili (Qinghai/Tibetan antelope, rare plant species)
Huaping Forest (Guangxi/silver pines)
Hulu Lake (Inner Mongolia/wetlands)
Huzhong (Heilongjiang/boreal mountain forests)
Jianfenling (Guangdong/long-armed monkeys)
Jigong (Henan/subtropical forests, rare wildlife)
Jingfu Mountain (Sichuan/giant panda bears)
Kanas (Xinjiang/lakes, snow mountains, glaciers, forests, "pure land of the world")
Laohegou Land Trust (Sichuan/snub-nosed monkeys, Asian cats, takin)
Leigong Mountain (Guizhou/virgin forests, Taiwan firs, ten peaks)
Momoge (Jilin/swamps, rivers, grasslands, white cranes and migratory birds)
Nanhaizi (Beijing Municipality/*Milu* deer)
Poyang Lake Migratory Birds (Jiangxi/birdlife)
Puducao National Park (Yunnan/white water stoned terraces)
Qiang Tang (Tibet/Tibetan antelopes, wolves, blue bears, pikas)

Qianshan (Jiangxi/forests)
Qilian Mountains (Gansu/forests)
Qomolangma (Tibet/Mt. Everest, glaciers)
Sanjiaozhou (Shandong/Yellow River)
Shanwang Paleo-Organism Fossils (Shandong/fossil remains)
Shapotou (Ningxia/deserts)
Shengnongjia (Hubei/dense forests, glaciers, golden-hair monkeys)
Tarim Poplar Forest (Xinjiang/"living fossil" poplar trees)
Tian-E Oxbow Lake (Hubei/*baiji* and finless porpoise conservation)
Tianmu Mountains (Zhejiang/lush forests, natural botanical garden)
Wolong (Sichuan/giant panda bear)
Wulingshan (Hebei/oak forests, birch forests, natural botanic garden)
Wuyi Mountains (Fujian/subtropical forests)
Xishuangbanna (Yunnan/rainforests, elephants, tigers)
Xuancheng Chinese Alligator Reserve (Anhui/alligators)
Yading (Sichuan/alpine forests)
Yancheng Littoral Mudflats and Rare Birds (Jiangsu/red-crowned cranes, golden eagles)
Yanchiwan Nature Reserve (Gansu/snow leopards)
Yiwulü Mountain (Liaoning/"one of three greatest mountains in northeast China")
Yongzhao Dupangling (Hunan/subtropical forests)
Zhalong (Heilongjiang/red-crowned cranes; marsh lands)

NATIONAL PARKS (PROVINCE/GEOLOGY AND WILDLIFE)

Changbai Mountain (Jilin/dormant volcano)
Cave of Stone Flower (Beijing/caves)
Dali Lake (Inner Mongolia/migratory birds)
Emei Mountain (Sichuan/highest of four sacred Buddhist mountains, macaques)
Guilin Li River (Guangxi/karst landscape)
Huangguoshu Waterfall (Guizhou/waterfalls, caves)
Huanglong (Sichuan/calcified, colorful water pools and waterfalls)
Huangshan (Anhui/granite peaks and gorges)
Huashan (Shaanxi/Daoist sacred mountains)
Jiuzhai Valley (Sichuan/waterfalls, multicolored mountain lakes)
Lijiang Yulong Snow Mountain (Yunnan/Tiger Leaping Gorge, glaciers)
Lushan (Jiangxi/gorges, valleys, caves, rivers, forests)
Mingsha Mountain and Crescent Lake (Gansu/"singing sands," desert oasis)
Nyainyêntanglha Mountain (Tibet/Namtso Lake, mountain peaks)

Potatso (Tibet/lakes, rolling pastures, forests, marsh lands)
Sanqing Mountain (Jiangxi/granite peaks, pillars, waterfalls, rhododendrons)
Shilin (Yunnan/"stone forest" rock formations, caves, lakes)
Taishan (Shandong/rock formations, Buddhist sacred mountain)
West Lake (Hangzhou, Zhejiang/gardens)
Wudang Mountain (Hubei/geopark, Daoist sacred mountain)
Wutai Mountain (Shanxi/Buddhist sacred peaks)
Zhangjiajie (Hunan/sandstone pillars)
Zhangye Danxia (Gansu/geopark, orangey-red and yellow-striped cliffs)

Appendix G
Wetland Nature Reserves and Parks

Aibi Lake (Xinjiang)
Argun River (Inner Mongolia)
Baiyangdian (Hebei)
Bayanbulak (Xinjiang)
Beihai (Yunnan)
Chongming Dongtan (Shanghai)
Dafeng (Jiangsu)
Dalai Lake (Inner Mongolia)
Dalian National Spotted Seal (Liaoning)
Danjiangshidi (Henan)
Dongting Lake (Hunan)
Dongzhai Port (Hainan)
Erguna Wetland (Heilongjiang)
Hainan Mangrove Forest (Hainan)
Lhalu (Tibet)
Liao River Delta Wetland (Liaoning)
Longgan Lake (Hubei)
Mangrove Nature Reserve (Guangdong)
Nuoshui River (Sichuan)
Ordos (Inner Mongolia)
Poyang Lake (Jiangxi)
Ruoergai (Sichuan)
Sanjiang Plain (Heilongjiang)
Shengjin Lake (Anhui)
Three Rivers Area/*Sanjiangyuan* (Qinghai)
Weining Wetland (Guizhou)
Wulagai Wetland (Inner Mongolia)
Xing Lake Wetland (Jiangsu)
Xixi Wetland (Zhejiang)
Yancheng Littoral Mudflats and Rare Birds (Jiangsu)
Yangchengmang River (Shanxi)
Yangzi River Delta (Shanghai)
Yellow River Delta (Shandong)
Zhalong Wetland (Heilongjiang)
Zoige (Sichuan)

Bibliography

CONTENTS

I. Introduction 325
II. Books and Articles 326
  A. English 326
  B. Chinese and Other Languages 334
III. Major Chinese and International Journals on the Environment 336
IV. Films 336
  A. China and International (Director/Topic) 336
V. Websites 337

I. INTRODUCTION

Major works on the environment in the People's Republic of China (PRC) in English, Chinese, and other languages cover important topics, including environmental history, government policies, and the range of environmental problems, from air and water pollution to soil erosion and degradation of plant species and wildlife. Limited by the relatively recent emergence of environmental protection and restoration as an issue in the PRC beginning in the 1990s, this bibliography contains sources most easily amenable to the general reader, while works dealing with more technical and complex subjects available from Chinese and international authors and scholars can be found in such outlets as *Science Direct*, *Research Gate*, *Science*, and *Nature Geoscience*. Major reports by international and Chinese institutions like the World Bank and the PRC's Ministry of Environmental Protection are also included, along with major Chinese and foreign journals on environmental topics, and Chinese and foreign films, both documentary and dramatic, on relevant topics. Publishers of works on the environment in the PRC include China Environmental Press and China Environmental Sciences Press, with such major government reports as the *China Environmental Yearbook* and frequent coverage by independent newspapers and magazines, for example, *South China Morning Post* and the *Diplomat*.

II. BOOKS AND ARTICLES

A. English

1.General Information and Policies

Brown, Lester. *Who Will Feed China? Wake-up Call for a Small Planet*. Washington, DC: Worldwatch Institute, 1995.

Brundtland, G. H. *Our Common Future: The Report of the World Commission on Environment and Development*. Oxford, U.K.: Oxford University Press, 1987.

Cannon, Terry, and Alan Jenkins, eds. *The Geography of Contemporary China*. New York: Routledge, 1990.

Carter, Niel T., and Arthur P. J. Mol, eds. *Environmental Governance in China*. New York: Routledge, 2007.

Chen Gang. *Politics of China's Environmental Protection: Problems and Progress*. Singapore: World Scientific Publishing, 2009.

China's Agenda 21: White Paper on China's Population, Environment, and Development in the 21st Century. Beijing: China Environmental Sciences Press, 1994.

Day, Kristen A., ed. *China's Environment and the Challenge of Sustainable Development*. Armonk, NY: M. E. Sharpe, 2005.

Economy, Elizabeth C. *Environmental Scarcities, State Capacities, Civil Violence: The Case of China*. Cambridge, MA: American Academy of Arts and Sciences, 1997.

Edmonds, Richard Louis, ed. *Managing the Chinese Environment*. Oxford, U.K.: Oxford University Press, 1998.

Feshbach, Murray. *Ecological Disaster: Cleaning Up the Hidden Legacy of the Soviet Regime*. New York: Twentieth-Century Fund Press, 1995.

Ho, Peter, and Eduard B. Vermeer, eds. *China's Limits to Growth: Prospects for Greening State and Society*. Oxford, U.K.: Blackwell, 2006.

Jahiel, Abigail R. "The Contradictory Impact of Environmental Protection in China." *China Quarterly* 149 (March 1997): 81–103.

———. "The Organization of Environmental Protection in China." *China Quarterly* 156 (December 1998): 757–87.

Komarov, Boris. *The Destruction of Nature in the Soviet Union*. White Plains, NY: M. E. Sharpe, 1980.

Lampton, David M., ed. *Policy Implementation in Post-Mao China*. Berkeley: University of California Press, 1987.

Law of the People's Republic of China on the Prevention and Control of Air Pollution. Beijing: China Environmental Press, 1989.

Law of the People's Republic of China on the Prevention and Control of Water Pollution. Beijing: China Environmental Press, 1989.

Leach, Beryl. "Long-Term Ecological Research in China: CAS Establishes a Network." *China Exchange News* 18, no. 4 (1990): 23–27.

Lora-Wainwright, Anna. *Resigned Activism: Living with Pollution in Rural China*. Cambridge, MA: MIT Press, 2017.

Ma Xiaoying, and Leonard Ortolano. *Environmental Regulation in China*. Lanham, MD: Rowman & Littlefield, 2000.

MacKinnon, John, and Wang Haibin. *The Green Gold of China*. London: EU–China Biodiversity Programme, 2008.

MacKinnon, John, Meng Sha, C. Cheung, G. Carey, Zhu Xiang, and D. Melville. *A Biodiversity Review of China*. Gland, Switzerland: World Wildlife Fund International, 1996.

Marine Environment Protection Law of the People's Republic of China. Beijing: China Environmental Press, 1989.

McElwee, Charles R. *Environmental Law in China: Mitigating Risk and Ensuring Compliance*. Oxford, U.K.: Oxford University Press, 2011.

Ministry of Science and Technology, China Meteorology Administration, and Chinese Academy of Sciences. *First National Climate Change Assessment*, 2006.

National Statistical Yearbook. Beijing: China National Bureau of Statistics, 2002– .

Ning, Datong, H. D. Wang, and J. T. Whitney. "Environmental Impact Assessment in China: Present Practice and Future Developments." *Environmental Impact Assessment Review* 8 (1985): 85–95.

Qu Geping. *Environmental Management in China*. Beijing: China Environmental Science Press, 1999.

———, and Lin Jinchang. *Population and the Environment in China*. Trans. Jiang Baozhong and Gu Ran. Boulder, CO: Lynne Rienner, 1994.

———, and Woyen Lee, eds. *Managing the Environment in China*. Dublin, Ireland: Tycooly International Publishing, Ltd., 1984.

Ross, Lester. *Environmental Policy in China*. Bloomington: Indiana University Press, 1988.

———, and Mitchell A. Silk. *Environmental Law and Policy in the People's Republic of China*. New York: Quorum Books, 1987.

Schreurs, Amanda, and Elim Papadakis. *Historical Dictionary of the Green Movement*, 2nd ed. Lanham, MD: Rowman & Littlefield, 2018.

Shapiro, Judith. *China's Environmental Challenges*. Cambridge, U.K.: Polity, 2012.

———. *Mao's War against Nature: Politics and the Environment in Revolutionary China*. Cambridge, U.K.: Cambridge University Press, 2001.

Sinkule, Barbara J., and Leonard Ortolano. *Implementing Environmental Policy in China*. Westport, CT: Praeger, 1995.

Smil, Vaclav. *The Bad Earth: Environmental Degradation in China*. Armonk, NY: M. E. Sharpe, 1984.

———. *China's Environmental Crisis: An Inquiry into the Limits of National Development*. Armonk, NY: M. E. Sharpe, 1993.

———. *Environmental Problems in China: Estimates of Economic Costs*. East–West Center Special Report, No. 5. Honolulu, HI: East–West Center, 1996.

Stalley, Philip. *Foreign Firms, Investment, and Environmental Regulation in the People's Republic of China*. Stanford, CA: Stanford University Press, 2010.

Stern, Rachel E. *Environmental Litigation in China: A Study in Political Ambivalence*. Cambridge, U.K.: Cambridge University Press, 2013.

Tang Xiyang, and Marcia Marks. *A Green World Tour*. Beijing: New World Press, 1999.

Tilt, Bryan. "The Political Ecology of Pollution Enforcement in China: A Case from Sichuan's Industrial Rural Sector." *China Quarterly* 192 (December 2007): 915–32.

———. *The Struggle for Sustainability in Rural China: Environmental Values and Civil Society*. New York: Columbia University Press, 2010.

Van Rooij, Benjamin, and Carlos Wing-Hung Lo. "Fragile Convergence: Understanding Variation in the Enforcement of China's Industrial Pollution Law." *Law & Policy* 32, no. 1 (January 2010): 14–37.

———. *Regulating Land and Pollution in China: Law-Making, Compliance, and Enforcement: Theory and Cases*. Leiden, Netherlands: Leiden University Press, 2006.

Vermeer, Eduard B. "Management of Environmental Pollution in China: Problems and Abatement Policies." *China Information* 5, no. 1 (1990): 32–65.

Wagner, Rudolf G. "Agriculture and Environmental Protection in China." In *Learning from China?* London: Allen and Unwin, 1987.

Wang, Alex L. "In Search of a Sustainable Legitimacy: Environmental Law and Bureaucracy in China." *Harvard Environmental Law Review* 37 (2013): 71–96.

Watts, Jonathan. *When a Billion Chinese Jump: How China Will Save Mankind—Or Destroy It*. New York: Scribner, 2010.

Wei Yiming, et al. *China Energy Report (2008)*. Beijing: CO2 Emissions Research, Science Press, 2008.

World Bank and State Environmental Protection Agency (PRC). *The Cost of Pollution in China: Economic Estimates of Physical Damages*. Washington, DC: World Bank, 2007.

World Bank. *Clear Water, Blue Skies: China's Environment in the New Century*. Washington DC: World Bank, 1997.

Xu Jianchu, and Andreas Wilkes. "Biodiversity Impact Analysis in Northwest Yunnan, Southwest China." *Biodiversity and Conservation* 13, no. 5 (May 2004): 959–83.

Xue Jinjun et al. *Green Low-Carbon Development in China*. New York: Springer, 2013.

Yi Xiao, Xuemei Bai, Zhiyun Ouyang, Hua Zheng, and Fangfang Xing. "The Composition, Trend, and Impact of Urban Solid Waste in Beijing." *Environmental Monitoring and Assessment* 135, no 1 (December 2007): 21–30.

Zhan Jiang. "Environmental Journalism in China." In *Changing Media, Changing China*, ed. Susan Shirk (pp. 115–27). Oxford, U.K.: Oxford University Press, 2011.

2. Air, Soils, and Water

Alford, William P., and Benjamin A. Liebman. "Clean Air, Clean Process? The Struggle over Air Pollution in the People's Republic of China." *Hastings Law Journal* 52 (2002–2003): 703–43.

Buckley, Michael. *Meltdown in Tibet: China's Reckless Destruction of Ecosystems from the Highlands of Tibet to the Deltas of Asia*. New York: St. Martin's, 2014.

Cao Hongfa. "Air Pollution and Its Effects on Plants in China." *Journal of Applied Ecology* 26 (1989): 763–73.

Chang, William Y. B. "Large Lakes of China." *Journal of Great Lakes Research* 13, no. 3 (1987): 235–49.

Dai Qing. *The River Dragon Has Come! The Three Gorges Dam and the Fate of China's Yangtze River and Its People*. Armonk, NY: M. E. Sharpe, 1998.

———. *Yangtze! Yangtze!* Toronto: Earthscan Publications, 1994.

Delang, Claudio O. *China's Air Pollution Problems*. London: Routledge, 2016.

———, and Yuan Zhen. *China's Grain for Green Program*. New York: Springer International, 2014.

Economy, Elizabeth C. *The River Runs Black: The Environmental Challenge to China's Future*. Ithaca, NY: Cornell University Press, 2004.

Foster, Simon. *China's Pearl River Delta, Guangzhou, and Shenzhen*. Phoenix, AZ: Hunter Books, 2010.

Fritz, Jack, et al. *Urbanization, Energy, and Air Pollution in China: The Challenge Ahead*. Washington, DC: National Academies Press, 2004.

Guo Huancheng, Wu Dengru, and Zhu Hongxing. "Land Restoration in China." *Journal of Applied Ecology* 26, no. 3 (December 1989): 787–92.

Grumbine, Edward R. *Where the Dragon Meets the Angry River: Nature and Power in the People's Republic of China.* Washington, DC: Island Press, 2010.

Hessler, Peter. *River Town: Two Years on the Yangtze*. New York: Harper-Collins, 2001.

Hong Songhe. "Survey of Environmental Protection in the Lead and Zinc Minds of China." *Industry and Environment* 8, no. 1 (1985): 8–14.

Hu Angang, and Wang Yi. "Current Status, Causes, and Remedial Strategies of China's Ecology and Environment." *Chinese Geographical Sciences* 1, no. 2 (1991): 97–108.

Huang Bingwei. "River Conservancy and Agricultural Development of the North China Plain and Loess Highlands: Strategies and Research." *Great Plains Quarterly* 6 (1986): 218–24.

Institute of Soil Science. *Soils of China*. Beijing: Science Press, 1990.

Knup, Elizabeth. *Environmental NGOs in China: An Overview*. Washington, DC: Woodrow Wilson Center Press, 1997.

Li Guoying. "Keep Healthy Lives of Rivers: A Case Study of the Yellow River." *Yellow River Conservancy Commission*, November 15, 2007, http://isi.irtces.org/isi/Publication/BooksandReports/webinfo/2007/11/1487239390233278.htm.

Lindert, Peter. *Shifting Ground: The Changing Agricultural Soils of China and Indonesia*. Cambridge, MA: MIT Press, 2000.

Liu Changming. "Environmental Issues and the South-to-North Water Transfer Scheme." *China Quarterly* 156 (December 1998): 899–910.

Luk, Shiu-hung, and Joseph B. R. Whitney, eds. *Megaproject: A Case Study of China's Three Gorges Project*. Armonk, NY: M. E. Sharpe, 1993.

Lynn, Madeline. *Yangtze River: The Wildest, Wickedest River on Earth*. Oxford, U.K.: Oxford University Press, 2007.

Ma Jun. *China's Water Crisis*. Trans. Nancy Yang Liu and Lawrence R. Sullivan. Norwalk, CT: EastBridge, 2004.

McBeath, Jennifer Huang, and Jerry McBeath. *Environmental Change and Food Security in China*. New York: Springer, 2010.

Mertha, Andrew. *China's Water Warriors: Citizen Action and Policy Change*. Ithaca, NY: Cornell University Press, 2008.

Mun Ho, and Chris Nielsen. *Clearing the Air: The Health and Environmental Dangers of Air Pollution in China*. Cambridge, MA: MIT Press, 2007.

National Research Council. *Grasslands and Grassland Sciences in Northern China*. Washington, DC: National Academy Press, 1992.

Nickum, James E. *Dam Lies and Other Statistics: Taking Measure of Irrigation in China, 1931–1991*. East–West Center Occasional Paper. Honolulu, HI: East–West Center, 1995.

———. *Irrigation in the People's Republic of China*. Washington, DC: International Food Policy Institute, 1990.

People's Republic of China, Ministry of Water Resources, Department of Hydrology. *Water Resources Assessment for China*. Beijing: China Water and Power Press, 1992.

Pietz, David A. *The Yellow River: The Problem of Water in Modern China*. Cambridge, MA: Harvard University Press, 2015.

Reardon-Anderson, James, and James Ellis. "Whither China's Grasslands." *China Exchange News* 8, no. 4 (1990): 13–17.

Steinfield, Edward S., and Edward A. Cunningham. *Greener Plants, Grayer Skies? A Report from the Front Lines of China's Energy Sector*. Cambridge, MA: MIT Industrial Performance Center, China Energy Group, 2008.

Van Slyke, Lyman P. *Yangtze: Nature, History, and the River*. New York: Addison-Wesley, 1988.

Wang Jusi. "Water Pollution and Water Shortage Problems in China." *Journal of Applied Ecology* 2 (1989): 851–57.

Wang, Mark, et al. "Rural Industries and Water Pollution in China." *Journal of Environmental Management* 8, no. 4 (March 2008): 648–59.

Winchester, Simon. *The River at the Centre of the World: A Journey up the Yangtze and Back in Chinese Time*. London: Viking, 1997.

Xu Weidong, Shirasaka Shigeru, and Ichikawa Takeo. "Farming Systems and Settlements in Xishuangbanna, Yunnan Province, China." *Geographical Review of Japan* 62, no. 2 (1989): 104–15.

Zhang, Andy. *Hu Jintao: Facing China's Challenges Ahead*. Bloomington, IN: iUniverse, 2002.

Zhang Xuexun, et al. "Studies on the Heavy Metal Pollution of Soils and Plants in Tianjin Waste Water Area." *China Environmental Science* 10, no. 1 (1990): 87–97.

Zhao Qiguo. "Soil Resources and Their Utilization in China." In *Soils and Their Management: A Sino-European Perspective*, ed. E. Malby and T. Wollerson (pp. 87–97). London: Elsevier Applied Science, 1989.

3. Deserts, Forests, Grasslands, and Wetlands

Cyranoski, David. "Putting China's Wetlands on the Map." *Nature*, March 11, 2009, https://www.nature.com/news/2009/090311/full/458134a.html.

Gao Qinghu, et al. "Grassland Degradation in Northern Tibet Based on Remote Sensing Data." *Journal of Geographical Sciences* 16, no. 2 (2006): 289–96.

Guo Kai, and Li Ling. *How Tibet's Water Will Save China*. Beijing: Changan Publishing House, 2005.

Hyde, William F., Brian Belcher, and Jintao Xu, eds. *China's Forests: Global Lessons from Market Reforms*. Washington, DC: Resources for the Future, 2003.
Meckelein, Wolfgang. "Land Use Problems in the Chinese Deserts." *Applied Geography and Development: A Biannual Collection of Recent German Contributions* 30 (1987): 7–29.
Raven, Peter. "Biodiversity and the Future of China." *Pacific Science Association Information Bulletin* 47, nos. 1–2 (1995): 1–8.
Richardson, S. D. *Forests and Forestry in China: Changing Patterns of Resource Development*. Washington, DC: Island Press, 1990.
Tyler, Christian. *Wild West China: The Untold Story of a Frontier Land.* London: John Murray, 2003.
Wang, C. W. *The Forests of China with a Survey of Grassland and Desert Vegetation*. Cambridge, MA: Maria Moors Cabot Foundation, 1961.
Wang Tao. *The Development Process of Aeolian Desertification in Typical Areas of Northern China during the Last 50 Years*. Lanzhou: Key Laboratory of Desert and Desertification, Cold and Arid Regions Environmental and Engineering Research Institute, Chinese Academy of Sciences, 2008.
Williams, Dee Mack. *Beyond Great Walls: Environmental Identity and Development on the Chinese Grasslands of Inner Mongolia*. Stanford, CA: Stanford University Press, 2002.
Zhu Zhenda, et al. *Deserts in China*. Lanzhou: Lanzhou Institute of Desert Research, 1986.

4. Fisheries, Wildlife, and Nature Reserves

Enderton, Catherine Shurr. "Nature Preserves and Protected Wildlife in the People's Republic of China." *China Geographer* 12 (1985): 117–40.
Harris, Richard B. *Wildlife Conservation in China: Preserving the Habitat of China's Wild West*. Norwalk, CT: EastBridge, 2008.
Santiapillai, Charles, and Li Yang. "Conservation of Elephants in Xishuangbanna Nature Reserve in the People's Republic of China." *Wildlife* 8, no. 2 (1990): 3.
Schaller, George B. *The Last Panda*. Chicago: University of Chicago Press, 1987.
Yang Xiyang. *Living Treasures: An Odyssey through China's Extraordinary Nature Reserves*. New York: Bantam, 1987.
Young, Llewellyn. "Pheasant Conservation in China." *Wildlife* 8, no. 1 (1990): 4.
Zweig, Ronald D. "Freshwater Aquaculture in China: Ecosystem Management for Survival." *Ambio* 14, no. 2 (1985): 66–74.

5. Economy, Energy, and International Affairs

Hewitt, Duncan. *Getting Rich First: Life in a Changing China*. London: Chatto & Windus, 2007.

Kynge, James. *China Shakes the World: A Titan's Rise and Troubled Future—and the Challenge for America*. Boston: Houghton Mifflin, 2006.

Liu Jianguo, and Jared Diamond. "China's Environment in a Globalizing World." *Nature*, June 30, 2005, pp. 1,179–86.

Mars, Neville, and Adrian Hornsby, eds. *The Chinese Dream: A Society under Construction*. Rotterdam, Netherlands: 010 Publishers, 2008.

Polonsko, Karen K. *The Technology–Energy–Environment–Health (TEEH) Chain in China: A Case Study of Cokemaking: 8*. New York: Springer, 2016.

Smil, Vaclav. *China's Past, China's Future: Energy, Food, and Environment*. London: Routledge, 2014.

Song Ligang, and Wing Thy Woo, eds. *China's Dilemma: Economic Growth, the Environment, and Climate Change*. Washington, DC: Brookings Institution, 2008.

Tibet Information Network. *Mining Tibet: Mineral Exploration in Tibetan Areas of the PRC*. Washington, DC: Tibet Information Network, 2002.

Whitney, Joseph B. R. "The Waste Economy and the Dispersed Metropolis in China." In *The Extended Metropolis: Settlement Transition in China*, ed. Norton Ginsburg et al. (pp. 177–92). Honolulu: University of Hawaii Press, 1991.

Wu Xiaoyu. *Coal History Review*. Beijing: China Coal Industry Publishing House, 2000.

Yi Qi. *Annual Review of Low-Carbon Development in China: 2010*. Singapore: World Scientific Publishing, 2013.

6. Environmental History: China and Russia

Anderson, E. N. *Food and Environment in Early and Medieval China*. Philadelphia: University of Pennsylvania Press, 2014.

Bello, David A. *Across Forest, Steppe, and Mountains: Environment, Identity, and Empire in Qing China's Borderlands*. Cambridge, U.K.: Cambridge University Press, 2016.

Edmonds, Richard Louis. *Patterns of China's Lost Harmony: A Survey of the Country's Environmental Degradation and Protection*. London: Routledge, 1994.

Elvin, Mark. *Retreat of the Elephants: An Environmental History of China*. New Haven, CT: Yale University Press, 2004.

———, and Liu Ts'ui-jung, eds. *Sediments of Time: Environment and Society in Chinese History*. New York: Cambridge University Press, 1997.
Hou Wenhui. "The Environmental Crisis in China and the Case for Environmental History Studies." *Environmental History Review* 14, nos. 1–2 (1990): 151–58.
———. "Reflections on Traditional Chinese Ideas of Nature." *Environmental History* 8, no. 4 (1997): 482–93.
Josephson, Paul, et al. *An Environmental History of Russia*. Cambridge, U.K.: Cambridge University Press, 2013.
Marks, Robert B. *China: Its Environment and History*. Lanham, MD: Rowman & Littlefield, 2013.
———. *Tigers, Rice, Silk, and Silt*. New York: Cambridge University Press, 1997.
Miller, James. *China's Green Religion: Daoism and the Quest for a Sustainable Future*. New York: Columbia University Press, 2017.
Murphy, Rhoads. "Man and Nature in China." *Modern Asian Studies* 1, no. 4 (1967): 313–33.
Needham, Joseph L. *Science and Civilization in China, Vol. 4: Physics and Physical Technology, Part 3: Civil Engineering and Nautics*. Cambridge, U.K.: Cambridge University Press, 1996.
Tucker, Mary Evelyn, and John Berthrong, eds. *Confucianism and Ecology: The Interrelationship of Heaven, Earth, and Humans*. Cambridge, MA: Harvard University Press, 1998.
Weller, Robert P. *Discovering Nature: Globalization and Environmental Culture in China and Taiwan*. Cambridge, U.K.: Cambridge University Press, 2006.

B. Chinese and Other Languages

Bie, Tao, ed. *Huanjing Hongyi Susong* (Environmental Public Interest Litigation). Beijing: Law Press, 2007.
Boyle, Alan, and Patricia Birnie. *Guojiafa yu Huanjing* (International Law and the Environment). Beijing: Higher Education Press, 2007.
Feng Yongfeng. *Meiyou Dashu de Guojia* (A Country without Big Trees). Beijing: China Law Press, 2008.
Han Depei, and Xiao Longan, eds. *Huanjing Baohu Fa Jiben Zhishi* (Basic Knowledge of Environmental Law). Beijing: Zhongguo Huanjing Kexue Chubanshe, 1990.
Isaia, Henri. *La Protection de L'Environnement en Chine* (Environmental Protection in China). Paris: Presses Universitaires de France, 1981.
Jin Suilin. *Huanjing Faxue* (Environmental Law). Beijing: Beijing Daxue Chubanshe, 1990.

Li Chuanlong. *Zhongguo Tielei Tupu* (An Atlas of Chinese Butterflies). Shanghai: Yuandong Publishing House, 1992.

Lin Chengkun. *Changjiang Sanxia yu Gezhouba de Nisha ji Huanjing* (Sediment and Environment of the Three Gorges and the Gezhouba Dam). Nanjing: Nanjing Daxue Chubanshe, 1989.

Luo Huihan. *Huanjing Faxue* (Environmental Law). Guangdong: Zhongshan Daxue Chubanshe, 1986.

Nian Zhongguo Huanjing Zhuangkuang Gongbao, 2003 (Report on the State of the Environment in China, 2003). Beijing: State Environmental Protection Administration, 2004.

State Environmental Protection Administration, Policy and Law Section, ed. *Zhongguo Huanjing Baohu Fagui Quanshu, 1982–1997* (Complete Book of China's Environmental Protection Laws and Regulations, 1982–1997). Beijing: Chemical Industry Press, 1997.

Wang Cangfa, ed. *Huanjing yu Ziyuan Baohufa Anli* (Cases in Environmental and Natural Resource Process Law). Beijing: Renmin University Press, 2005.

Wang Lili. *Lü Meiti* (Green Media). Beijing: Tsinghua University Press, 2005.

Wu Chen. *Huabei Pingyuan Siwan Nian lai Ziran Huanjing Yanbian* (Changes in the Natural Environment of the North China Plain during the Last 40,000 Years). Beijing: Zhongguo Kexue Jishu Chubanshe, 1992.

Yang Chaofei. *Huanjing Baohu yu Huanjing Wenhua* (Environmental Protection and Environmental Culture). Beijing: Zhongguo Zhengfa Daxue Chubanshe, 1994.

Yang Jianyi. *Ma Yinchu*. Shijiazhuang: Huashan Wenyi Chubanshe, 1997.

Zhang Kunmin, and Jin Suilin. *Huanjing Baohu Fa Jianghua* (A Guide to Environmental Law). Beijing: Qinghua Daxue Chubanshe, 1990.

Zhang Zitai. *Huanjing Baohu Fa* (Environmental Protection Law). Nanjing: Hehai Daxue Chubanshe, 1994.

Zheng Yisheng, and Qian Yihong. *Shendu Youhuan: Dangdai Zhongguo de Kechixu Fazhan Wenti* (Deep Hardships: Problems of Sustainable Development in China). Beijing: Jinti Zhongguo Chubanshe, 1998.

Zhongguo Huanjing Nianjian (China Environment Yearbook). Beijing: China Environmental Press, 2001, 2002, 2010.

Zhongguo Huanjing Tongji Nianjian (China Environmental Statistical Yearbook). Beijing: China Statistics Press, 2001, 2009, 2010.

III. MAJOR CHINESE AND INTERNATIONAL JOURNALS ON THE ENVIRONMENT

Acta Ecological Sinica
China Environmental Law Review
China Environmental News
China Forestry Daily
China Geothermal Environmental Bulletin
China Green Times
Chinese Journal of Applied and Environmental Biology
China Journal of Environmental Engineering (Chinese Academy of Sciences)
China Journal of Environmental Law
Chinese Journal of Population Resources and Environment
Chinese Journal of Urban and Environmental Studies (Singapore)
Cropwatch Bulletin
Great Nature Magazine
Journal of Ecology and Rural Environment (China Environmental Sciences Press)
Journal of Environmental Sciences (Chinese Academy of Sciences)
Lake Studies
Plateau Meteorology
Tropical Geography

IV. FILMS

A. China and International (Director/Topic)

Battle for the Elephants (National Geographic Special)
Behemoth (Zhao Liang/mining)
Beijing Besieged by Waste (Wang Jiuliang/urban waste)
Blind Shaft (Li Yang/corruption in the environmentally damaging coal industry)
Born in China (Disney Nature/wildlife)
China Megaprojects: Energy (China Global Television Network [CGTN])
China Megaprojects: Food (CGTN)
Damming the Nu River (Zhou Xiaoli/popular opposition to dam construction)
Dirty Business: What Really Happens to Your Recycling (Sky News, United Kingdom)
Gambling on Extinction (National Geographic Documentary)
Green China Rising (Mandarin Films/strategy of green development)

The Ivory Game (National Geographic Documentary)
Lessons of the Loess Plateau (John D. Liu/destruction and restoration of the Loess)
Mountain Patrol: Kekexili (Lu Chuan/wildlife poaching)
Plastic China (Wang Jiuliang/plastic waste)
The Road (Zhang Zangbo/environmental impact of construction projects)
Sea of Shadows (National Geographic Documentary)
Searching for Sacred Mountain (Shi Lihong/Tibet)
Still Life (Jia Zhangke/impact of the Three Gorges Dam)
Under the Dome (Chai Jing/impact of air pollution)
Up the Yangtze (Yung Chang/environmental impact of the Three Gorges Dam)
Voice of an Angry River (Shi Lihong/documentary on the Nu River)
Waking the Green Tiger (Gary Marcuse/local opposition to hydropower)
Warriors of Qiugang (Ruby Yang/villager opposition to industrial pollution)
Wild China (BBC Earth/human impact on the natural world and wildlife)
Wolf Totem (Jean-Jacques Annaud/wildlife extinction)

V. WEBSITES

Chinadialogue.org
China.org.cn
Ecowatch.com
Mdpo.com
Wri.org

About the Authors

Nancy Liu-Sullivan received a Ph.D. in pharmacology (State University of New York, Stony Brook, 2006), M.S. in biology (Adelphi University), M.S. in speech pathology (Columbia University), and B.A. in English literature (Foreign Language Teachers' College, Beijing, China). Liu-Sullivan has conducted research at Sloan-Kettering Institute, Memorial Sloan-Kettering Cancer Center, New York City, and is a lecturer in the Department of Biology, College of Staten Island, City University of New York, where she also conducts cancer research. She is coeditor and cotranslator of *China's Water Crisis* (2004), by Ma Jun, and *The River Dragon Has Come! The Three Gorges Dam and the Fate of China's Yangtze River and Its People* (1998), by Dai Qing, and coauthor of *Historical Dictionary of Science and Technology in Modern China* (2015).

Lawrence R. Sullivan is professor emeritus of political science, Adelphi University, Garden City, New York, and a research associate, East Asian Institute, Columbia University, New York City. He received a Ph.D. in political science at the University of Michigan and is coeditor and cotranslator of several works on the Chinese environment, including *China's Water Crisis* (2004), by Ma Jun, and *Yangtze! Yangtze!* (1989) and *The River Dragon Has Come! The Three Gorges Dam and the Fate of China's Yangtze River and Its People* (1998), both by Dai Qing. Sullivan is also author and coauthor of several books on contemporary China, including *Leadership and Authority in China: 1895–1976* (2012), *Historical Dictionary of the Chinese Communist Party* (2012), *Historical Dictionary of the Chinese Economy* (2018, with Paul Curcio), and *Historical Dictionary of Chinese Foreign Affairs* (2018, with Robert L. Paarlberg).

www.ingramcontent.com/pod-product-compliance
Ingram Content Group UK Ltd.
Pitfield, Milton Keynes, MK11 3LW, UK
UKHW041128071026
15587UKWH00009B/287

9 781538 120354